The Science of Algorithmic Trading and Portfolio Management

The Science of Algorithmic Trading and Portfolio Management

Robert Kissell Ph.D

ELSEVIER

AMSTERDAM • BOSTON • HEIDELBERG • LONDON
NEW YORK • OXFORD • PARIS • SAN DIEGO
SAN FRANCISCO • SINGAPORE • SYDNEY • TOKYO
Academic Press is an imprint of Elsevier

Academic Press is an imprint of Elsevier
525 B Street, Suite 1800, San Diego, CA 92101–4495, USA
The Boulevard, Langford Lane, Kidlington, Oxford, OX5 1GB, UK
225 Wyman Street, Waltham, MA 02451, USA

First published 2014

British Library Cataloguing-in-Publication Data
A catalogue record for this book is available from the British Library

Library of Congress Cataloging-in-Publication Data
A catalog record for this book is available from the Library of Congress

ISBN: 978-0-12-401689-7

For information on all Academic Press publications
visit our website at elsevierdirect.com

Printed and bound in United States of America

14 15 16 17   10 9 8 7 6 5 4 3 2 1

Landon and Mason

A continuous source of joy and inspiration

And a reminder to keep asking why why why....

Contents

Preface

If we knew what it was we were doing, it would not be called research, would it?

Albert Einstein

The Science of Algorithmic Trading and Portfolio Management is a reference book intended to provide traders, portfolio managers, analysts, students, practitioners, and financial executives with an overview of the electronic trading environment, and insight into how algorithms can be utilized to improve execution quality and fund performance.

We provide a discussion of the current state of the market and advanced modeling techniques for trading algorithms, stock selection, and portfolio construction.

This reference book will provide readers with:

- An understanding of the new electronic trading environment.
- An understanding of transaction cost analysis (TCA) and proper metrics for cost measurement and performance evaluation.
- A thorough understanding of the different types of trading algorithms: liquidity seeking, dark pools, arrival price, implementation shortfall (IS), volume weighted average price (VWAP), arrival price, and portfolio implementation shortfall.
- Proven market impact modeling techniques.
- An understanding of algorithmic trading across various asset classes: equities, futures, fixed income, foreign exchange, and commodities.
- Advanced algorithmic forecasting techniques to estimate daily liquidity and monthly volumes.
- An algorithmic decision making framework to ensure consistency between investment and trading objectives.
- A best execution process.

Readers will subsequently be prepared to:

- Develop real-time trading algorithms customized to specific institutional needs.
- Design systems to manage algorithmic risk and dark pool uncertainty.
- Evaluate market impact models and assess performance across algorithms, traders, and brokers.
- Implement electronic trading systems.

For the first time, portfolio managers are not forgotten and will be provided with proven techniques to better construct portfolios through:

- Stock Selection
- Portfolio Optimization
- Asset Allocation
- MI Factor Scores

- Multi-Asset Investing
- Factor Exposure Investing

The book is categorized in three parts. Part I focuses on the current electronic market environment where we discuss trading algorithms, market microstructure research, and transaction cost analysis. Part II focuses on the necessary mathematical models that are used to construct, calibrate, and test market impact models, as well as to develop single stock and portfolio trading algorithms. The section further discusses volatility and factor models, as well as advanced algorithmic forecasting techniques. Part III focuses on portfolio management techniques and how TCA and market impact can be incorporated into the investment decisions, stock selection, and portfolio construction to improve portfolio performance. We introduce readers to an advanced portfolio optimization process that incorporates market impact and transaction costs directly into the portfolio optimization. We provide insight into how MI factor scores can be used to improve stock selection, as well as a technique that can be used by portfolio managers to decipher broker-dealer black box models. This section concludes with an overview of high frequency trading, and the necessary mathematical knowledge required to develop black box trading models.

Acknowledgments

There are several people who made significant contributions to the concepts introduced throughout the text. Without their insights, comments, suggestions, and criticism, the final version of this book and these models would not have been possible. They are:

Roberto Malamut, Ph.D., was instrumental in the development of the methodologies and framework introduced in this book. His keen mathematical insight and market knowledge helped advance many of the theories presented throughout the text. Morton Glantz, my coauthor from Optimal Trading Strategies, provided invaluable guidance and direction, and helped turn many of our original ideas into formulations that have since been put into practice by traders and portfolio managers, and have now become mainstream in the industry.

The All-Universe Algorithmic Team: Roberto Malamut (again), Andrew Xia, Hernan Otero, Deepak Nautiyal, Don Sun, Kevin Li, Peter Tannenbaum, Arun Rajasekhar, and Mustaq Ali, and Tom M. Kane and Dan Keegan too! And to complete the All-Universe team: Pierre Miasnikof, Agustin Leon, and Alexis Kirke for all of their early contribution in developing and testing many of the ideas and models that have now become ingrained into the algorithmic trading landscape. Their contribution to algorithmic trading is second to none.

Wayne Wagner provided valuable direction and support over the years. His early research has since evolved into its own science and discipline known as transaction costs analysis (TCA). His early vision and research has helped pave the way for making our financial markets more efficient and investor portfolios more profitable. Robert Almgren and Neil Chriss provided the ground breaking work on the efficient trading frontier, and introduced the appropriate mathematical trading concepts to the trading side of the industry. Their seminal paper on Optimal Liquidation Strategies is the reason that trading desks have embraced mathematical models and algorithmic trading.

Victoria Averbukh Kulikov, Director of Cornell Financial Engineering Manhattan (CFEM), allowed me to lecture on Algorithmic Trading (Fall 2009 & Fall 2010) and test many of my theories and ideas in a class setting. I have a great deal of gratitude to her and to all the students for correcting my many mistakes before they could become part of this book. They provided more answers to me than I am sure I provided to them during the semester.

Connie Li, Quantitative Analyst at Numeric Investments (and M.S. in Financial Engineering from Cornell University), provided invaluable comments and suggestions throughout the writing of the book. And most importantly, corrected the errors in my math, the grammar in my writing, and helped simplify the many concepts discussed throughout the book. Scott Wilson, Ph.D., Analyst at Cornerstone Research, provided invaluable insight and direction for modeling trading costs across the various asset classes, and was influential in helping to structure the concepts behind the factor exposure allocation scheme.

Ayub Hanif, Ph.D. Researcher, Financial Computing and Applied Computational Science, University College London, for his extraordinary contribution to the book as the author of Chapter 13: High Frequency Trading and Black Box Models. This chapter has provided more insight into the secretive word of black box modeling and high frequency trading than has been disseminated in all the seminars and conferences I have attended put together. It is a must read for any investor seeking to manage a portfolio and earn a profit in the ultracompetitive high frequency and high velocity trading space.

Additionally, Dan Dibartolomeo, Jon Anderson, John Carillo, Sebastian Ceria, Curt Engler, Marc Gresack, Kingsley Jones, Scott Wilson, Eldar Nigmatullin, Bojan Petrovich, Mike Rodgers, Deborah Berebichez, Jim Poserina, Mike Blake, and Diana Muzan for providing valuable insight, suggestions, comments, during some of the early drafts of this manuscript. This has ultimately lead to a better text. The team at Institutional Investor and Journal of Trading, Allison Adams, Brian Bruce, and Debra Trask for ongoing encouragement and support on the research side of the business.

A special thanks to Richard Rudden, Stephen Marron, John Little, Cheryl Beach, Russ Feingold, Kevin Harper, William Hederman, John Wile, and Kyle Rudden, from my first job out of college at R.J. Rudden Associates (now part of Black and Veatch) for teaching the true benefits of thinking outside of the box, and showing that many times a non-traditional approach could often prove to be the most insightful.

Finally, Hans Lie, Richard Duan, Trista Rose, Alisher Khussainov, Thomas Yang, Joesph Gahtan, Fabienne Wilmes, Erik Sulzbach, Charlie Behette, Min Moon, Kapil Dhingra, Harry Rana, Michael Lee, John Mackie, Nigel Lucas, Steve Paridis, Thomas Reif, Steve Malin, Marco Dion, Michael Coyle, Anna-Marie Monette, Mal Selver, Ryan Crane, Matt Laird, Charlotte Reid, Ignor Kantor, Aleksandra Radakovic, Deng Zhang, Shu Lin, Ken Weston, Andrew Freyre-Sanders, Mike Schultz, Lisa Sarris, Joe Gresia, Mike Keigher, Thomas Rucinski, Alan Rubenfeld, John Palazzo, Jens Soerensen, Adam Denny, Diane Neligan, Rahul Grover, Rana Chammaa, Stefan Balderach, Chris Sinclaire, James Rubinstein, Frank Bigelow, Rob Chechilo, Carl DeFelice, Kurt Burger, Brian McGinn, Dan Wilson, Kieran Kilkenny, Kendal Beer, Edna Addo, Israel Moljo, Peter Krase, Emil Terazi, Emerson Wu, Trevor McDonough, Simon, Jim Heaney, Emilee Deutchman, Seth Weingram, and Jared Anderson.

Best Regards,

Robert Kissell, Ph.D.

Algorithmic Trading

INTRODUCTION

Algorithmic trading represents the computerized executions of financial instruments. Algorithms trade stocks, bonds, currencies, and a plethora of financial derivatives. Algorithms are also fundamental to investment strategies and trading goals. The new era of trading provides investors with more efficient executions while lowering transaction costs—the result, improved portfolio performance. Algorithmic trading has been referred to as "automated," "black box" and "robo" trading.

Trading via algorithms requires investors to first specify their investing and/or trading goals in terms of mathematical instructions. Dependent upon investors' needs, customized instructions range from simple to highly sophisticated. After instructions are specified, computers implement those trades following the prescribed instructions.

Managers use algorithms in a variety of ways. Money management funds—mutual and index funds, pension plans, quantitative funds and even hedge funds—use algorithms to implement investment decisions. In these cases, money managers use different stock selection and portfolio construction techniques to determine their preferred holdings, and then employ algorithms to implement those decisions. Algorithms determine the best way to slice orders and trade over time. They determine appropriate price, time, and quantity of shares (size) to enter the market. Often, these algorithms make decisions independent of any human interaction.

Similar to a more antiquated, manual market-making approach, broker dealers and market makers now use automated algorithms to provide liquidity to the marketplace. As such, these parties are able to make markets in a broader spectrum of securities electronically rather than manually, cutting costs of hiring additional traders.

Aside from improving liquidity to the marketplace, broker dealers are using algorithms to transact for investor clients. Once investment decisions are made, buy-side trading desks pass orders to their brokers for execution

The Science of Algorithmic Trading and Portfolio Management. DOI: http://dx.doi.org/10.1016/B978-0-12-401689-7.00001-5

using algorithms. The buy-side may specify which broker algorithms to use to trade single or basket orders, or rely on the expertise of sell-side brokers to select the proper algorithms and algorithmic parameters. It is important for the sell-side to precisely communicate to the buy-side expectations regarding expected transaction costs (usually via pre-trade analysis) and potential issues that may arise during trading. The buy-side will need to ensure these implementation goals are consistent with the fund's investment objectives. Furthermore, it is crucial for the buy-side to determine future implementation decisions (usually via post-trade analysis) to continuously evaluate broker performance and algorithms under various scenarios.

Quantitative, statistical arbitrage traders, sophisticated hedge funds, and the newly emerged class of investors known as high frequency traders will also program buying/selling rules directly into the trading algorithm. The program rules allows algorithms to determine instruments and how they should be bought and sold. These types of algorithms are referred to as "blackbox" or "profit and loss" algorithms.

For years, financial research has focused on the investment side of a business. Funds have invested copious dollars and research hours on the quest for superior investment opportunities and risk management techniques, with very little research on the implementation side. However, over the last decade, much of this initiative has shifted towards capturing hidden value during implementation. Treynor (1981), Perold (1988), Berkowitz, Logue, and Noser (1988), Wagner (1990), and Edwards and Wagner (1993) were among the first to report the quantity of alpha lost during implementation of the investment idea due to transaction costs. More recently, Bertsimas and Lo (1996), Almgren and Chriss (1999, 2000), Kissell, Glantz, and Malamut (2004) introduced a framework to minimize market impact and transaction costs, as well as a process to determine appropriate optimal execution strategies. These efforts have helped provide efficient implementation—the process known as algorithmic trading[1].

While empirical evidence has shown that when properly specified, algorithms result in lower transaction costs, the process necessitates investors be more proactive during implementation than they were previously utilizing manual execution. Algorithms must be able to manage price, size, and timing of the trades, while continuously reacting to market condition changes.

---

[1]A review of market microstructure and transaction cost literature is provided in Chapter 2, Market Microstructure.

Advantages

Algorithmic trading provides investors with many benefits such as:

- *Lower Commissions*. Commissions are usually lower than traditional commission fees since algorithmic trading only provides investors with execution and execution-related services (such as risk management and order management). Algorithmic commissions typically do not compensate brokers for research activities, although some funds pay a higher rate for research access.
- *Anonymity*. Orders are entered into the system and traded automatically by the computer across all execution venues. The buy-side trader either manages the order from within his firm or requests that the order is managed by the sell-side traders. Orders are not shopped or across trading floor as they once were.
- *Control*. Buy-side traders have full control over orders. Traders determine the venues (displayed/dark), order submission rules such as market/limit prices, share quantities, wait and refresh times, as well as when to accelerate or decelerate trading based on the investment objective of the fund and actual market conditions. Traders can cancel the order or modify the trading instructions almost instantaneously.
- *Minimum Information Leakage*. Information leakage is minimized since the broker does not receive any information about the order or trading intentions of the investor. The buy-side trader is able to specify their trading instructions and investment needs simply by the selection of the algorithm and specifications of the algorithmic parameters.
- *Transparency*. Investors are provided with a higher degree of transparency surrounding how the order will be executed. Since the underlying execution rules for each algorithm are provided to investors in advance, investors will know exactly how the algorithm will execute shares in the market, as algorithms will do exactly what they are programmed to do.
- *Access*. Algorithms are able to provide fast and efficient access to the different markets and dark pool. They also provide co-location, low latency connections, which provides investors with the benefits of high speed connections.
- *Competition*. The evolution of algorithmic trading has seen competition from various market participants such as independent vendors, order management and execution management software firms, exchanges, third party providers, and in-house development teams in addition to the traditional sell-side broker dealers. Investors have received the benefits of this increased competition in the form of better execution services and lower costs. Given the ease and flexibility of choosing and switching between providers, investors are not locked into any one

selection. In turn, algo providers are required to be more proactive in continually improving their offerings and efficiencies.

- *Reduced Transaction Costs.* Computers are better equipped and faster to react to changing market conditions and unplanned events. They are better capable to ensure consistency between the investment decision and trading instructions, which results in decreased market impact cost, less timing risk, and a higher percentage of completed orders (lower opportunity cost).

Disadvantages

Algorithmic trading has been around only since the early 2000s and it is still evolving at an amazing rate. Unfortunately, algorithms are not the be all and end all for our trading needs. Deficiencies and limitations include:

- Users can become complacent and use the same algorithms regardless of the order characteristics and market conditions simply because they are familiar with the algorithm.
- Users need to continuously test and evaluate algorithms to ensure they are using the algorithms properly and that the algorithms are doing what they are advertised to do. Users need to measure and monitor performance across brokers, algorithms and market conditions to understand what algorithms are most appropriate given the type of market environment.
- Algorithms perform exactly as they are specified, which is nice when the trading environment is what has been expected. However, in the case that unplanned events occur, the algorithm may not be properly trained or programmed for that particular market, which may lead to sub-par performance and higher costs.
- Users need to ensure consistency across the algorithm and their investment needs. Ensuring consistency is becoming increasingly difficult in times where the actual algorithmic trading rule is not as transparent as it could be or when the algorithms are given non-descriptive names that do not provide any insight into what they are trying to do.
- Too many algos and too many names. VWAP, volume weighted average price, is an example of a fairly descriptive algorithmic name and is fairly consistent across brokers. However, an algorithm such as Tarzan is not descriptive and does not provide insights into how it will trade during the day. Investors may need to understand and differentiate between hundreds of algorithms, and keep track of the changes that occur in these codebases. For example, a large institution may use twenty different brokers with five to ten different algorithms each, and with at least half of those names being non-descriptive.

■ *Price Discovery.* As we discuss in Chapter 2 (Market Microstructure) the growth of algorithms and decline of traditional specialists and market marker roles has led to a more difficult price discovery process at the open. While algorithms are well versed at incorporating price information to determine the proper slicing strategy, they are not yet well versed at quickly determining the fair market price for a security.

CHANGING TRADING ENVIRONMENT

The US equity markets have experienced sweeping changes in market microstructure, rapid growth in program trading, and a large shift to electronic trading. In 2001, both the New York Stock Exchange (NYSE) and NASDAQ moved to a system of quoting stocks in decimals (e.g., cents per share) from a system of quoting stocks in fractions (e.g., 1/16th of a dollar or "teenies"). As a consequence, the minimum quote increment reduced from $0.0625/share to $0.01/share. While this provides investors with a much larger array of potential market prices and transactions closer to true intrinsic values, it has also been criticized for interfering with the main role of financial markets, namely, liquidity and price discovery.

The decrease in liquidity shortly after decimalization has been documented by Bacidore, Battalio, and Jennings (2001), NASDAQ Economic Research (2001), and Beesembinder (2003). This was also observed in the US equity markets after moving from a quoting system of 1/8th ($0.125/share) to 1/16th (0.0625/share) and in Canada when the Toronto Stock Exchange moved from a system of 1/8th ($0.125/share) to nickels ($0.05/share). For example, see Harris (1994, 2003), Jones and Lipson (1999), and Goldstein and Kavajecz (2000).

Some market participants argued that actual market liquidity did in fact remain stable after decimalization, although it was spread out over a larger number of price points. For example, suppose that prior to decimalization the best offer for a stock was 5000 shares at $30.00. However, after decimalization the market offers were 500 shares at $29.98, 1000 shares at $29.99, 2000 shares at $30.00, 1000 shares at $30.01, and 500 shares at $30.02. In this example, neither the total liquidity or average offered price for 5000 shares has changed, but the measured liquidity at the best market ask has decreased from 5000 shares pre-decimalization to 500 shares post-decimalization. So, while market depth (e.g., "transaction liquidity") measured as the total quantity of shares at the national best bid and offer (NBBO) has decreased, actual market liquidity may be unaffected. What has changed in this example is that now it takes five times as many transactions to fill the 5000 share order. Thus, even if liquidity has remained stable, trading difficulty, as measured by the number of transactions required to complete the order, has in fact increased.

Another major structural event that occurred with dramatic effects on trading was Reg NMS, regulation of national market systems. Reg NMS was a series of initiatives to modernize our national securities markets. The main goals were to promote competition across market centers and orders, and provide fair and efficient pricing mechanisms. The rules that had the greatest effect on the markets were:

- Order Protection (Rule 611)—Establish, maintain, and enforce written policies and procedures to prevent "trade-throughs." Trade-throughs are the execution of trades at prices inferior to protected quotes. A protected quote is an order that is immediately and automatically accessible.
- Access (Rule 610)—Requires fair and non-discriminatory access to quotes and routing of orders to those market centers with the best prices.
- Sub Penny Pricing (Rule 612)—minimum pricing increments.

Analysis of market volumes has found that while visible liquidity at the best bid and ask may have decreased, actual traded volume has increased dramatically. Figure 1.1 shows the increase in volume for NYSE listed securities from 2000 to 2012. Total consolidated volume peaked during the financial crisis in 2008–2009 where it was more than 6× the average in 2000. Volume has succumbed since 2009, but in 2012 it is still 3.3× the 2000 average. While volumes increased in the market, the issue of fragmentation transpired. Traded volume in the NYSE was between 80 and 90% on average for 2000 through 2005, but then started decreasing

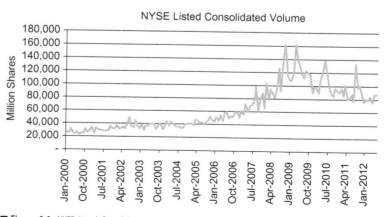

■ **Figure 1.1** NYSE Listed Consolidated Volume.

(partly due to Reg NMS in 2007) through 2008/2009. Currently, the NYSE group accounts for only 33% and the traditional NYSE floor exchange accounts for 20%−23% of total volume in NYSE listed stocks. The NYSE group includes trades in the NYSE, American Express (AMEX), and Archipelago Exchange (ARCA). This is illustrated in Figure 1.2.

Empirical evidence from the NYSE confirms that decimalization made trading more difficult by increasing the total number of transactions required to fill an order. Average trade size on the NYSE has decreased 82% since 2000 when the average trade size was 1222 shares compared to only 222 shares in 2012. Investors with large positions now require more than five times the number of trades to complete the order than was required prior to decimalization. This does not even include the increase in quote revisions and cancellations for price discovery purposes (and fishing expeditions). Figure 1.3 shows the rapid decrease in trade size over the period 2000 through about 2005, with the decline continuing more gradually until about 2009. Since 2009, the average trade size has been just over 200 shares per trade. Figure 1.4 shows the decline in average trade size for the NYSE group. The difference in this analysis is that in 2004 the NYSE combined with AMEX and ARCA started to include crossing session volumes, auctions, etc., into its reported volume statistics. As a result, higher volume values were reported on the NYSE group reports than to the consolidated public tape, and average trade sizes were much larger than the median or the average trade sizes excluding auctions and crossing session volume. Even under this new definition we can see that the average trade sizes (computed by dividing the NYSE group reported volume by the total number of reported trades) dramatically declined. These average sizes

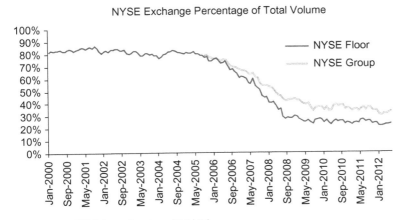

■ **Figure 1.2** NYSE Exchange Percentage of Total Volume.

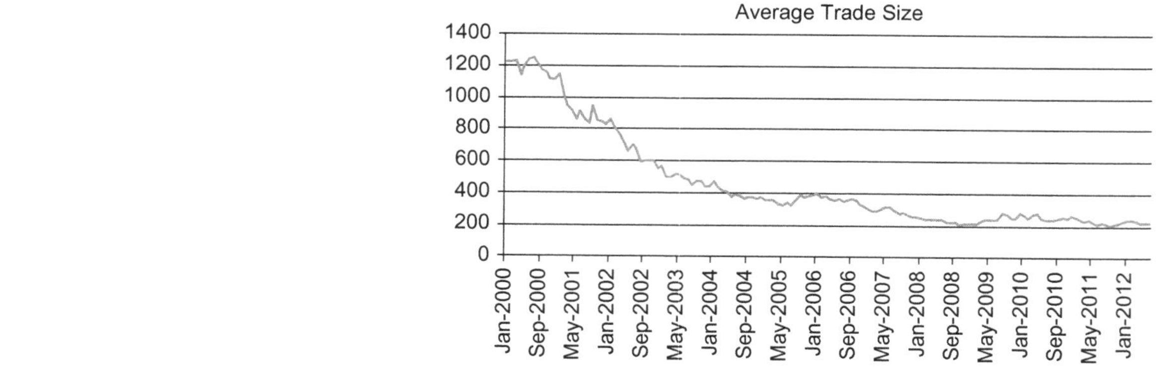

Average Trade Size

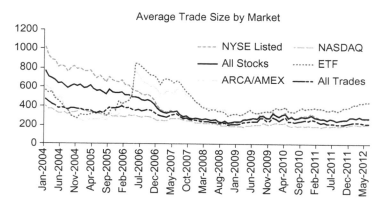

■ **Figure** 1.3 Average Trade Size.

Average Trade Size by Market

■ **Figure** 1.4 Average Trade Size by Market.

show the different equity listings (NYSE, AMEX, ARCA, and NASDAQ), and exchange traded funds (ETFs). In May and June of 2006, the NYSE group started reporting all crossing session volumes in its ETF and AMEX/ ARCA volume statistics resulting in what appears to be an increase in trade size. The increase was due to the reporting data and not any increase in trade size. The most representative average trade size figures in this chart are the all trades average (computed from market data) and the NYSE reported NASDAQ values, since all NASDAQ trades and volume were reported to the tape. The NYSE does not run any crossing or auctions for NASDAQ listed stocks. One interesting observation that follows is that the average ETF trade size (441 shares in 2012) is almost double the average equity size

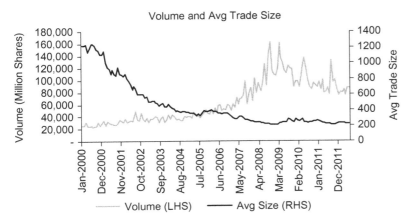

■ **Figure 1.5** Volume and Average Trade Size.

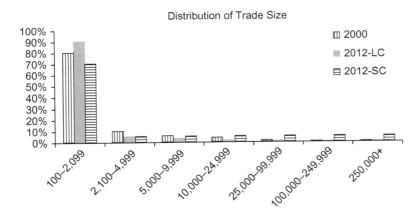

■ **Figure 1.6** Distribution of Trade Size.

(222 shares in 2012). A comparison of volume and trade size is shown in Figure 1.5. Notice the inverse relationship between the two. While volumes have increased, average trade size has declined, thus causing a higher number of trades and more work to complete an order. A comparison of the distribution of trade sizes in 2000 and in 2012 is shown in Figure 1.6.

The quantity of block trading activity on the NYSE has also decreased considerably (Figure 1.7). The decreased activity has greatly contributed to the smaller trade sizes and increased difficulty in completing an order. However, many of the challenges can be attributed to algorithmic trading. The percentage of block volume (10,000 shares or more) has decreased dramatically from 52% of total volume in 2000 to fewer than 20%

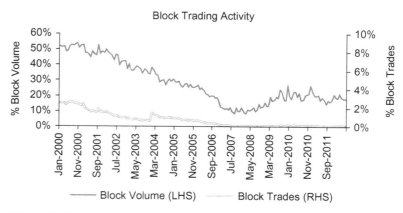

■ Figure 1.7 Block Trading Activity.

in 2012. The number of block transactions has decreased 95% from about 2.6% of total trades to 0.1% of total trades in 2012. Clearly, trading has become more difficult, measured from the number of transactions required to fill an order.

Our analysis period has seen a dramatic surge in program trading activity over 2000 through 2007. During this time, program trading increased from about 10% of total volume to almost 40%. Then in August 2007, there was a market correction that is believed to have been due to a high correlation across different quantitative strategies and which caused many quantitative funds to incur losses and lose assets under management. Subsequently, program trading activity declined through about September 2008. Moreover, due to the financial crisis many funds turned to program trading to facilitate their trading needs from a risk management perspective, which caused program trading to again increase through 2010 where it seems to have since leveled off at about 30%−35% of total market volume (Figure 1.8).

It is easy to see that the increase in program trading has also helped open the door for algorithmic trading. Program trading is defined as trading a basket of 15 or more stocks with a total market value of $1 million or more. Since 2000, program trading on the NYSE has increased 273% from 10% in 2000 to 36% in 2012. This more than threefold increase in program trading has been attributable to many factors. For example, institutional investors have shifted towards to embracing quantitative investment models (e.g., mean-variance optimization, long-short strategies, minimal tracking error portfolios, stock ranking model, etc.) where the model results are a list of stocks and corresponding share quantities, compared to the more traditional fundamental analysis that only recommends whether a stock should be

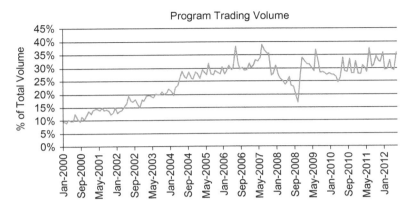

■ Figure 1.8 Program Trading Volume.

bought or sold without any insight into the number of shares to transact. The recent increase in program trading activity is also due to large broker-dealers offering program trading at lower commission rates than traditional block executions. Technological improvements (e.g., Super DOT on the NYSE, Super Montage on NASDAQ, and in-house trading systems) have made it much easier for brokers to execute large lists of stocks more efficiently. Now, combine the large increase in program trading with the large reduction in block volume and we begin to appreciate the recent difficultly in executing large positions and the need for more efficient implementation mechanisms. Execution and management of program trades have become much more efficient through the use of trading algorithms, and these program trading groups have pushed for the advancement and improvement of trading algorithms as a means to increase productivity.

RECENT GROWTH IN ALGORITHMIC TRADING

To best position themselves to address the changing market environment, investors have turned to algorithmic trading. Since computers are more efficient at digesting large quantities of information and data, more adept at performing complex calculations, and better able to react quickly to changing market conditions, they are extremely well suited for real-time trading in today's challenging market climate. Algorithmic trading became popular in the early 2000s. By 2005, it accounted for about 25% of total volume. The industry faced an acceleration of algorithmic trading (as well as a proliferation of actual trading algorithms) where volumes increased threefold to 75% in 2009. The rapid increase in activity was largely due to the increased difficulty investors faced executing orders. During the financial crisis, it was not uncommon to see stock price swings of 5−10% during the

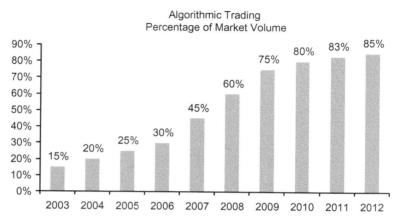

Algorithmic Trading
Percentage of Market Volume

■ Figure 1.9 Algorithmic Trading Percentage of Market Volume.

day, as well as the changing market environment (discussed above). These trends are shown in Figure 1.9. Over the years there have been various sources providing algorithmic trading estimates. For example, Tabb Group and Aite Group have published participation rates for buy-side algorithmic trading usage that are lower than our figures reported in this book. Our estimates include the execution's end product. So, even if the investor did not trade directly with an algorithm but did in fact route the algorithm to a broker who ultimately transacted the shares with an algorithm, those shares are included with the algorithmic trading volume figures.

The decade 2000–2010 was also associated with changing investor styles and market participants. We analyzed market participant order flow by several different categories of investors: traditional asset managers (including mutual funds, indexers, quantitative funds, and pension funds), retail investors, hedge funds (including statistical arbitrage and proprietary trading funds), market makers, and high frequency traders. In our definition, the high frequency trader only consisted of those investors considered liquid or rebate traders. We discuss the different types of high frequency trading below.

In 2003–2004 market volumes were led by asset managers, accounting for 40% of total volume. High frequency traders had almost negligible percentages in 2003 but grew to about 10% of the total market volumes in 2006. During the financial crisis, high frequency/rebate traders accounted for about 33% of volumes followed by hedge funds (21%). The biggest change we have observed over 2000–2012 is the decrease in asset manager volumes from 40% (2003) to about 23% (2012), and the increase in high frequency trading from 1–3% to about 30% of total volumes.

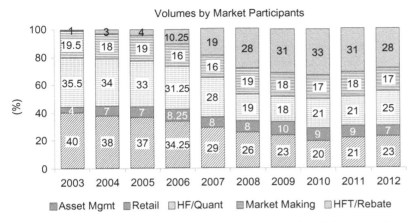

■ **Figure 1.10** Volumes by Market Participants (values in the graph may not add to 100 due to rounding).

Hedge fund trading volumes as a percentage of the total decreased slightly, but this is due to the increased competition across hedge funds, withdrawal of dollars, and a shift for some of these funds and managers from traditional hedge fund strategies into the ultra-short-term high frequency strategies. Hedge fund volume percentage also increased slightly in 2012. These changing volume percentages are illustrated in Figure 1.10.

Opponents of high frequency trading will often argue that it is disadvantageous to institutional and retail investors because it is causing trading costs to spike and liquidity to dryup. We have always found it humorous that whenever someone is trying to make a point or effect change to something that will benefit them, they pull retail into the mix and imply that the current structure is harming those retail investors in particular, so we must make changes in order to protect these mom and pop investors. Otherwise, these retail investors are almost entirely ignored. What is more likely in these situations is that the market has become increasingly competitive and high frequency traders have achieved higher levels of profitability than traditional investors. As such, they are pushing for change that will either favor their particular trading needs or investment style, or help them, albeit in an indirect way, by taking an advantage away from a competitor.

For example, during the period of increased high frequency trading activity (2006−2009) these funds were reported to be highly profitable. Conversely, over the same period the more traditional funds were not as profitable as they had been previously, especially in comparison to the high frequency traders. As a result, the belief (either correct or not) is that high frequency traders must be doing something that is harmful and detrimental to the general markets. Their trading must be toxic. A similar situation occurred back in the early 2000s when traditional hedge funds were highly profitable. They were greatly criticized at the time for being "fast money" and causing price dislocations due to their buying and selling pressure, which wasn't good for, you've guessed it, retail investors.

As a follow-up to the high frequency criticism, many financial pundits are stating that high frequency trading accounts for upwards of 50−70% of total market volumes. These are much different values than what our research has found. We estimated high frequency trading to account for only 33% of total volume. What is behind this difference? It just so happens that this difference is due to the definition of "high frequency" trader. These parties have grouped high frequency traders (our rebate trader definition) with market maker participants and some hedge fund traditional statistical arbitrage traders (including pairs trading, index arbitrage, and market neutral strategies).

Figure 1.11 shows a plot of the percentage of market volume for high frequency trading (HFT), market making (MM), and hedge funds (HF). Notice the increasing trend of HFT and the initial decrease in HF trading percentage, but with a recent increase. Market making (MM) appears to have been relatively steady, but has decreased from about 20% down to 17% of market volumes. Notice our maximum HFT market volume percentage of 33% in 2010. We additionally plot trends consisting of high frequency trading and market making (HFT + MM) and high frequency trading, market making, and hedge funds (HFT + MM + HF). Notice that at the peak of the financial crisis, 2008−2009, high frequency trading was only about 33% of total market share—a large discrepancy from the widely reported 50−70%. But the combination of HFT and MM

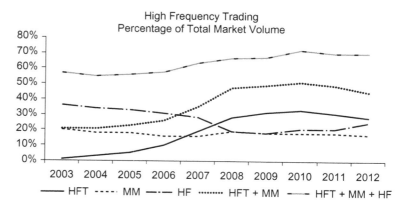

■ **Figure 1.11** High Frequency Trading Percentage of Total Market Volume.

percentage peaks at about 50% during the period and then tapers off in 2011–2012 when markets have become increasingly competitive and some high frequency traders have exited the business. Furthermore, as we add the hedge fund trader percentage to HFT and MM, we start approaching values of about 60–70%. Thus, the classification of the different types of traders is what accounts for the disparity in statistical figures.

An important issue to point out is that the actual market volume percentage of HFT + MM + HF has only increased slightly over 2003–2012 from 57 to 70%. So, it is more likely that the more nimble market makers and hedge fund traders turned to high frequency trading strategies rather than this class of investors just appearing overnight.

Algorithmic trading is currently one of the hottest areas of capital expenditure for Wall Street firms (both buy-side and sell-side). There are numerous conferences and seminars dedicated to algorithmic trading throughout the US, Europe, Asia, and Australia. Unfortunately, the amount of academic research has not kept pace with the surge in algorithmic trading. Most industry awareness regarding algorithmic trading has come from broker-dealers whose marketing information is mainly self-serving, with the main purpose being to increase order flow and business. There is a strong need for unbiased academic research and a well-tested decision making methodology. We seek to bridge the gap between academia and Wall Street.

INVESTMENT CYCLE

The investment cycle consists of four distinct phases: asset allocation, portfolio construction, implementation, and portfolio attribution. *Asset allocation* consists primarily of distributing investment dollars across stocks, bonds, cash, and other investment vehicles in order to achieve a target level of return within a specified level of risk exposure and tolerance. *Portfolio construction* consists primarily of selecting the actual instruments to hold in each asset class. *Implementation* has historically consisted of selecting an appropriate broker-dealer, type of execution (e.g., agency transaction or principal bid), and now includes specification algorithms and algorithmic trading rules. Portfolio managers evaluate the performance of the fund to distinguish between market movement and skilled decision making ability in the *portfolio attribution* phase of the cycle.

Until more recently, the vast majority of research (academic and practitioner) has focused on improved investment decisions. Investors have a large array of investment models to assist in asset allocation and portfolio construction. Unfortunately, investors do not have nearly the same quantity of trading tools to analyze implementation decisions. The quality of trading tools has changed significantly with the rise of portfolio trading tools and transition management. With the advent of algorithmic trading these tools are being developed further and gaining greater traction.

CLASSIFICATIONS OF ALGORITHMS

One of the more unfortunate events in the financial industry is the proliferation of the algorithmic nomenclature used to name trading algorithms. Brokers have used catchy names and phrases for the algorithms to have them stand out from competitors rather than using naming conventions that provide insight into what it is that the algorithm is trying to accomplish. While some of the industry algorithms do have logical, descriptive names, such as "VWAP," "TWAP," "Arrival Price," and "Implementation Shortfall," there are many others such as "Tarzan," "Bomber," "Lock and Load," and one of the all-time favorites "The Goods," although this name is soon to be replaced. None of these catchy names offer any insight into what it is that the algorithm is trying to accomplish or the actual underlying trading strategy.

As a way to shed some light on the naming convention used, we suggest classifying algorithms into one of three categories: Aggressive, Working Order, and Passive. These are as follows:

Aggressive: The aggressive family of algorithms (and sometimes hyper-aggressive strategies) are designed to complete the order with a high level of urgency and capture as much liquidity as possible at a specified price or better. These algorithms often use terminology such as "get me done," "sweep all at my price or better," "grab it," etc.

Working Order: The working order algorithms are the group of algorithms that look to balance the trade-off between cost and risk, as well as the management of appropriate order placement strategies through appropriate usage of limit/market orders. These algorithms consist of VWAP/TWAP, POV, implementation shortfall (IS), arrival price, etc.

Passive: The passive family of algorithms consists of those algorithms that seek to make large usage of crossing systems and dark pools. These algorithms are mostly designed to interact with order flow without leaving a market footprint. They execute a majority of their orders in the dark pools and crossing networks.

TYPES OF ALGORITHMS

Single Stock Algorithms: Single stock algorithms interact with the market based on user specified settings and will take advantage of favorable market conditions only when it is in the best interest of the order and the investor. Single stock algorithms are independent of one another while trading in the market and make decisions based solely on how those decisions will affect the individual order.

VWAP: Volume weighted average price. These algorithms participate in proportion with the intraday volume curve. If 5% of the day's volume trades in any specified period then the VWAP algorithm will transact 5% of the order in that period. The intraday volume profile used to follow a U-shaped pattern with more volume traded at the open and close than mid-day. But recently, intraday volume profiles have become more back-loaded and resemble more of a J-shaped pattern than U-shaped pattern. A VWAP strategy is a static strategy and will remain constant throughout the day.

TWAP: Time weighted average price. These algorithms execute orders following a constant participation rate through the entire day. A full day order will trade approximately 1/390th of the order in each 1 minute bucket (there are 390 minutes in the trading day in the US). It is important to note that many TWAP algorithms do not participate with volume in the opening and closing auctions since there is no mathematical method to determine the quantity of shares to enter into these auctions. In *Optimal Trading Strategies*, the TWAP curve was referred to as the uniform distribution or uniform strategy and was used for comparison purposes. A TWAP strategy is a static strategy and will remain constant throughout the day.

Volume: These strategies are referred to volume inline, percentage of volume (POV), of participation rate algorithms. These algorithms participate with market volume at a pre-specified rate such as 20% and will continue to trade until the entire order is completed. The algorithms will trade more shares in times of higher liquidity and fewer shares in times of lower liquidity, and thus react to market conditions (at least to changing volume profiles). One drawback to these volume strategies is that they do not guarantee completion of the order by the end of the time horizon. For example, if we are trading an order that comprises 20% of the day's volume at a POV = 20% rate but the actual volume on the day is only half of its normal volume, the order would not complete by the end of the day. As a safety around potential uncompleted orders, some brokers have offered a parameter to ensure completion by the end of the period. This parameter serves as a minimum POV rate and adjusts in real-time to ensure order completion by the designated end time.

Arrival Price: The arrival price algorithm has different meanings across different brokers and vendors. So it is important to speak with those parties to understand the exact specifications of these algorithms. But in general, the arrival price algorithm is a cost minimization strategy that is determined from an optimization that balances the trade-off between cost and risk (e.g., Almgren and Chriss, 1997). Users specify their level of risk aversion or trading urgency. The resulting solution to the optimization is known as the trade schedule or trade trajectory and is usually front-loaded. However, some parties solve this optimization based on a POV rate rather than a static schedule in order to take advantage of changing liquidity patterns.

Implementation Shortfall: The implementation shortfall algorithm is similar to the arrival price algorithm in many ways. First, its meaning varies across the different brokers and different vendors and so it is important to speak with those parties to understand their exact specifications. Second, we base the implementation shortfall algorithm on Perold's (1988) paper and seek to minimize cost through an optimization that balances the trade-off between cost and risk at a user specified level of risk aversion. In the early days of algorithms trading, the arrival price and implementation shortfall algorithms were identical across different brokers. Thus, to distinguish implementation shortfall from arrival price, brokers began to incorporate real-time adaptation tactics into the implementation shortfall logic. These rules specify how the initial solution will deviate from the optimally pre-scribed strategy in times of changing market liquidity patterns and market prices. Thus, while arrival price and implementation shortfall still do not have a standard definition across the industry, the general consensus is that the arrival price algorithm is constant while the implementation shortfall algorithm incorporates a second level of adaptation tactics based on market volumes and market prices.

Basket Algorithms: Basket algorithms, also known as portfolio algorithms, are algorithms that manage the trade-off between cost and total basket risk based on a user specified level of risk aversion. These algorithms will manage risk throughout the trading day and adapt to the changing market conditions based on user specifics. The algorithms are usually based on a multi-trade period optimization process. They make real-time trading decisions based on how those decisions will affect the overall performance of the basket. For example, a basket algorithm may choose to not accelerate trading in an order even when faced with available liquidity and favorable prices if doing so would increase the residual risk of the basket. Furthermore, the basket algorithm may be more aggressive in an order even in times of illiquidity and adverse price movement if doing so would result in a significant reduction of residual basket risk. The biggest difference between single stock and basket algorithms is that the basket algorithm will manage cost and total basket risk (correlation and covariance) whereas the single stock algorithm will seek to manage the cost and individual risk of the stock. Important basket trading constraints include cash balancing, self-financing, minimum and maximum participation rate.

Risk Aversion Parameter: The meaning of the risk aversion parameter used across the different brokers will vary. First, the optimization technique is not constant. For example, some parties will optimize the trade-off between cost and variance since it fits a straightforward quadratic optimization formulation. Others optimize based on the trade-off between cost and standard deviation (square root of variance) which results in a non-linear optimization formulation. Second, the definition of the risk aversion parameter, usually denoted by λ, varies. Some brokers specify $\lambda = dCost/dRisk$ where $\lambda > 0$. Some map λ to be between 0 and 1 (0 = most passive and 1 = most aggressive), and still others map risk aversion to be between 1 and 3 or 1 and 10. Thus selecting a value of $\lambda = 1$ could mean the most aggressive strategy, the most passive strategy, or somewhere in the middle. Still others use a qualitative measure such as passive, medium, aggressive, etc., rather than a specified value of λ. Investors need to discuss the meaning of the risk aversion parameter with their providers in order to determine how it should be specified in the optimization process in order to ensure consistency across trading goal and investment objective.

Black Box Algorithms: The family of black box trading algorithms are commonly referred to as profit and loss algorithms and/or robo trading algorithms. These include all algorithms that make investment decisions based on market signals and execute decisions in the marketplace. Unlike the implementation algorithms that are tasked with liquidating a predetermined position within some specified guidelines or rules, the black box algorithms monitor market events, prices, trading quantities, etc., for a profiting opportunity search. Once profiting opportunity appears in the market, the algorithm instantaneously buys/sells the shares. Many black box algorithms have time horizons varying from seconds to minutes, and some longer time horizons run from hours to days. While many investors use blackbox algorithms, they are still primarily tools of the quants, and especially when it comes to high frequency trading. Some black box trading algorithms are pairs-trading, auto market making, and statistical arbitrage. Black box trading strategies and the corresponding mathematics are discussed in detail in Chapter 13.

ALGORITHMIC TRADING TRENDS

Algorithmic usage patterns have also changed with the evolution of trading algorithms. In the beginning, algorithmic trading was mostly dominated by VWAP/TWAP trading that utilized a schedule to execute orders. The advantage: investors acquired a sound performance benchmark, VWAP, to use for comparison purposes. The improvement in algorithms and their ability to source liquidity and manage micro order placement strategies more efficiently led the way for price-based algorithms such as arrival price, implementation shortfall, and the Aggressive in the Money (AIM) and Passive in the Money (PIM) tactics. During the financial crisis, investors were more concerned with urgent trading and sourcing liquidity and many turned to "Liquidity Seeking" algorithms to avoid the high market exposure present during these times. The financial crisis resulted in higher market fragmentation owing to numerous venues and pricing strategies and a proliferation of dark pools. However, the industry is highly resourceful and quick to adapt. Firms developed internal crossing networks to match orders before being exposed to markets, providing cost benefits to investors, and incorporating much of the pricing logic and smart order capabilities into the liquidity seeking algorithms. Thus usage in these algos has remained relatively constant. Currently, liquidity seeking algos account for 36% of volumes, VWAP/TWAP 25%, volume 16%, arrival price/IS 10%, and portfolio algos 7%. This is shown in Figure 1.12.

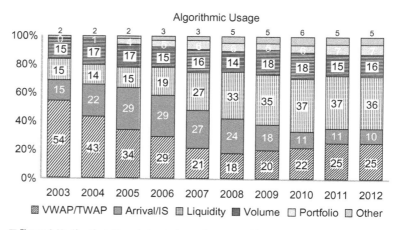

■ **Figure 1.12** Algorithmic Usage (values in the graph may not add to 100 due to rounding).

TRADING VENUE CLASSIFICATION
Displayed Market

A displayed exchange is a trading venue that discloses order book information. This consists of bid and offer prices, share quantities, and depth of book. Investors transacting in a displayed venue are able to see exactly how many shares are available at each price increment. This allows investors to compute expected transaction prices for a specified number of shares and also the expected wait time for a limit order to transact since they have knowledge where the order would sit in the queue and how many orders would need to transact ahead of them before their order will execute. Examples of displayed exchanges are the NYSE, NASDAQ, Chicago, CBOE, etc.

Dark Pool

A dark pool is a crossing network or other type of matching system that does not display or disseminate order information such as bids and offers, depth of book, number of orders, buy/sell imbalances, etc. Customers enter buy or sell orders into the dark pool. The order is executed only if there is a match. Dark pools do have drawbacks, however. These include no prior knowledge of order execution or where orders will sit in the order book queue. The dark pool's opaque/non-transparent nature makes it difficult for customers to determine if a market order or marketable limit order will execute at the specified prices. In addition it is problematic to calculate the likelihood that a limit order will be executed at a specified price increment since the customer does not know where it will sit in the queue. An advantage of the dark pool is that since order details are not disseminated there is no information leakage. Investors can enter large block orders without showing their hand to market participants. Dark pools also allow investors to cross at the midpoint of the bid-ask spread, and are maintained by brokers and third party vendors. Broker "internal dark pools" are used for matching internal and client orders away from the displayed exchanges. Third party dark pools, such as Liquidnet, Bids Trading, and Level ATS, provide investors with the opportunity to trade large block positions anonymously, thus reducing information leakage and market impact.

Grey Pool

The term grey pool denotes a displayed venue that allows investors to enter hidden orders and view displayed orders, prices, depth of book, etc., similar to the displayed exchanges. However, there may be another level of hidden orders on the order book transacting with incoming orders providing there is a match. For example, an investor sending a market order to a "grey pool" may transact at a price better than the NBBO if there is a hidden order in the venue at a better price.

Dark Pool Controversies

Historically there has been a large amount of debate surrounding dark pool executions, adverse selection, and toxic order flow. Adverse selection refers to situations when you use a dark pool and have the order executed fully (100%). Subsequent price movement is in your favor (e.g., buys become cheaper and sells become higher) so you would have been better off waiting to make the trade. And when you do not execute in the dark pool or execute less than the full order (<100%) the subsequent price movement is away from your order (e.g., buys become more expensive and sells become lower in value). The belief is that there is either some information leakage occurring in the dark pool or the interaction with high frequency orders is toxic, meaning that the high frequency traders are able to learn information about the order, such as the urgency of the investor or the number of shares that still need to be executed. In turn, they adjust their prices based on leaked knowledge. However, we have not yet found evidence of adverse selection in dark pools to confirm these suspicions.

But let us evaluate the above situation from the order level. Suppose we have a buy order for 100,000 shares and there is a seller with an order for 200,000 shares. Thus, there is a sell imbalance of −100,000 shares. If both parties enter the order into the crossing network (dark pool or other type of matching system) there will be 100,000 shares matched with the buy order being 100% filled and the sell order being only 50% filled. The seller will then need to transact another 100,000 shares in the market and the incrementing selling pressure will likely push the price down further due to the market impact cost of their order. So the downward price movement is caused by the market imbalance, not by the dark pool. Next, suppose that the seller only has 50,000 shares. Thus, there is a +50,000 buy imbalance. If these orders are entered into the crossing network, 50,000 shares of the buy order will match. The buyer will then need to transact the additional 50,000 shares in the displayed market where the buying pressure will likely push the price up further. Thus we can see that the adverse price movement is caused by the market imbalance and not the dark pool. This type of price movement is commonly observed in times of market imbalances.

To be fair, there was a time when dark pools and venues allowed flash orders to be entered into their systems. These flash orders would provide some market participants with a preview of whether there would be a match before the entire marketplace. Many believed that this provided an unfair advantage to those privileged to these flash orders. Flash trading is no longer allowed in any of the market venues.

TYPES OF ORDERS

The market allows numerous different types of orders such as market, limit, stop loss, etc. But the three most important order types for algorithmic trading are market, limit, and marketable limit orders.

Market Order: A market order specifies to the algorithm to buy or sell at the best available market price. This order is most likely to be executed because there are no restrictions on its price and it will not be placed into an order book. The disadvantage is that in today's markets, prices can move away so quickly that the best ask or best bid could in effect be much higher or much lower than they were at the time the order was routed for execution. Market order will "take" liquidity.

Limit Order: A limit order specifies to the algorithm to buy or sell at the specified limit price or better. In most cases the limit order will be entered into the order book of the exchange or venue and is subject to the queue before it is eligible to be executed. For example, in price-time priority, existing orders at that price or better will need to transact before that order with an offsetting buyer. A limit order is not guaranteed to execute, but provides some safety surrounding the execution price and ensures that the execution will not be worse than the pre-specified limit price. A limit order will "provide" liquidity and is also referred to as posting an order.

Marketable Limit Order: A marketable limit order is an order that specifies to the algorithm to buy or sell at a specified price or better. This order will either be executed in the market at the specified price or better, or be cancelled if there are no existing orders at that price or better in the market.

Rebates: Depending upon the exchange or venue, investors may receive a rebate for posting liquidity to the exchange and others may provide a rebate for taking liquidity from the exchange. Different rebate models are discussed further in Chapter 2.

EXECUTION OPTIONS

Here the investor provides the broker with the order or basket to trade in the market on their behalf. The broker exerts "best efforts" to achieve the best prices for the investor. They receive a commission for their role in the execution. The investor, however, incurs all market risk and price uncertainty. For example, if prices for a buy order increase during trading investors will receive less favorable prices, but if prices decrease investors will receive more favorable prices. Investors do not know what the prices will be in advance. The broker's profit in an agency execution will be the commission received less any applicable fees incurred during trading.

Principal bid. A principal bid, also known as a capital commitment or risk bid, is when the investor provides the broker with the order or basket, and the broker provides the investor with immediate executions at specified prices such as the day's closing price or the midpoint of the bid-ask spread at some agreed upon point in time. The broker charges the investor a premium (e.g., the principal bid) which is more than the standard agency commission fee. In this case, the investor transfers all risk and price uncertainty to the broker. If the broker is able to transact the acquired position or basket in the market at a lower cost than the principal bid premium they will make a profit, but if they incur a cost higher than the principal bid premium they will incur a loss. The advantage that brokers often have over investors in a principal bid transaction is that they have an inventory of customer order flow that could be used to offset the acquired position, or they may have potential hedging vehicles such as futures, ETFs, etc., that will allow them to trade more passively and incur lower costs. Quite often investors need to implement an investment decision within some specified time constraint which may lead to higher transactions costs. Brokers are not necessarily tied to these requirements.

A principal bid for an order can occur for a single stock order or a basket of stock. For a single stock order the broker will be provided with the name of the stock and shares to trade. Depending on the relationship between broker and investor, the broker may or may not be provided with the order side. The broker will then provide the investor with the principal bid for the order. If they are not provided with the side they may provide the investor with a two way market. Since the broker knows the exact stock, they are able to incorporate actual market events and company specific risk into the principal bid premium. For a basket principal bid, investors will often solicit bids from multiple brokers. To keep their trading intentions and actual orders hidden until they select the winning broker, they only provide the brokers with a snapshot of the trade list: includes average order size, trade list value, volatility, risk, tracking error, and depending upon their relationship, a breakdown by side, although the sides may simply be listed as side A and side B. Since the broker is not privileged to the actual names in the trade list, they incur a second level of risk.

Thus, they often factor in a buffer to their principal bid premium to account for not knowing the exact names or position sizes.

THE TRADING FLOOR

The equity trading operation at a broker dealer is primarily broken into three trading desks: cash, program, and electronic. Investors utilize these desks in different manners and for many different reasons. An overview of the primary functions is provided below.

Cash Trading: The cash trading desk, also known as the single stock or block desk, is utilized by investors who have orders subject to potential adverse price momentum, or when they have a strong alpha conviction or directional view of the stock. Traditionally, the block trading desk was used to transact large block orders and for capital commitment in favorable and adverse market conditions. Nowadays, investors additionally use block desks to transact single stock and multi-stock orders, large and small order sizes, in times of potential price movement. In these cases, investors rely on the expertise of block traders and their understanding of the stock, sector, and market, to determine the appropriate implementation strategy and timing of order placement. The cash desk has also historically been the desk where investors would route orders to pay for research and to accumulate credits for future investment banking allocations from IPOs and secondary offerings. Fundamental portfolio managers (e.g., stock pickers) who transact single stock positions are primary clients of the cash desk. We can summarize their trading goal as to minimize the combination of market impact cost and adverse price movement.

Program Trading: The program trading desk, also known as the portfolio trading desk, is used by investors to trade baskets of stocks. These baskets are also known as lists, programs, or portfolios. Investors will utilize a program trading desk primarily for risk management and cash balancing. In these cases, the portfolio manager does not typically have a strong short-term view of a stock and is concerned with the overall performance of the basket. They seek the expertise of program traders to determine the best way to manage the overall risk of the basket so that they can trade in a more passive manner and minimize market impact cost. In times of a two sided basket consisting of buys and sells, the program trader will trade into a hedged position to protect the investor from market movement. In times of a one-sided basket, the program trader will seek to offset orders and partial orders with the highest marginal contribution to risk. Very often these are the names with high idiosyncratic or company specific risk, pending news, or otherwise deemed as toxic due to liquidity or unstable trading patterns. Investors will transact with a program desk either via an agency execution or capital commitment. Other investors will solicit the expertise of a program trader when they are trading a basket where the sell orders will be financing the buy orders and wish to keep cash position balanced throughout the day so that they are not short cash at the end of the day. For program trades, the capital commitment is also known as a principal trade or risk bid. Quantitative portfolio managers are the primary clients of the program desk since these are the investors who more often trade baskets. Their primary trading objective is to minimize market impact and timing risk.

Electronic Trading: The electronic trading desk, also known as the algorithmic or "algo" desk, is the primary destination for investors who are seeking to capture liquidity, retain full control of the trading decision, remain anonymous, and minimize information leakage. Investors will often utilize an electronic desk when they are not anticipating any type of short-term price momentum. Here the primary goal of the investor is to gain access to the numerous market venues and be positioned to capture as much liquidity as they can within their price targets. Traditionally, the electronic trading desk was utilized for smaller orders, e.g., $\leq 1-3\%$ ADV, or what were believed to be "easy" trades. Now, investors use algorithms to trade both large and small orders, single stock orders and portfolios consisting of hundreds of names or more. Many investors do in fact use algorithms for their block and portfolio program trading needs, providing they have ample control over the execution of the algorithm and the algorithm is customizable to the investment objective of the fund. Electronic trading is performed on an agency basis only. The primary trading objective of these clients is to minimize market impact and opportunity cost—that is, to complete the entire order without adversely affecting market prices.

Research Function

The research function on the equity side also has three main segments and each is closely interconnected with each of the trading desks. These research roles are equity analyst, quantitative analyst, and transaction cost analyst.

Equity analysts evaluate individual companies using primarily fundamental data and balance sheet information. These analysts then provide ratings on the company such as buy, sell, hold, or short, or provide price targets or expected levels of return, based on their earnings and growth expectations. If a highly regarded analyst changes their rating on a stock, such as changing a sell rating to a buy rating, it is pretty likely that the stock price will move and move quickly right after the analyst's report is made public. Equity analysts do move stock prices and are considered the "rock stars" of investment research.

Quantitative analysts evaluate the relationship between various factors (both company and economic) and company returns. They use these factors to determine what is driving market returns (as opposed to company specific returns)—e.g., growth, value, quality, etc. Quantitative analysts determine optimal portfolios based on these relationships and their expectations of future market conditions. They also rely on optimization techniques, statistical analysis, and principal component analysis. However, unlike their equity analyst brethren, quantitative analysts do not move the market or cause volumes to increase. Portfolio managers do not typically incorporate recommendations from quantitative analysts directly into their portfolio. Instead, managers will use quantitative analysis for independent verification of their own findings, and as an idea generation group. Managers tend to rerun quantitative analyst studies to verify their results and to see if there is potential from their suggestions. Quantitative analysts are also used at times to run specified studies, evaluate specific factors, etc. In this role, they serve as an outsource consultant.

Transaction cost analysts are tasked with evaluating the performance of algorithms and making changes to the algorithms when appropriate. These analysts study actual market conditions, intraday trading patterns, and market impact cost. They perform market microstructure studies. The results and findings of these studies are incorporated into the underlying trading algorithms and pre-trade models that assist investors in determining appropriate trading algorithms. Unlike equity and quantitative analysts, transaction cost analysts do not make any stock or investment recommendations, and their research findings do not move stock prices. Buy-side traders rely on transaction cost analysts to understand current market conditions and the suite of trading algorithms.

Sales Function

The role of the sales person on the trading floor is to connect the buy-side client with sell-side research. There are three main areas of the selling function, which follows the research offerings described above. First, *equity sales*, also known as research or institutional sales, is responsible for providing the portfolio manager client with all company research. However, since the primary concern of the majority of portfolio managers is stock specific company research, the equity sales person focuses on providing their portfolio manager clients with equity analyst research. Since this is the research that could potentially move stock prices immediately, it has a high level of urgency. The *program sales trader* for the most part takes the lead in connecting their clients with quantitative research. Since they deal with these quant managers on a daily basis, they are well aware of their clients' research interests. Quant managers do not have the same sense of urgency in reviewing quant research, since this research is not company specific and will not move stock prices. Again, they are interested in quant research to verify their own findings, to gain insight into what is affecting the market, what approaches are working and not working, and for additional investment ideas. Buy-side quant managers will often re-check and verify the results of the sell-side quant research teams before they incorporate any of the findings into their portfolio. Transaction cost analysis (TCA) research, as mentioned, is not intended to provide managers with stock specific information, stock recommendations, or price targets. TCA research is performed to gain an understanding of the market. This information is then incorporated into the underlying trading algorithms and pre-trade analytics that are intended to assist investors in determining the appropriate algorithm for their order.

Consequently, electronic trading desks usually have a team of analysts that provide buy-side traders with transaction cost analysis research. This research will also provide insight into what algorithms or trading strategies are best suited for various market conditions. The primary client of TCA research is the buy-side trader, although recently a trend has emerged where portfolio managers (both fundamental and quantitative) are becoming interested in learning how to incorporate transaction costs into the portfolio construction phase of the investment cycle and uncover hidden value and performance. TCA is beginning to target managers as well as traders.

Table 1.1 Trading Floor Function

Desk	Primary Research	Sales
Cash	Equity	Institutional Sales Team
Program	Quant	Program Sales Traders
Electronic	TCA	Algo Sales Team

ALGORITHMIC TRADING DECISIONS

As the trading environment has become more complex and competitive, investors have turned to "efficient" algorithms for order execution and navigation. However, utilization of algorithms alone does not guarantee better performance. Investors need to become more proactive than a simple "set and forget" mindset. They need to specify an appropriate set of algorithmic trading rules and corresponding parameters, and most important, ensure that the implementation parameters are consistent with the overall investment objectives of the fund. Otherwise, it is unlikely that best execution will be achieved.

Proper specification of algorithmic parameters requires rules to be specified on a macro- and microscale. The macro-level decisions are specified by users prior to trading and consist of selecting the appropriate optimal trading strategy that is consistent with the investment objectives, and defining how the algorithms are to adapt in real-time to changing market conditions. The micro-level decisions consist of order submission rules and are made at the trade level by the algorithm level. These decisions are made through usage of limit order models and smart order routers.

To ensure "best execution," investors need to select those brokers and algorithms that can best align the micro-level trading with the user specified macro-level goals.

Macro-Level Strategies

The macro-level strategy decision rules consist of specifying an appropriate optimal trading strategy (e.g., order slicing schedule or percentage of volume rate) and real-time adaptation tactics that will take advantage of real-time market conditions such as liquidity and prices when appropriate. This type of decision making process is consistent with the framework introduced by Kyle (1985), Bertsimas and Lo (1998), Almgren and Chriss (1999, 2000), Kissell and Glantz (2003), and Kissell, Glantz, and Malamut (2004). For investors, macro-level trading specification consists of a three-step process:

1. Choose implementation benchmark
2. Select optimal execution strategy
3. Specify adaptation tactics

To best address these questions, investors need a thorough understanding of market impact, timing risk, and efficient intraday optimization. A detailed explanation of the algorithmic decision making process is provided in Chapter 8.

Step 1—Choose Implementation Benchmark

The first step of the macro decision process is selection of the implementation price benchmark. The more common price benchmarks include implementation shortfall (IS), decision price, price at order entry ("inline"), opening price, prior night's close, future closing price, and VWAP. Another common implementation goal is to minimize tracking error compared to some benchmark index. It is essential that the implementation goal be consistent with the manager's investment decision. For example, a value manager may desire execution at their decision price (i.e., the price used in the portfolio construction phase), a mutual fund manager may desire execution at the closing price to coincide with valuation of the fund, and an indexer may desire execution that achieves VWAP (e.g., to minimize market impact) or one that minimizes tracking error to their benchmark index.

Step 2—Select Optimal Execution Strategy

The second step of the process consists of determining the appropriate optimal execution strategy. This step is most often based on transaction cost analysis. Investors typically spend enormous resources estimating stock alphas and consider these models proprietary. Market impact estimates, however, remains the holy grail of transaction cost analysis and are usually provided by brokers due to the large quantity of data required for robust estimation. Risk estimates, on the other hand, can be supplied by investors, brokers, or a third party firm.

The selected optimal execution strategy could be defined in terms of a trade schedule (also referred to as slicing strategy, trade trajectory, waves), a percentage of volume ("POV"), as well as other types of liquidity participation or price target strategies. For example, trade as much as possible at a specified price or better.

The more advanced investor's will perform TCA optimization. This consists of running a cost-risk optimization where the cost component consist of price trend, and market impact cost. That is,

$$Min \quad (MI + Trend) + \lambda \cdot Risk$$

where λ is the investor's specified level of risk aversion. Investors who are more risk averse will set $\lambda > 1$ and investors who are less risk averse will set $\lambda < 1$. In situations where the trader does not have any expectations regarding price trend, our TCA optimization is written in terms of market impact cost and risk as follows:

$$Min \quad MI + \lambda \cdot Risk$$

Depending upon expected price trend, optimization may determine an appropriate front- and/or back-loading algorithm to take advantage of better prices. For example, an algorithm may call for a 15% POV rate in the morning, increasing to a POV of 25% beginning midday (back-loading) to take advantage of expected better prices in the afternoon while still balancing the trade-off between market impact and timing risk. Furthermore, an algorithm may call for a POV of 30% in the morning, falling to 10% in the afternoon (front-loading) as a means to reduce risk and hedge the trade list.

Solving the optimization problem described above for various levels of risk will result in numerous optimal trading strategies. Each has the lowest cost for the specified level of risk and the lowest risk for the specific amount of cost. The set of all these optimal strategies comprises the efficient trading frontier (ETF) first introduced by Almgren and Chriss (1999).

After computing the ETF, investors will determine the most appropriate "optimal" strategy for their implementation goal. For example, informed traders with expectations regarding future price movement are likely to select an aggressive strategy (e.g., POV = 30%) with higher cost but more certainty surrounding expected transaction prices. Indexers are likely to select a passive strategy (e.g., POV = 5%) or a risk neutral strategy to reduce cost. Some investors may select a strategy that balances the trade-off between cost and risk depending upon their level of risk aversion, and others may elect to participate with volume throughout the day (e.g., VWAP strategy).

It is important that investors thoroughly evaluate alternative strategies to determine the one that provides the highest likelihood of achieving their investment goal. Figure 1.13 shows four different "optimal" strategies.

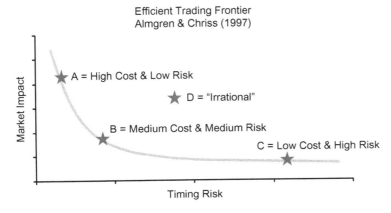

■ **Figure 1.13** Efficient Trading Frontier.
Almgren and Chriss (1997)

The aggressive strategy (A) corresponds to higher cost but low risk. The normal strategy (B) is associated with a mid-level of cost and a mid-level of risk. The passive strategy (C) is a low cost strategy but does have higher risk. Selection of a strategy such as D would not be appropriate because there are alternative strategies with the same risk but lower cost, lower risk at the same cost, or both lower cost and lower risk. A strategy such as D is deemed an irrational trading strategy and will never provide best execution.

Step 3—Specify Adaptation Tactic

The next step in the decision process consists of specifying how the algorithm is to adapt to changing market conditions. Algorithms may also include multiple types of adaptation tactics. Below are some common adaptation settings. Mathematical techniques behind these adaptation tactics are provided in Chapter 8.

Volume-based. Adjust the trading schedule based on market liquidity. A POV/participation rate algorithm is an example of a volume-based adaptation tactic. While these are often constant volume rates, they result in faster trading in times of higher market volume and slower trading in times of lower market volume.

Price-based. Adjust the trading schedule based on market prices. Aggressive-in-the-Money (AIM) algorithms that trade faster in times of favorable prices and slower in times of adverse price movement, and Passive-in-the-Money (PIM) algorithms that trade slower in times of favorable prices and faster in times of adverse price movement are types of price-based scaling algorithms.

Time-based. The algorithm adjusts its trading rate to ensure executions by a specified time, such as no later than the close. This algorithm may well finish sooner than specified but will not finish any later.

Probabilistic. Determines the appropriate trading rate to always provide the best chances (highest likelihood) of achieving the investment objective. It is based on a non-linear optimization technique, such as maximizing Sharpe ratio or minimizing tracking error.

Optimization Technique. The trade schedule is continuously adjusted so that its expected finishing price will be within a specified tolerance. These types of algorithms will often be based on a variation of a z-score measure (see Chapter 9) and incorporate realized costs (past), sunk cost or savings (dependent upon price increase or decrease since commencement of trading), and expected future price (based on actual market conditions and specified trading strategy).

Cash Balancing. In times of trading investors often select cash balancing adaptation techniques. The most common variations of cash balancing are risk management and self-financing. Risk management adaptation techniques will manage the unexecuted positions to keep the risk within a specified tolerance level. Self-financing adaptation techniques are used when the sells will be used to pay for the buys. Here we are managing the shares that have already traded, and depending upon market prices and movement, may require the unexecuted buy shares to be revised (either increasing or decreasing) depending on actual prices.

Dark Pool Utilization. Investors may elect to use dark pools in a different manner than the displayed venues. For example, investors may choose to try to maximize trading in dark pools but keep trading in the displayed markets at a different rate. Furthermore, the participation in dark pools and displayed venues may also be determined by stock prices or market movement.

Micro-Level Decisions

The micro-level pricing decisions, as stated above, consist of the order submission rules. The actual decision models are embedded directly into the trading algorithms and utilize limit order models and smart order routers. These decision points are not entered by the user directly as with the macro decisions, but the algorithms and decision path are to ensure that executions adhere to the higher level macro goals entered by the investor. The goal of a micro-level scheme is threefold. First, to ensure the executions follow the optimally prescribed strategy entered by the user. Second, to ensure that the algorithms deviate from the optimally prescribed strategy only when it is in the best interest of and defined by the investor. Third, to achieve fair and reasonable prices without incurring unnecessary market impact cost. In situations where the fixed costs and exchange costs are high, optimizing, crossing, and micromanagement on each exchange can also lead to substantial cost savings.

It is essential that the micro pricing strategy ensures consistency with the macro-level objective and ensures transactions adhere to the specified implementation goal. For example, it would not be in the best interest of the fund to execute an aggressive strategy (e.g., POV = 40%) using solely limit orders, because execution with limit orders is not guaranteed execution and there is a high likelihood that this type of strategy may fall behind the targeted rate. But it would be appropriate to transact a passive macro strategy (e.g., POV = 5%) utilizing a larger number of limit orders to avoid crossing the bid-ask spread, since there is ample time for these limit orders to be lifted in the market.

In most situations it will be appropriate to use a combination of limit, market, floats and reserve orders. For example, suppose the specified macro-level optimal strategy is a POV rate of 15%. Here a micro-level algorithm may submit limit orders to the market for execution better than the mid-quote for as long as the actual POV rate is consistent with 15% of market volume, but once the algorithm starts lagging behind the specified rate or some specified tolerance level it would submit appropriately sized and spaced market orders to be more aggressive and adhere to the 15% rate. A reserve order could also be used to automatically replenish limit orders at favorable prices. Some of the more advanced micro pricing strategies utilize real-time data, prices and quotes, order book, and recent trading activity to forecast very short-term price trends and provide probabilistic estimates surrounding the likelihood that a limit order will execute within a certain period of time.

Limit Order Models

Limit order models determine the appropriate mix of limit and market orders to best adhere to the higher level macro goals. The limit order model is a probabilistic model that takes into account current market conditions, price momentum, order book information, macro goal and timing. Traditionally, limit order models will determine the probability that an order will execute in the market at a stated price and within a stated amount of time. The limit order model here is a modified limit order model with the output being a mix of prices and share quantities to ensure completion by the end of the time period (or at least a high enough likelihood of completion) rather than a probability estimate of executing at a specified price point.

For example, if the optimal trading rate specified in the macro-level decision for a buy order is POV = 10% and we forecast 10,000 shares will be traded in the next 1 minute, then we will need to execute 1000 shares in the next 1 minute to adhere to our 10% POV rate. If the current market is $30.00–$30.10 the limit order model may determine that the most cost effective mix of prices, order type, and share quantity to trade these 1000 shares in the next 1 minute is:

- Limit order, 200 shares at $29.95
- Limit order, 300 shares at $30.00
- Limit order, 300 shares at $30.05
- Market order, 200 shares at $30.10

Smart Order Routers

The smart order router (SOR) is responsible for routing the child orders to the different exchanges, venues, and dark pools. The SOR will collect, monitor, and maintain trading activity data at the different venues and dark pools throughout the day using market/public data and in-house transactions. The SOR determines the likelihood of executing an order at each of the different venues based on frequency of trading and where the order would reside in the order book queue at that price. If the trading frequency and all else is equal across two venues, the SOR will route the limit order to the venue where it will sit highest in the queue. If one venue has 10,000 shares at the desired price and another venue has 5000 shares at the desired price, the SOR will route the shares to the venue with only 5000 shares since it is more likely to execute at that location quicker. The SOR determines on an expected value basis the likelihood of trading so it may route the order to a venue where it initially sits lower in the queue if that venue has higher trading activity or it may route the order to a venue that does not trade as frequently as others if it sits higher in the queue. For the most part this is an expected value calculation.

Revisiting our example above, where we are tasked with trading 1000 shares in the next 1 minute with the best mix of limit and market prices being: 200 @ $29.95, 300 @ $39.00, 300 @ $30.05, and 200 @ $30.10 (the offer), the SOR may decide to enter 200 shares in a dark pool at the midpoint of the bid-ask spread ($30.05), 300 shares at the primary exchange at the best bid price of $30.00. Furthermore, 200 shares may be entered into a non-traditional exchange @ $29.95 where it sits first in the queue and perhaps even having to pay a rebate for posting (inverted pricing model).

Finally, the SOR may determine that to avoid potentially falling behind the schedule and having to possibly trade 200 shares at the market at the end of the 1 minute period where the price may move away (increase for the buy order) it would be best to trade 100 shares at the market immediately and 100 shares after perhaps 30–45 seconds, thus increasing the likelihood that they will not fall behind and have to trade in an aggressive and costly manner to catch up.

One reason why someone might decide to pay to post an order is to ensure that they will be the first 200 shares to trade at that price (if the market falls). Rational investors would route an order to a venue where they will receive a rebate to trade over a venue where they have to pay to trade. Here the savings achieved by transacting at the better price will more than make up for the rebate that has to be paid for posting liquidity.

Another reason why someone may pay a rebate to post is that rather than have to increment the bid price to be first in the queue (at a 1 cent increment), the investor may decide to enter the order to an exchange where they pay to post (inverted pricing model) to almost ensure themselves top of the queue because the counterparty would rather transact with their order and receive a rebate than transact with another order on a different venue and pay a fee. Here, the rebate cost of posting the order will be lower than the 1 cent that they would need to increment the bid even after accounting for any rebate they may receive.

In times of the NYSE and NASDAQ (well really ARCA and INET for algorithms) order routing was a much easier problem. But now with multiple destinations, venues, dark pools, etc., this is a much more complex problem.

In addition to maintaining order routing data across the various exchanges and computing the probabilities of executions, the SOR is also responsible for all order submission rules. The more common pricing rules as mentioned are market, limit, marketable limit, and floating prices that are pegged to a reference price such as the bid, ask, or midpoint and change with the reference price, etc. Varying these order types allows the algorithm to adhere to the optimally prescribed strategy by executing aggressively (i.e., market orders) and/or passively (i.e., limit orders) when needed. Order sizes are set in quantities that can be easily absorbed by the market. The order type variation disguises the actual trading intentions (i.e., limit orders) and minimizes potential adverse price impact (i.e., market orders). A reserve (iceberg) order is another technique commonly used in micro pricing strategies and refers to a continuously replenishing order at a stated size. For example, a 10,000 share order could be entered as a 1000 share reserve book order where 1000 shares would be displayed and immediately replenished each time it transacts until the order is filled. Finally, the SOR maintains, randomizes, and varies wait and cancellation times to help disguise trading intentions. Some orders may remain in the market for longer periods of time while other orders remain on the book for a shorter period of time before cancellation. Additionally, randomizing the time between orders and waves helps hide trading intentions, minimizes information leakage, and helps improve the likelihood of achieving more favorable prices.

An important note that is often overlooked in the algorithmic trading arena is that a smart order router should only be used as a smart order router. Many vendors have made the mistake of forcing the smart order router to provide a combination of services such as limit order model, macro strategy selection model, etc. These parties have tried to market these services as an all-in-one algorithmic smart order solution. Without understanding the macro-level needs of the investor or their trading goals, it is simply not possible to provide this type of all-in-one solution. The best in class solutions to these algorithmic issues have followed our algorithmic decision making process: trading goal, adaptation tactic, limit order model, order submission rules, not the all-in-one type of solution.

ALGORITHMIC ANALYSIS TOOLS
Pre-Trade Analysis

The first step in developing an algorithmic trading strategy is to perform pre-trade analysis. This provides investors with the necessary data to make informed trading decisions on both the macro- and micro-levels, and serves as input into the algorithms. Pre-trade analysis provides investors with liquidity summaries, cost and risk estimates, and trading difficulty indicators as a means to screen which orders can be successfully implemented via algorithmic trading and which orders require manual intervention. It also provides potential risk reduction and hedging opportunities to further improve algorithmic execution strategies. Pre-trade data is comprised of current prices and quotes, liquidity and risk statistics, momentum, and an account of recent trading activity. This also provides investors with necessary data to develop short-term alpha models.

Intraday Analysis

Intraday analysis is used to monitor trading performance during trading. These systems will commonly provide in real-time the number of shares executed, the realized costs for those executed shares, the price movement since trading began (which translates to either a sunk cost or savings), and the expected market impact cost and timing risk for the remaining shares based on the implementation strategy and expected market conditions (which could be different from those expected at the beginning of trading). Some of the more advanced intraday analysis systems will provide z-score estimates, which are the projected risk adjusted trading costs for all shares (based on strategy and market conditions), as well as comparisons to projected final trading costs for various different algorithms and strategies. The intraday analysis systems are used by traders to evaluate market conditions and make revisions to their algorithms.

Post-Trade Analysis

Algorithmic post-trade analysis is a two part process that consists of cost measurement and algorithm performance analysis. First, cost is measured as the difference between the actual realized execution price and the specified benchmark price. This allows investors to critique the accuracy of the trading cost model to improve future cost estimates and macro strategy decisions, and provide managers with higher quality price information to improve investment decisions. Second, algorithmic performance is analyzed to assess the ability of the algorithm to adhere to the optimally prescribed strategy, its ability to achieve fair and reasonable prices, and determine if the algorithm deviates from the optimally specified strategy in an appropriate manner. Investors must continuously perform post-trade analysis to ensure brokers are delivering as advertised, and question those executions that are out of line with pre-trade cost estimates.

Rule-Based Trading

Rule-based trading algorithms control the macro-level decisions such as order slicing strategies that break larger orders into smaller pieces to trade over time in order to reduce market impact cost. These instructions, however, are based on simple logic and heuristics, with many of the guidelines being completely arbitrary (e.g., participate with volume over a specified period, trade faster when prices are favorable, trade slower in times of adverse trends, etc.). Furthermore, many of these so-called rule-based algorithms do not provide insight into potential costs or associated risk. Furthermore, they do not provide necessary transparency to determine the most appropriate algorithm given the objectives of the fund. For the most part, rule-based trading algorithms are a "black box" approach to trading.

Quantitative Techniques

Quantitative algorithmic trading also controls the macro-level decisions but differs from rule-based trading in that all decisions are based on robust statistical models and a sound mathematical framework. For example, quantitative models serve as the basis for developing strategies to minimize the difference between expected execution price and a specified benchmark price, to minimize cost subject to a maximum level of risk exposure, or to maximize the probability of transacting more favorably than a specified benchmark price. Quantitative algorithms manage overall transaction costs (e.g., market impact, price momentum, and timing risk), and provide investors with necessary transparency regarding cost and risk, and insight into how the algorithm will behave in times of changing prices or varying market conditions. This transparency also allows investors to evaluate alternative strategies (algorithms) and determine the most appropriate "optimal" strategy (algorithm) given the underlying trading goal and investment objective.

HIGH FREQUENCY TRADING

High frequency trading (HFT) is the usage of sophisticated mathematical techniques and high speed computers to trade stocks, bonds, or options with the goal to earn a profit. This differs from the execution trading algorithms that are tasked with implementing an investment decision that has previously been determined. In other words, the HFT system makes both the investment and trading decisions simultaneously. High frequency trading in this sense is also called "black box" and "robo" trading.

HFT strategies can be classified into three different styles: Auto Market Making (AMM), Quant Trading/Statistical Arbitrage, and Rebate/Liquidity Trading. Donefer (2010) provides a similar classification in "Algos Gone Wild," Journal of Trading (Spring 2010). There is often some overlap across these styles as we point out below, but for the most part, each of these styles has completely different goals and objectives. In short, high frequency trading has these features:

- *Automated trading.* Algorithms determine what to buy and what to sell, as well as the micro order placement strategies such as price, size, and timing of the trade. These decisions are determined from actual real-time market data including price signals, momentum, index or sector movement, volatility, liquidity, and order book information. These decisions are made independent of human interaction.
- *No net investment.* HFT does not require a large cash inflow since the inventory imbalances are netted out by the close each day. HFT strategies take both long and short positions in different names and close these position before the end of day so that they do not take on any overnight risk. In cases where the HFT holds overnight positions they will mostly likely use the proceeds from short sales to pay for the buys.
- *Short trading horizons.* Depending upon the strategies, HFT time horizons can vary from seconds to minutes, but also up to hours.

Auto Market Making

Auto market making (AMM) provides the financial community with the same services as the traditional market makers or specialists. The main difference, however, is that rather than employing human market makers the AMM system uses advanced computer systems to enter quotes and facilitate trades. The registered AMM still has an obligation to maintain a fair and orderly market, provide liquidity when needed, and provide market quotes a specified percentage of the time.

AMM systems automatically enter bids and offers into the market. After the AMM system transacts with a market participant they become either long or short shares and they will seek to offset any acquired position through further usage of limit orders. The AMM systems look to profit from buying at the bid and selling at the offer and earning the full spread. And as an incremental incentive, registered auto market maker firms are often provided an incremental rebate for providing liquidity. Therefore, they can profit on the spread plus rebates provided by the exchange. This is also causing some difficulty for portfolio managers seeking to navigate the price discovery process and determine fair value market prices.

AMM black box trading models will also include an alpha model to help forecast short-term price movement to assist them in determining the optimal holding period before they are forced to liquidate an acquired position to avoid a loss. For example, suppose the bid-ask spread is $30.00 to $30.05 and the AMM system bought 10,000 shares of stock RLK at the bid price of $30.00. If the alpha forecast expects prices to rise the AMM will offer the shares at the ask price of $30.05 or possibly higher in order to earn the full spread of $0.05/share or possibly more. However, if the alpha forecast expects prices to fall, the AMM system may offer the shares at a lower price such as $30.04 to move to the top of the queue or if the signal is very strong the AMM systems may cross the spread and sell the shares at $30.00 and thus not earn a profit, but not incur a loss either.

Most AMM traders prefer to net out all their positions by the end of the day so that they do not hold any overnight risk. But they are not under any obligation to do so. They may keep positions open (overnight) if they are properly managing the overall risk of their book or if they anticipate future offsetting trades/orders (e.g., they will maintain an inventory of stock for future trading). Traditional AMM participants continue to be concerned about transacting with an informed investor, as always, but it has become more problematic with electronic trading since it is more difficult to infer if the other side is informed (has strong alpha or directional view) or uninformed (e.g., they could be a passive indexer required to hold those number of shares), since the counterparty's identity is unknown.

Some of the main differences between AMM and traditional MM are that AMM maintains a much smaller inventory position, executes smaller sizes, and auto market makers are not committing capital for large trades as the traditional market makers once did.

Quantitative Trading/Statistical Arbitrage

Traditional statistical arbitrage trading is trying to profit between a mispricing in different markets, in indexes, or even ETFs. Additionally, statistical arbitrage trading strategies in the high frequency sense will try to determine profiting opportunities from stocks that are expected to increase or decrease in value, or at least increase or decrease in value compared to another stock or group of stocks (e.g., relative returns). Utilizing short time frame "long-short" strategies relies on real-time market data and quote information, as well as other statistical models (such as PCA, probit and logit models, etc.). These traders do not necessarily seek to close out all positions by the end of the day in order to limit overnight risk, because they are based on alpha expectations and the profit and loss is expected to be derived from the alpha strategies, not entirely from the bid-offer spread. This is the traditional statistical arbitrage strategy in the past, but the time horizon could be much shorter now due to potential opportunity, better real-time data, and faster connectivity and computational speeds. This category of trading could also include technical analysis based strategies as well as quant models (pairs, cointegration). These types of trading strategies have traditionally been considered as short-term or medium-term strategies, but due to algorithmic and electronic trading, and access to an abundance of real-time data and faster computers, these strategies have become much more short-term, reduced to hours or minutes, and are now also considered as HFT strategies. However, they do not necessarily need to be that short-term or an HFT strategy. These participants are less constrained by the holding period of the positions (time) and most concerned by the expected alpha of the strategy.

Rebate/Liquidity Trading

This is the type of trading strategy that relies primarily on market order flow information and other information that can be inferred or perceived from market order flow and real-time pricing, including trades, quotes, depth of book, etc. These strategies include "pinging" and/or "flash" orders, and a strong utilization of dark pools and crossing venues (e.g., non-traditional trading venues). Many of these non-traditional trading venues have structures (such as the usage of flash orders) that may allow certain parties to have access to some information before other parties. These traders seek to infer buying and selling pressure in the market based on expected order flow and hope to profit from this information. The liquidity trading strategies can be summarized as those strategies that seek to profit through inefficient market information. What is meant by this is the information that can be inferred, retrieved, processed, computed, compiled, etc., from market data to generate a buy or sell signal, through the use of quick systems and better computers, infrastructure, location of servers, etc., co-location, pinging, indications of interest (IOIs), flash orders. The "liquidity trading" HFT is often the category of HFT that is subject to the most scrutiny and questions in the market. Market participants are worried that these strategies have an unfair advantage through the co-location, available order types, ability to decipher signals, etc. These participants counter argue that they adhere to the same market rules and have an advantage due to their programming skills or mathematical skills, better computers, connectivity (e.g., supercomputers) and co-location of servers, which are available to all market participants (albeit for a cost).

Another variation of the rebate trader is an opportunistic AMM. This is again similar to the AMM and the traditional market-making role, but the opportunistic trader is not under any obligations to provide liquidity or maintain a fair and orderly market. These market participants will provide or take liquidity at their determined price levels, as they are not required to continuously post bids and offers, or maintain an orderly market. Since they are not registered or under any obligations to provide liquidity, these parties do not receive any special rebates that are made available to the registered AMM. This party tends to employ alpha models to determine the best price for the stocks (e.g., theoretical fair value models) and corresponding bids and offers to take advantage of market prices—they only tend to provide quotes when it is in their best interest to do so and when there is sufficient opportunity to achieve a profit. If prices are moving away from them, they may no longer keep a market quote. As a result, they may only have a quote on one side of the market, or will quickly close the position via a market order to avoid potential adverse price movement. These parties expect to profit via the bid-ask spread (similar to the tradition AMM participants) as well as via market rebates and alpha signals. But unlike traditional AMM participants, the rebates and alpha signals are a primary P/L opportunity. They only perform the AMM function when these signals are in their favor, and they do not have any obligation to continuously provide market quotes. The opportunistic AMM participants are more likely to net and close their positions by the end of the day because they do not want to hold any overnight risk even if they are well hedged. Furthermore, the opportunistic AMM participants are not willing to hold any inventory of stock in anticipation of future order flow. But they will hold an inventory (usually small) of stock (either long or short) based on their alpha signal—which is usually very short-term (before the end of the day). They often close or net their positions through market orders, and do so especially when they can lock in a profit. Additionally, some of the opportunistic AMM may continuously net positions throughout the day so that they keep very little cash exposure. These parties also try to profit via rebates, and utilize limit order models (and other statistical models relying on real-time data) to infer buying and selling pressure and their preferred prices.

DIRECT MARKET ACCESS

Direct market access or "DMA" is a term used in the financial industry to describe the situation where the trader utilizes the broker's technology and infrastructure to connect to the various exchanges, trading venues, and dark pools. The buy-side trader is responsible for programming all algorithmic trading rules on their end when utilizing the broker for direct market access. Often funds combine DMA services with broker algorithms to have a larger number of execution options at their disposal.

Brokers typically provide DMA to their clients for a reduced commission rate but do not provide the buy-side trader with any guidance on structuring the macro- or micro-level strategies (limit order strategies and smart order routing decisions). Investors utilizing DMA are required to specify all slicing and pricing schemes, as well as the selection of appropriate pools of liquidity on their own.

In the DMA arena, the buy-side investor is responsible for specifying:

1. *Macro trading rules.* Specify the optimal trading time and/or trading rate of the order.
2. *Adaptation tactics.* Rules to determine when to accelerate or decelerate trading, based on market prices, volume levels, realized costs, etc.
3. *Limit order strategies.* How should the order be sliced into small pieces and traded in the market, e.g., market or limit order, and if limit order, at what price and how many shares.
4. *Smart order routing logic.* Where should orders be posted, displayed or dark, how long should we wait before revising the price or changing destination, how to best take advantage of rebates.

The investor then takes advantage of the broker's DMA connectivity to route the orders and child orders based on these sets of rules. Under DMA, the investor is in a way renting the broker's advanced trading platforms, exchange connectivity, and market gateways.

Many broker networks have been developed with the high frequency trader in mind and are well equipped to handle large amounts of data, messages, and volume. The infrastructure is built on a flexible ultra-low latency FIX platform. Some of these brokers also provide smart order routing access, as they are often better prepared to monitor and evaluate level II data, order book queues, and trading flows and executions by venue in real-time.

Advantages

- *Lower Commissions.* Brokers are paid a fee by the fund to compensate them for their infrastructure and connectivity to exchanges, trading venues, dark pools, etc. This fee is usually lower than the standard commission fee and the fund does not receive any additional benefit from the broker such as order management services, risk management controls, etc.
- *Anonymity.* Orders are entered into the system and managed by the trader. Brokers do not see or have access to the orders.
- *Control.* Traders have full control over the order. Traders determine the venues (displayed/dark), order submission rules such as market/limit prices, share quantities, wait and refresh times, as well as when to accelerate or decelerate trading based on the investment objective of the fund and actual market conditions. Information leakage is minimized since the broker does not receive any information about the order or trading intentions of the investor.
- *Access.* Access to the markets via the broker's technology and infrastructure. This includes co-location, low latency connections, etc.
- *Perfectly Customized Strategies.* Since the investor defines the exact algorithmic trading rules, they are positioned to ensure the strategy is exactly consistent with their underlying investment and alpha expectations. Funds rarely (if ever) provide brokers with proprietary alpha estimates.

Disadvantages

- *Increased Work.* Funds need to continuously test and evaluate their algorithms, write and rewrite codes, develop their own limit order models and smart order routers.
- *Lack of Economies of Scale.* Most funds do not have access to the large number and breadth of orders entered by all customers. Therefore, they do not have as large a data sample to test new and alternative algorithms. Brokers can invest substantial resources in an algorithmic

undertaking since the investment cost will be recovered over numerous investors. Funds incur the entire development cost themselves.

- *Research Requirements.* Need to continuously perform their own research to determine what works well and under what types of market conditions.
- *Locked-Into Existing Systems.* Difficult and time consuming to rewrite code and redefine algorithms rules for all the potential market conditions and whenever there is a structural change in the market or to a trading venue. However, many traders who utilize DMA also have the option of utilizing broker suites of algorithms (for a higher commission rate). The main exception in this case is the high frequency traders.
- *Monitoring.* Need to continuously monitor market conditions, order book, prices, etc., which could be extremely data intensive.

Market Microstructure

INTRODUCTION

Market microstructure is the study of financial markets and how they operate. Market microstructure research primarily focuses on the structure of exchanges and trading venues (e.g. displayed and dark), the price discovery process, determinants of spreads and quotes, intraday trading behavior, and transaction costs. Market microstructure continues to be one of the fastest growing fields of financial research due to the rapid development of algorithmic and electronic trading.

As a result of the high speed evolution of financial markets in the last decade, today's number of trading venues and exchanges have mushroomed, with more on the horizon, and trading processes have evolved exponentially in complexity and sophistication. Traditional trading functions and participants have been surpassed by computers and electronic trading agents. Human intervention in the trading process has expanded from the traditional matching and routing of orders to one requiring complex analysis and sophisticated real-time decision making. The market microstructure analyst is now tasked with understanding all issues surrounding the ever-changing marketplace.

Most of the academic research up until now has focused on valuation techniques to uncover the fair market price of an instrument, expected return forecasts, and risk modeling techniques. Analysts may employ a bottom-up approach to determine the fair value price of a company by examining the company's balance sheet, fundamentals, sales figures, year-over-year growth, and revenue forecasts. Analysts may also utilize a top-down approach to determine which factors, sectors, or other subgroups will likely under- or overperform going forward. Additionally, analysts may utilize a quantitative process, such as the capital asset pricing model (CAPM) or arbitrage pricing theory (APT) to perform asset allocation and determine optimal portfolio mixes.

The Science of Algorithmic Trading and Portfolio Management. DOI: http://dx.doi.org/10.1016/B978-0-12-401689-7.00002-7

While the bottom-up, top-down, and quantitative approaches provide valuable insight into the market's assessment of fair value and appropriate stocks to hold in the portfolio, these techniques do not provide insight into how to measure and incorporate investors' subjective assessment and preference of the securities or how markets will react to the arrival of "new" information. Without taking into account these preferences, portfolio managers could transact at unfavorable prices, causing a drag on performance; a common reason why funds underperform their peers. As will be a theme in this chapter, it is the role of the market microstructure researcher/analyst to provide insight and valuable information to traders and portfolio managers so that the best possible market prices are captured—thus reducing the performance drag and improving portfolio returns.

One of the main roles of financial markets is to provide investors with a fair and transparent price discovery and liquidity snapshot. Unlike formal economic theory suggests, financial markets are far from frictionless. In fact, it is this inefficiency that often leads to costly implementation. As such, market microstructure analysts are tasked with understanding not only the price discovery process and market liquidity, but also how prices will change with the arrival of new information and competing customer orders.

Historically, companies would either list on a traditional exchange (listed) such as the New York Stock Exchange (NYSE) or trade in an over-the-counter-market (OTC) such as the National Association of Securities Dealers Automatic Quotation market (NASDAQ). Stocks listed on the NYSE would trade via a system utilizing a sole specialist, whose job was to match customer orders, keep a fair and orderly market, provide liquidity, and maintain a central order book consisting of all customer buy and sell orders. Specialists would have access to all orders residing in the central order, allowing them to determine the overall buy-sell imbalance and establish a fair value price based on market information. Specialists would disseminate the best bid and best ask prices to the public marketplace.

Stocks listing in the OTC market would trade via NASDAQ's dealer-based system with multiple market makers. Each market maker provides their best bid and ask price to a central quotation system. Customers wishing to trade would route their order to the market maker with the best quoted price. Through competition with multiple parties vying for orders, customers were almost ensured that they would receive the best available fair prices. Similar to specialists, market makers were responsible for maintaining a fair and orderly market, providing liquidity, and disseminating prices. But unlike specialists, market makers were not privileged to the entire market buy-sell imbalance, as they only had access to their central order book consisting of their customer orders. The privilege differences provided some difficulties at times in establishing fair value market prices due to the fragmentation of orders. Still, the competitive system played a large role in determining fair market prices.

Times have changed, and the underlying trading process is far from the simple process of yesteryear. There are more than thirteen displayed trading venues including exchanges and alternative trading systems (ATSs) and from thirty to forty electronic crossing networks (ECNs) which are known as electronic communication networks. To make matters more difficult, the rules of engagement for each venue differ with several different pricing and queuing models. For example, maker-taker, inverted, commission based, price-time, price-size, etc.

To keep abreast of the rapidly changing market structure, many funds have created an in-house market microstructure analyst role and some have hired a team of analysts. In order to develop the most efficient and profitable trading algorithms and systems, market participants need to fully understand the inner workings of each venue and how it affects their orders, and ultimately their bottom line.

We provide an overview of the different venues and provide empirical data relating to intraday trading patterns such as spreads, volumes, volatility, and trading stability. The chapter concludes with a discussion of the market microstructure surrounding special event days, with special attention paid to the "Flash Crash" of May 6, 2010 and what really happened during the day.

In the remainder of this chapter, we describe the current state of the equity markets and review some of the important market microstructure research papers that have not only provided the groundwork for the evolution of exchanges, but also provided a foundation for trading algorithms, high frequency trading, and ultimately improved stock selection and portfolio construction.

Readers are encouraged to study these research works and empirical findings as if it were a final exam.

MARKET MICROSTRUCTURE LITERATURE

Academia has been performing market microstructure research since the 1970s. And while the research has expanded immensely, the current work

Table 2.1 A Brief History of Market Microstructure Research and TCA	
Era	**General Interest**
1970−1990	Spreads, Quotes, Price Evolution, Risk Premium
1990−2000	Transaction Costs, Slippage, Cost Measurement, Friction
2000−2010	Algorithms, Pre-Trade, Black Box Models, Optimal Trading Strategies
2010−	Market Fragmentation, High Frequency, Multi-Assets, and Portfolio Construction

that is being incorporated into trading algorithms and black box models is based on many of the earlier groundbreaking works.

How exactly has market microstructure research evolved over time? First, research in the 1970s and 1980s focused mainly on spreads, quotes, and price evolution. In the following decade, 1990−2000, interest turned to transaction cost analysis but focused on slippage and cost measurement. The following period, 2000−2010, was marked with decimalization and Reg-NMS, and research turned to algorithmic trading. Studies focused on pre-trade analysis in order to improve trading decision and computer execution rules, and the development of optimal trading strategies for single stocks and portfolios. The 2010 era so far has been marked with market fragmentation, high frequency trading, algorithmic trading in multi-asset classes, and exchange traded fund research. And many portfolio managers are studying how best to incorporate transaction costs into stock selection and portfolio construction. These eras of market microstructure research are shown in Table 2.1.

To help analysts navigate the vast amount of market microstructure research and find an appropriate starting point, we have highlighted some sources that we have found particularly useful. First, the gold standard is the research by Ananth Madhavan in two papers: "Market Microstructure: A Survey" (2000) and "Market Microstructure: A Practitioner's Guide" (2002). *Algorithmic Trading Strategies* (Kissell, 2006) provides a literature review in Chapter 1, and *Algorithmic Trading & DMA* by Barry Johnson (2010) provides in-depth summaries of research techniques being incorporated into today's trading algorithms for both order execution and black box high frequency trading. Finally, Institutional Investor's Journal of Trading has become the standard for cutting edge academic and practitioner research.

Table 2.2 Spread Cost Analysis

Category	Payment	Study
Order Processing	Order processing fee Service provided	Tinic (1972) Demetz (1968)
Inventory Cost	Risk-reward for holding inventory	Garmen (1976) Stoll (1978) Ho and Stoll (1981) Madhavan and Sofianos (1998) Amidhud and Mendelson (1980) Treynor (1971) Copeland and Galai (1983) Gosten and Harris (1988) Huang and Stoll (1997) Easley and O'Hara (1982, 1987) Kyle (1985)
Adverse Selection	Payment for transacting with informed investors Payment for transacting with investor with private information	

Some of the more important research papers for various transaction cost topics and algorithmic trading needs are shown in the following tables. Table 2.2 provides important insight into spread cost components. These papers provide the fundamental needs for order placement and smart order routing logic. Table 2.3 provides important findings pertaining to trade cost measurement and trading cost analysis. These papers provide a framework for measuring, evaluating, and comparing trading costs across brokers, trading, and algorithms. Table 2.4 provides insight into pre-trade cost estimation. These papers have provided the groundwork for trading cost and market impact models across multi-assets as well as providing insight into developing dynamic algorithmic trading rules. Finally, Table 2.5 provides an overview of different trade cost optimization techniques for algorithmic trading and portfolio construction.

THE NEW MARKET STRUCTURE

Over the last few years, there have been some dramatic and significant changes to the structure of the equities markets. Some of the changes were in response to changing regulations and the electronic environment, while some have been due to changing investor preference and investment style. We are certain these will continue to evolve over time.

Table 2.3 Cost Measurement Analysis

Study	Type	Observation
Perold (1988)	Value Line Fund	17.5% Annual
Wagner and Edwards (1993)	Total Order Cost	
	Liquidity Demanders	5.78%
	Neutral Investors	2.19%
	Liquidity Suppliers	−1.31%
Beebower and Priest (1980)	Trades	
	Buys	−0.12%
	Sells	0.15%
Loeb (1983)	Blocks	
	Small Sizes	1.1−17.3%
	Large Sizes	2.1−25.4%
Holthausen, Leftwich and Mayer (1999)	Trades	
	Buys	0.33%
	Sells	0.40%
Chan and Lakonishok (1993)	Trades	
	Buys	0.34%
	Sells	−0.04%
Chan and Lakonishok (1995)	Orders	
	Buys	0.98%
	Sells	0.35%
Chan and Lakonishok (1997)	Orders	
	NYSE	1.01−2.30%
	NASDAQ	0.77−2.45%
Keim and Madhavan (1995)	Block Orders	3−5%
Keim and Madhavan (1997)	Institutional Orders	0.20−2.57%
Plexus Group (2000)	Market Impact	0.33%
	Delay	0.53%
	Opportunity Cost	0.16%
Kraus and Stoll (1972)	Blocks	1.41%
Lakonishok, Shleifer, and Vishny (1992)		1.30−2.60%
Malkiel (1995)		0.43−1.83%
Haung and Stoll (1996)	NYSE	25.8 cents
	NASDAQ	49.2 cents
Wagner (2003)		0.28−1.07%
Conrad, Johnson and Wahal (2003)		0.28−0.66%
Berkowitz, Logue and Noser (1988)	VWAP Cost	5 bp
Wagner (1975)	Trading Costs	0.25 − 2.00% +

Note: Positive value indicates a cost and negative value indicates a saving.

Table 2.4 Cost Estimation

Study	Structure	Factors
Chan and Lakonishok (1997)	Linear	Size, Volatility, Trade Time, Log(Price)
Keim and Madhavan (1997)	Log-Linear	Size, Mkt Cap, Style, Price
Barra (1997)	Non-Linear	Size, Volume, Trading Intensity, Elasticity, etc.
Bertsimas and Lo (1998)	Linear	Size, Market Conditions, Private Information
Almgren and Chriss (2000)	Non-Linear	Size, Volume, Sequence of Trade
Breen, Hoodrisk, and Korajczyk (2002)	Linear	14 Factors - Size, Volatility, Volume, etc.
Kissell and Glantz (2003)	Non-Linear	Size, Volatility, Mkt Conditions, Seq. of Trades
Lillo, Farmer, and Mantegna (2004)	Non-Linear	Order Size, Mkt Cap

Table 2.5 Trade Cost Optimization

Study	Optimization Technique
Bertsimas and Lo (1998)	Price Impact and Private Information
Almgren and Chriss (2000)	Risk Aversion, Value at Risk
Kissell and Glantz (2003)	Risk Aversion, Value at Risk, Price Improvement
Kissell and Malamut (2006)	Trading and Investing Consistency

Historically, companies were either listed on an exchange such as the New York Stock Exchange (NYSE) and traded primarily on that stock exchange floor or listed on the over-the-counter (OTC) market and traded primarily in the National Association of Securities Dealers Automated Quotation (NASDAQ) market. The NYSE market was primarily comprised of established blue chip companies; companies with strong balance sheets, low volatility, and a long history of earnings and dividends. They were household names such as GE, IBM, Ford, AT&T, etc. The NASDAQ market was primarily comprised of newer and smaller companies and attracted technology companies such as MSFT, WorldCom, Red Hat, Ebay, Amazon, etc. Many of these smaller companies at the time of their IPO did not have the large balance sheets or a long history of earnings or dividends, thus making their stocks more volatile in general.

Listed securities transacted more than 90% of their volume on the NYSE. Customers would use brokers to help route orders to the specialist. Limit orders would be entered into the specialist's central order book which included all customer limit orders from all member brokers. They would execute based on priority determined by price and time (price-time priority). Market orders routed to the specialist would transact with the best market prices in the limit order book. Buy market orders would transact with the lowest priced ask or offer in the limit book. Sell market orders would transact with the highest priced bid in the limit order book. If there were multiple "Bids" or "Asks" at the same prices, the orders which have been in the order book the longest amount of time transacted first.

NASDAQ listed (OTC) stocks traded in a market maker system. Each market maker would provide the market with their best bid and ask price and stand ready to transact the specified number of shares at the specified prices. Market makers, like the specialist, were required to maintain a stable and orderly market. Customers would enter orders with their brokers, who could also display these orders to the market as a customer limit order and have the shares and price shown in the quote. While price-time priority was still the rule for routed orders, brokers could transact with the market maker of their choice. Due to the fragmentation of the market surrounding multiple market makers, a market maker who just increased their bid price could receive an order over a market maker who has displayed that better price for a significantly longer amount of time. However, the competition surrounding multiple market makers led to better prices and a more efficient market—competition at its best. The equity markets are much different now compared to the days of the NYSE and a single specialist model, or NASDAQ with multiple market makers, even though parts of each still exist today. Now, instead of two trading exchanges, there are numerous trading venues. Currently, there are thirteen trading venues consisting of exchanges, alternative trading systems (ATS), and electronic communication networks (ECNs) also known as electronic crossing networks. At current count, there are in excess of thirty dark pools including third party and broker-owned internal dark pools which are used to match internal order flow.

Looking at the sheer number of potential venues shows that there could be an enormous amount of fragmentation and inefficient pricing if analysts and traders do not have a full understanding of how these venues behave and operate. This has further created the need for market microstructure analysts.

Going back to the circa 1995−2005 era, an important distinguishing characteristic between listed and NASDAQ companies was that listed companies were usually the larger, more established companies. They had strong balance sheets and a long history of earnings and dividends, they had lower price volatility, and more stable trading patterns, partly due to the role played by a single intermediary—the specialist. NASDAQ companies were usually the smaller and less established companies. They did not have as strong balance sheets as the listed companies and they certainly did not have as long a corporate earnings and dividend paying history (some did not even have any history of profits at all). In addition, these companies had higher price volatility, less stable intraday trading patterns, and higher company specific risk.

At that time fewer analysts covered NASDAQ companies. Many of these companies were also new era technology stocks so techniques used to value them were not as developed as they are today. Analysts at times would force the traditional valuation techniques for companies with historical earnings, profits, sales figures, etc., such as GE, AT&T, IBM, etc. But these approaches would break down for companies that did not have any profits or pay dividends. How could we value a company using a discounted dividend model if it had never paid out any dividends? How could we forecast sales if the company is based on an idea without any previous sales, or forecast performance in a new industry such as internet shopping, where there was no history whether or not consumers would embrace the approach? While at the time we all knew these concepts would take off and be embraced, we did not know which companies would excel and which would falter.

What we did observe during that time was that company characteristics had a great effect on the underlying market impact cost of an order. And they still do!

1995–2005 triggered a great deal of discussion surrounding which exchange (NYSE or NASDAQ) was more efficient and provided investors with lower costs. But the reason was not because of the NYSE structure having lower costs than NASDAQ. In actuality, it was due to the majority of NYSE companies with trading characteristics that caused lower trading costs, such as lower volatility, larger market cap, more stable intraday trading patterns, lower company specific risk, and were followed by a larger number of equity analysts—producing less potential for informed trader surprises. As mentioned, the majority of NASDAQ companies at that time were more volatile, with smaller market cap, less stable trading patterns, a greater quantity of company specific risk—creating greater potential for informed trader surprises than NYSE companies. However, when we isolated these characteristics across exchanges, and held trade size and strategy constant, we did not find any statistical difference across costs. Stocks on both exchanges had very similar cost structures and we concluded that costs were similar across the NYSE and NASDAQ (after adjusting for trading characteristics). In today's markets, with numerous exchanges and trading venues, we find the same patterns; costs are related to trading characteristics. Different exchanges provide different needs for investors. We did not find that any one structure was less efficient or more costly than another. One factor, however, that stuck out as having a very large correlation with cost is liquidity. Both historically and in today's markets, we have found that those venues with greater liquidity were associated with lower trading costs.

PRICING MODELS

Currently there are three different pricing models used by trading venues: maker-taker, taker-maker, also known as inverted pricing models, and commission based. All of the different models are put forth to attract as much liquidity as possible to their exchange.

Maker-Taker Model. In the maker-taker model, the maker is the investor posting and providing liquidity. In return, the maker is paid a rebate to provide liquidity to the exchange, and the taker of liquidity is charged a larger transaction fee than the rebate paid to the maker. The rebate is only provided if a transaction takes place. Thus, investors are incentivized to provide liquidity to the particular exchange or venue.

Taker-Maker (Inverted) Model. In the taker-maker model, the investor posting the order is charged a fee for supplying liquidity and the investor taking the liquidity is provided a rebate. Why would investors pay to provide liquidity when they could enter liquidity on another venue and receive a rebate for this service? The answer is simple. Suppose the investor is at the end of a long price-time priority queue. The investor could increment their bid or offer price costing them a full price increment. However, a better option that allows them to jump to the top of the queue is to place the order on a taker-maker exchange, where the rebate charged would still be less than the price increment. Hence, liquidity taking investors have two options; one being charged a fee to take liquidity and another being paid to take liquidity. At the same price, rational investors would always select the option where they would be paid a rebate as opposed to having to pay a rebate. The taker-maker model allows investors to jump to the front of the line for a small fee. As long as this fee is less than the full price increment (less any expected rebate the investor is expected to receive from the transaction) they would select the pay to post option. The option proves valuable in situations where investors are looking to improve the best market price or utilize a market order where they would pay the entire spread and rebate in addition to crossing the spread.

Commission. In the commission-based model, both the liquidity provider and liquidity supplier are charged a fee for transacting with liquidity. This was the original pricing model of exchanges, but currently attracts the least amount of interest. Investors could place an order on a commission-based venue and they could jump to the front of the queue (similar to with the taker-maker model). Although there is no incentive for the liquidity taker to transact with that particular exchange over a maker-taker exchange unless the commission fee is less than the rebate. Commission fee structures are popular with dark pools and crossing networks where investors are allowed to transact within the bid-ask spread, thus receiving better prices even after commissions paid.

ORDER PRIORITY

Currently there are two types of priority models in use: "price-time" priority and "price-size" priority. In price-time models, orders receive execution priority based on the time the order was entered into the venue. Orders are sorted based on price and then time so that the best priced and longest standing orders are executed first. In price-size models, orders are sorted based on price and then order size so that the largest orders are executed first. This incentivizes investors to enter larger orders as a way to move to the front of the line, rather than increase their price or submit orders to a taker-maker model. Investors with large orders are often encouraged to utilize the price-size priority model.

EQUITY EXCHANGES

The landscape of trading exchanges has changed dramatically since the early days of the NYSE and NASDAQ. There are four exchange groups operating ten exchanges with an additional three independent venues, and dark pools. These groups are (Table 2.6):

- NYSE/EuroNext: NYSE, ARCA, AMEX
- NASDAQ/OMX: NASDAQ, BSX, PSX
- Bats: BYX, BZX
- DirectEdge: EDGA, EDGX
- Independents: NSX, CBOE, CBSX
- Finra: TRF (trade reporting facilities)

NEW NYSE TRADING MODEL

The NYSE model has changed dramatically from what is depicted in television and movies with crowds of people huddled around a central figure yelling and screaming orders at one another and manically

Table 2.6 Trading Venue Pricing Rules

Parent	Name	Short	Pricing Rules			Priority Trading Rules	
			Maker-Taker	Taker-Maker	Per Share	Price-Time	Price-Size
NYSE EuroNext	New York Stock Exchange	NYSE	✓			✓	
	American Stock Exchange	AMEX	✓			✓	
	Archipelago	ARCA	✓			✓	
NASDAQ OMX	NASDAQ	NASDAQ	✓			✓	
	NASDAQ/Philly	PSX	✓				✓
	NASDAQ/Boston	BX		✓		✓	
Bats	Bats X	BZX	✓			✓	
	Bats Y	BYX		✓		✓	
DirectEdge	Direct Edge A	EDGA	✓			✓	
	Direct Edge X	EDGX	✓			✓	
Regionals	National Stock Exchange	NSX	✓			✓	
	Chicago Board Options Exchange	CBSX			✓	✓	
	Chicago Exchange	CHX	✓			✓	
Finra	Finra TRF	TRF			✓	✓	

Source: NYSE, as of 1Q-2012.

gesturing hand signals. Much of this change has been forced by the evolution algorithmic and electronic trading.

The NYSE antiquated specialist system has been replaced with a model that features a physical auction managed by designated market makers (DMMs) and a completely automated auction that includes algorithmic quotes from DMMs and other market participants. The old NYSE specialist system has evolved into a system comprised of three key market participants: DMMs Supplemental Liquidity Providers (SLPs), and Trading Floor Brokers. Each participant's role is described as follows:

Designated Market Makers

Each stock traded on the NYSE has a single designated market maker (DMM) whose role has superseded the specialist. The role of the DMM is to:

- Maintain fair and orderly markets for their assigned securities.
- Quote at the national best bid and offer (NBBO) for a specified percentage of the time.

Table 2.7 Comparison of Specialist vs. Designated Market Maker (DMM)

Role	Specialist (Old)	Designated Market Maker (New)
Trading Responsibility	Agency responsibility, visibility to all incoming orders	Market maker with quoting obligation, no advanced "look"
Priority	Yield on all orders	Parity
Obligations	Affirmative and negative obligations, open, close	Affirmative obligations, open, close
Technology	S-quotes, SAPI	Addition of Capital Commitment Schedule (CCS)
Economics	Economic incentive	Economic incentive tied to providing liquidity

- Facilitate price discovery during the day, market open and close, periods of trading imbalances, and times of high volatility.
- Provide liquidity and price improvement to the market.
- Match orders based on a pre-programmed capital commitment schedule.

DMMs will no longer be the agent for orders on the Display Book, and their algorithms do not receive a "look" at incoming orders. This ensures fairness and that the DMMs compete as a market participant. In exchange for these obligations and restrictions, DMMs will be:

- Parity with incoming orders.
- Permitted to integrate their floor-based trading operations into a related member firm, while subject to strict information barriers.

DMMs will have their performance periodically reviewed, and will receive transparent economic incentives based on performance (Table 2.7).

Supplemental Liquidity Providers

Supplemental liquidity providers (SLPs) are high volume trading members who provide liquidity to their NYSE stocks. Unlike the DMM, where there is only a single member designated to each stock, multiple SLPs may be assigned to a stock. The rules of engagement for the SLP are:

- Maintain a bid or offer at the NBBO in each assigned security for at least 10% of the trading day.
- Trade only for their proprietary accounts. They are prohibited from trading for public customers or on an agency basis.
- Cannot act as a DMM and SLP for the same stock.
- Will not receive any unique information (unlike a DMM).

The SLP program rewards aggressive liquidity suppliers who complement and add competition to existing quote providers. SLPs who post liquidity in their assigned security and execute against incoming orders will be awarded a financial rebate by the NYSE.

Trading Floor Brokers

Floor brokers of the NYSE exchange will continue to trade customer orders on parity with other orders. They are positioned at the point of sale during the market open and close, as well as during any intraday situation that requires unique handlings to execute orders (such as halts and restarts). Floor brokers are now equipped with state-of-the art computer systems, handheld order management devices, and trading algorithms that provide parity with DMMs and the NYSE Display Book. The revised role of the floor broker now provides:

- Trading customer orders parity with other market participants with the parity benefit being integrated into algorithms designed specifically for the NYSE environment.
- Ability to offer customers the competitive benefits of algorithmic speed and strategies benchmarked against the NBBO, directly from the NYSE point of sale.
- A new technology "Block Talk" enables NYSE floor brokers to more efficiently locate deep liquidity. Block Talk allows floor brokers to broadcast and subscribe to specific stocks, thus creating an opportunity to trade block-sized liquidity not otherwise accessible electronically.
- These messages do not contain any specific order information, thus minimizing information leakage.

Source:

http://www.nyse.com/pdfs/fact_sheet_dmm.pdfhttp://www.nyse.com/pdfs/03allocation_policy_instructions_wd.pdf

http://usequities.nyx.com/listings/dmms

NASDAQ SELECT MARKET MAKER PROGRAM

NASDAQ has also implemented a new market maker program, now known as the NASDAQ Select Market Maker Program (SMMP). The program has been implemented to encourage market making firms to provide liquidity at the NBBO, which in turn will improve the price discovery process and ensure stable or orderly markets. Firms that achieve select market maker designation status will be provided with increased visibility to the issuer community (CEO/CFO/IR Officers) through: NASDAQ Online, NASDAQ Market Intelligence Desk, and also through NASDAQ sponsored events.

Table 2.8 NASDAQ Select Market Maker Program

	ADV	Requirement
Tier 1	<1,000,000 Shares	NBBO 15% or more
Tier 2	≥1,000,000 Shares	NBBO 10% or more

Source: http://www.nasdaqtrader.com/content/ProductsServices/Trading/SelectMarketMaker/smm_factsheet.pdf.

NASDAQ select market maker designations are available for NASDAQ and non-NASDAQ listed securities. All registered market maker firms are eligible for select status and there is no limit to the number of select market makers in a particular stock. NASDAQ does not grant any special trading privileges to select market makers.

The program is as follows:

Securities are split into one of two tiers of stock. Tier 1 includes securities with an average daily trading volume (ADV) of less than 1 million shares per day. Tier 2 includes securities with an ADV greater than 1 million shares per day. For tier 1 stocks select market maker needs to display orders at the NBBO at least 15% of the time. For tier 2 stocks select market maker needs to display orders at the NBBO at least 10% of the time. Select status is based on the previous month's quoting statistics on a stock by stock basis (Table 2.8).

EMPIRICAL EVIDENCE

In this section we examine the underlying data to see if the market microstructure findings from a two exchange system (NYSE and NASDAQ) still hold true. Later in Chapter 7, Advanced Algorithmic Forecasting Techniques, we provide techniques needed to best forecast volumes and intraday profiles (among other things), as well as offer insight into how we can best utilize this information when developing algorithms and specifying trading strategies. Our analysis period was 1Q-2012.

Trading Volumes
Market Share

The equity markets are dominated by four major regional and crossing exchange groups. Of these, the NYSE has the largest market share with 24.5% of total share volume. NASDAQ is second with 21.6% of total

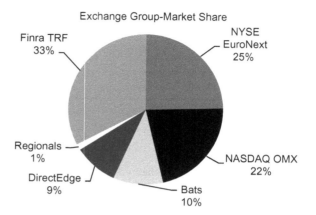

■ Figure 2.1 Market Share.

share volume. Bats and DirectEdge are third and fourth, respectively. For the period analyzed we found Bats with 10.4% and DirectEdge with 9.3% of stock share volume. DirectEdge and Bats have been pretty consistent with each having about 9%–10% of total market share. The regional exchanges, surprisingly, only account for about 1.3% of total share volume. Finra TRF volume consists of the dark pools and crossing networks matches including both third party networks and internal broker dealing crossing systems. These trades account for about 33% of total share volume traded in the market (Figure 2.1).

Large and Small Cap Trading

NYSE listed large cap (LC) stocks account for a much larger percentage of market share than NYSE listed small cap (SC) stocks. NYSE LC stocks account for 77% of market volumes and NYSE SC stocks account for about 23% of market volumes. The breakdown of NASDAQ market share is not as skewed for large cap as it is for NYSE stocks. NASDAQ large cap stocks account for 52% of the market volume and NASDAQ small cap stocks account for 48% of the market volume. It is important to point out that there is not a difference across exchange. The NYSE trades more volume in large cap compared to small cap, and vice versa for NASDAQ. Historically, the smaller, less established companies would list on the NASDAQ exchange and would often migrate to the NYSE exchange as they grew and matured. This trend, however, is beginning to change, as we see some large cap NYSE listed companies moving to the NASDAQ exchange. For example, Texas Instruments (Dec. 2011) and Kraft (Jun. 2012) moved from the NYSE to NASDAQ.

Do Stocks Trade Differently Across the Exchanges and Venues?

We next analyzed volumes across all trading venues. Stocks were segmented by exchange listing and then by market cap to determine if there is any material difference to how they trade.

For NYSE stocks we found very little difference between large and small cap stocks. Small cap stocks did trade slightly more in the NYSE/AMEX market (Specialist/DMM model) and in the crossing networks (Finra-TRF). Large caps traded a little more in the alternative exchanges, NASDAQ, and ARCA. For NASDAQ stocks, we once again found similar trading activity across large and small cap stocks. Large cap stocks traded a little more on NASDAQ and ARCA and small cap stocks traded slightly more in the crossing networks.

We next analyzed the difference across market cap stocks and how they trade in the different venues. We found that the listed large cap stocks traded much more on the NYSE/AMEX venue as expected, since these are primarily the stocks that are listed on the NYSE exchange, and the NASDAQ large cap stocks traded more in the NASDAQ venue as well as in the alternative exchanges and ARCA. Similar results were found for small cap stocks. Listed small cap stocks traded a larger percentage in NYSE/AMEX and NASDAQ small cap stocks traded a larger percentage in NASDAQ. The trading volumes across the different alternative exchanges were similar but with NASDAQ small caps trading slightly more in the alternative exchanges and in the dark pools (Figure 2.2).

Volume Distribution Statistics

Investigation into stock trading patterns uncovered some interesting volume distribution properties, namely, the difference in trading volume across market capitalization. As expected, large cap stocks trade a much higher number of shares than small cap stocks. Large cap stocks traded an average of 6.1 million shares per day and small cap stocks traded an average of 496,000 shares per day. NASDAQ large cap stocks (7.4 million) trade more daily volume than NYSE listed large cap stocks (5.8 million). NYSE listed small cap stocks 639,000 trade more daily volume than NASDAQ listed small cap stocks (400,000).

The difference across trading volume by large cap is likely to be driven by sector and information content. First, NASDAQ is more dominated by large cap technology and internet stocks than the NYSE. These companies are also more likely to be growth companies or companies with a

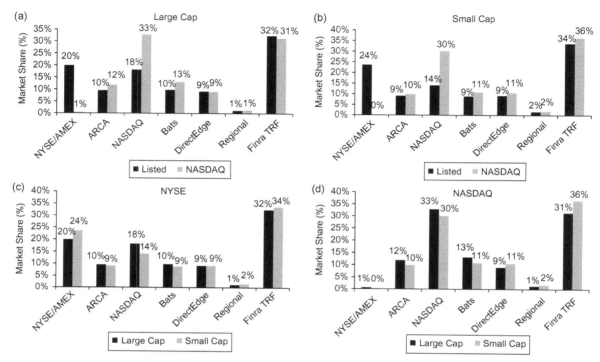

■ **Figure 2.2** Percentage of Share Volume.

high return potential, thus making them attractive to those seeking higher returns. Since NASDAQ stocks are traded more via market makers there is some double counting in the names. Additionally, the NASDAQ technology and internet stocks are becoming a favorite of retail investors, since they are known names and it is sexier to trade technology and internet stocks than, say, utilities. NYSE small cap stocks are believed to trade more volume than NASDAQ small cap stocks because these are the more established companies with a history of earnings and company specific fundamentals. Furthermore, NYSE small cap stocks are usually larger than NASDAQ small cap stocks. Finally, the designated market maker model has lower spreads that result in lower trading costs and hence make NYSE small cap stocks a more attractive investment opportunity than NASDAQ small cap stocks. Another interesting daily volume property is that the median daily volume is much lower than the average daily volume measure. The ratio of median volume to average volume is NYSE-LC = 0.48, NASDAQ-LC = 0.38, NYSE-SC = 0.40, and NASDAQ-SC = 0.37. This is an indication that NYSE stocks have much more stable day-to-day patterns than

NASDAQ stocks. Analysis of the skewness of absolute daily volumes finds NYSE = 4.33 and NASDAQ = 5.59, and large cap = 3.23 and small cap = 5.43. This shows that daily volumes are very positively skewed for all stocks (NYSE and NASDAQ, and large and small cap stocks). High kurtosis is also an indication of peaked means and fat tails, or in this case, an indication of outliers. Non-normal distribution patterns for asset returns and their repercussion on risk and return have been greatly researched and published. But the non-normal volume profiles and their repercussion on daily trading costs and ultimately portfolio returns have not been given the same level of attention as with asset prices. But it is just as important.

Another aspect of daily volume distribution and the use of "average" over "median" is that the positive skewness of the distribution causes the "average" volume metric to overestimate actual volumes the majority of times. Since volumes are highly skewed by positive outliers, it results in a value that could be much higher than the middle point of the distribution. In our data, the average is approximately in the 60−65% percentile, thus leading to an overestimation of daily volume 60−65% of the time. We have found (as we show in later chapters) that the median is a much more accurate predictor of daily volume than average. Why is this so? The reason is simple. When there are earnings announcements, company news, or macro events there will be more volume traded than on a normal day. Stock volume is highly skewed. As a fix, many practitioners compute the average by taking out the highest and lowest one or two data points, which results in a much more accurate measure. Interestingly, practitioners can compute a more accurate adjusted average by taking out the highest one or two data points since they are the most skewed outliers (Table 2.9).

Day of Week Effect

Historically there has always been a day of week pattern associated with market volumes. Stocks would trade the least amount on Monday, increasing on Tuesday and Wednesday and then decreasing on Thursday and Friday. Examination of large cap trading volume in 2012 found similar patterns. Mondays were the least active trading day with volume increasing on Tuesday−Thursday and falling slightly on Friday. We found similar patterns across NYSE and NASDAQ listed large cap stocks. Analysis of small cap stocks found a different pattern. Volumes were the lowest on Monday as expected, but highest on Tuesday and Friday. Similar patterns for small cap stocks were found for NYSE and NASDAQ listed stocks.

Next we examined the day of week effect for large and small cap stocks across time to see if the Friday effect for small cap stocks is relatively new

Table 2.9 Daily Volume Distribution Statistics

	Avg ADV	Avg Median	Avg St Dev	Avg Skew	Avg Kurt	Avg CoV
Exchange						
NYSE	2,450,075	592,194	1,360,792	4.33	38.98	74%
NASDAQ	991,463	177,048	689,431	5.59	58.01	99%
Market Cap						
LC	6,160,167	2,785,869	3,186,060	3.31	23.65	55%
SC	495,593	187,777	429,620	5.43	55.60	95%
NYSE						
LC	5,836,089	2,785,869	2,986,093	3.23	22.77	54%
SC	639,273	255,871	491,600	4.92	47.65	85%
NASDAQ						
LC	7,405,640	2,753,902	3,954,560	3.61	27.04	59%
SC	400,455	148,711	388,579	5.77	60.86	102%
Large Cap						
NYSE	5,836,089	2,785,869	2,986,093	3.23	22.77	54%
NASDAQ	7,405,640	2,753,902	3,954,560	3.61	27.04	59%
Small Cap						
NYSE	639,273	255,871	491,600	4.92	47.65	85%
NASDAQ	400,455	148,711	388,579	5.77	60.86	102%

or due to randomness. We evaluated volumes by day over 2009–2012 to address this question. For large cap stocks we found similar weekly patterns for each of the years except for 2010 where volumes increased on Friday. For small cap stocks we found that the current weekly pattern has existed since at least 2009. One of the common beliefs for why Friday volumes have increased for small cap stocks is that portfolio managers do not want to hold weekend risk for stocks with a large amount of company specific risk. Thus, where managers may have been willing to hold open positions over the weekend at one point in time, they may be less willing to hold open positions in these names in today's trading environment. In 2010 large cap stocks exhibited similar patterns with Friday volumes increasing. A belief by market participants is that this was the result of a hangover effect where managers were not willing to hold open positions for large as well as small cap stocks due to market environment and uncertainty at that point in time.

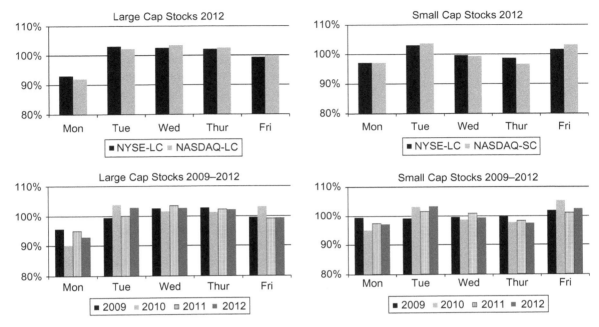

■ **Figure 2.3** Day of Week Effect.

Managers are still haunted by the overnight collapse of many large cap companies during the financial crisis (Figure 2.3).

Intraday Trading Profiles

We examined the intraday trading patterns for spreads, volume, volatility, and intraday trading stability. Historically, each of these measures followed a U-shaped pattern. For example, volume and volatility were both high at the open, decreasing into mid-day and then increasing again into the close. Spreads also followed a similar U-shaped intraday trading pattern. Examination of the same profiles in today's market found a distinctly different trading pattern. There were also distinct differences across market capitalization (LC/SC) and exchange listing (NYSE/NASDAQ). Our analysis period was 1Q-2012.

Spreads

■ Intraday spreads were measured as the average bid-ask spread in each 15 minute trading period.
■ Spreads are higher at the open than mid-day, but do not spike at the close. Spreads in fact decrease slightly into the close.

- Spreads decrease and level out after about the first 15–30 minutes for large cap stocks and after about 30–60 minutes for small cap stocks. The amount of time spreads persist at the higher values is longer than it was historically where both large and small cap spreads used to decline rather quickly and often within 15 minutes for both.
- Small cap spreads persist longer than large cap spreads.
- Small cap spreads are higher than large cap spreads due to the higher risk of each company, lower trading frequency, and higher potential for transacting with an informed investor.
- NYSE stocks have slightly lower spreads than NASDAQ stocks even after adjusting for market capitalization.

Analysis of intraday spreads found three observations worth noting. First, spreads are much higher in the beginning of trading and these higher spreads persist longer due to a difficult price discovery process. Specialists and market makers used to provide a very valuable service to the financial markets in terms of price discovery and determining the fair starting price for the stock at the market open. Now, the price discovery is often left to trading algorithms transacting a couple of hundred shares of stock at a time. While algorithms have greatly improved, they are still not as well equipped as the specialists and market makers in assisting price discovery. For example, specialists used to have access to the full order book and investor orders prior to the market open. So they could easily establish a fair value opening price by balancing these orders and customer preferences. Market makers also had a large inventory of positions and customer orders which allowed them to provide reasonable opening prices. The current electronic trading arena, where investors only have access to their individual orders, does not allow for an efficient price discovery process. Second, NASDAQ spreads are lower than NYSE spreads even after adjusting for market cap. The belief is that this is likely due to the NYSE designated market maker (DMM) system that has been established to encourage the DMM participants to participate in the process by providing liquidity when necessary. Third, spreads now decrease going into the close rather than increasing. This is likely due to greater transparency surrounding closing imbalances and investors' ability to offset any closing auction imbalance (Figure 2.4).

Volumes

- Intraday volume is measured as the percentage of the day's volume that traded in each 15 minute trading period.
- Intraday volume profiles are not following the traditional U-shaped trading patterns.

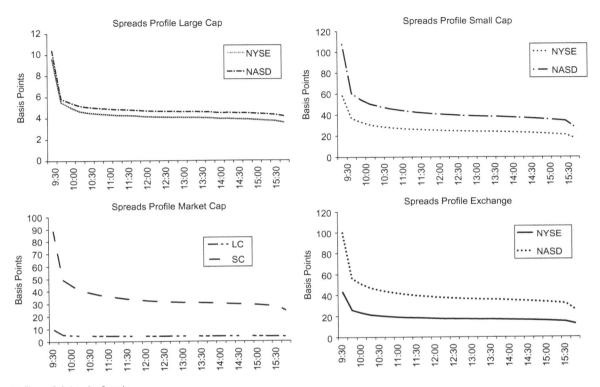

■ **Figure 2.4** Intraday Spreads.

- Intraday profiles are more "J" shaped, with only slightly more volume traded at the open and then decreasing and then increasing significantly into the close.
- NYSE listed companies trade a little less at the open and a little less at the close than NASDAQ listed companies.
- Small cap stocks trade more at the open and close than large caps.

There are five likely reasons why intraday trading patterns have shifted from a U-shaped trading pattern to a J-shaped pattern. First, similar to an increase in Friday daily volumes, where managers are less inclined to hold open positions over the weekend, traders are less inclined to hold open positions overnight and they make a more conscious effort to complete those positions before the market close. Second, there are currently fewer active investment strategies in today's markets than previously. Active managers typically trade at the open at known and preferred prices, rather than wait until towards the end of the day when prices are not known. Third, there are more index funds and closet index funds today than previously. These are the managers who more often trade at and into the close than in the morning hours. This is done in order to transact as close to the closing price as possible and to minimize tracking error compared to the closing price, since the fund will be valued based on the day's closing price. Fourth, there has been a dramatic increase in exchange traded funds (ETFs). ETFs are used for many different reasons. For example, ETFs are used to gain certain market exposures or to hedge very short-term risk. And very often investors who were hedging short-term market risk net out those positions at the end of the day, resulting in increased trading volume. ETF trading has also caused an increase in trading volume towards the end of the day due to the creation and redemption of these ETFs. While we have not found evidence suggesting correlation between overall ETF and stock volumes, we have found a statistically significant relationship between the shift in intraday volume trading and ETF volume. Thus, ETF trading has played a part in shifting the stock's intraday volume profile towards the close and away from the open. Fifth, the increase in closing imbalance data and the investor's ability to offset end of day imbalances (as well as easy and direct access to this information) has helped improve the price discovery process. This coupled with less end of day price volatility allows funds an easier time to achieve market prices at the close than previously and hence leads to more trading at the close. Finally, over the last few years there has been a decrease in quantitative investment strategies. Quant managers have reduced leverage of their portfolios and the general economic climate has not provided these managers with as much profiting opportunity as in years past. Since quant managers do not currently have a strong need to achieve particular market prices they are able to utilize a more passive execution style such as VWAP rather than utilizing a front-loading execution style such as IS or arrival price.

To the extent that there is another shift away from Index and ETF trading back to active management and quantitative styles, we are likely to see a corresponding shift in intraday volume profiles with more volume trading at the open and less dramatic spiking towards the close. Portfolio manager investment styles have a dramatic effect on when volumes occur throughout the day (Figure 2.5).

Volatility

- Measured as the average high-low percentage price range in each 15 minute trading period.
- Does not currently follow its historical U-shaped profile. Intraday volatility is higher at the open than mid-day and only increases slightly into the close.

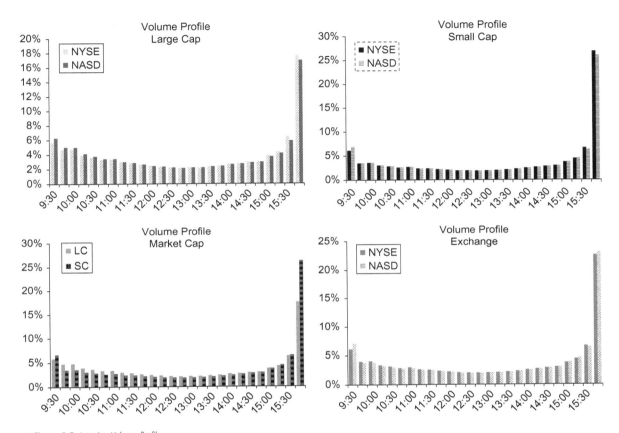

■ **Figure 2.5** Intraday Volume Profile.

- Higher at the open due to a more difficult price discovery process and the higher levels persist for a longer period of time than historically.
- Does not increase into the close to the same extent that it did historically.
- Slightly lower volatility at NYSE listed stocks than NASDAQ listed stocks.
- Small cap intraday volatility is much higher than large cap volatility as expected.

As mentioned, the old specialist and market maker system provided valuable price discovery. Today's market algorithms, however, are left to determine the fair value price by trading a couple of hundred shares at a time. It is not uncommon to look at the tape and see a price change of $0.50/share at the open with only a few hundred shares trading in the interval. Leaving algorithms to determine fair market prices by trading

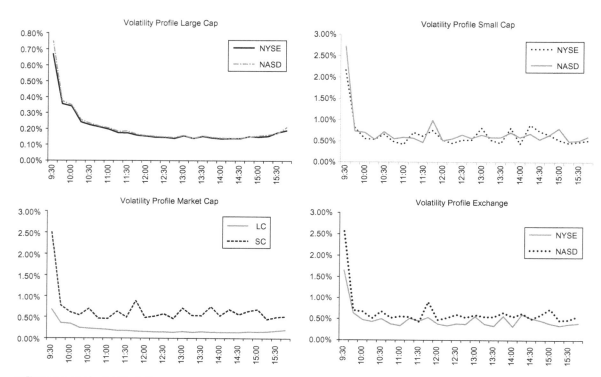

■ **Figure 2.6** Intraday Volatility Profile.

relatively small amounts causes opening period volatility to be higher than it was previously as well as to persist longer than it did before. It appears that the NYSE DMM system is providing value in terms of lower opening spreads and volatility levels. The decrease in end of day volatility is due to an improved and more transparent closing auction process (Figure 2.6).

Intraday Trading Stability—Coefficient of Variation
- Measured as the average standard deviation of interval volume.
- High variation in volumes at the open. Leveling off mid-day and then decreasing into the close.
- Small cap volume variation is about $2 \times$ large cap variation.
- No difference in volume variation across NYSE and NASDAQ listed stocks after adjusting for market cap.

Coefficient of variation is a rarely used risk statistic in the industry. As any trader will confirm, intraday liquidity risk is one of the most important aspects of trading and will be one of the primary reasons for a trader to

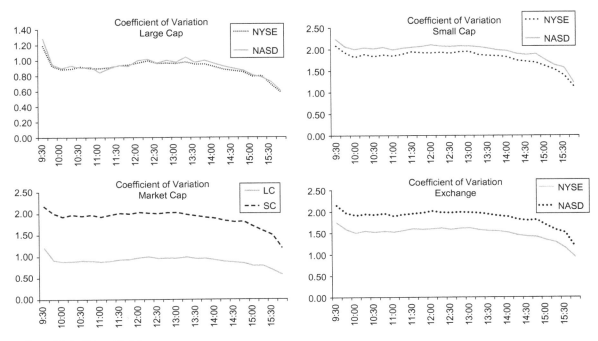

■ **Figure 2.7** Coefficient of Variation.

deviate from a prescribed strategy, change the algorithm, or adjust the algorithmic parameters and settings. The intraday coefficient of variation, when computed and used properly, will serve as a valuable liquidity risk measure and provide information that allows traders to improve overall performance. Additionally, as we show in later chapters, the coefficient of variation is a large determinant of stock specific trading cost and could be a valuable component of any market impact model (Figure 2.7).

Special Event Days

In the paper, "US Exchange Auction Trends: Recent Opening and Closing Auction Behavior, and the Implications on Order Management Strategies" (Kissell and Lie, 2011), the authors evaluated intraday trading profiles across exchanges, indexes, and market cap categories. The analysis found that trading behavior varied greatly on special event days including: FOMC, triple witching, company earnings, index changes, month and quarter end, day before and after holidays, and early closing days (Table 2.10).

Table 2.10 Special Event Day Volumes - Percentage of a Normal Day's Volume

	Normal Day	FOMC	Triple Witching	Company Earnings	Index Chg	Month End	Qtr End	Before/After Holidays	Early Close
Daily Volumes									
NYSE	100%	104%	120%	189%	187%	108%	104%	94%	37%
NASDAQ	100%	104%	147%	191%	281%	109%	105%	96%	46%
SP500	100%	104%	119%	184%	124%	107%	105%	92%	33%
R2000	100%	104%	145%	192%	285%	109%	106%	96%	48%
Intraday Volume									
NYSE	100%	105%	94%	191%	141%	103%	100%	93%	37%
NASDAQ	100%	105%	115%	193%	186%	103%	102%	95%	45%
SP500	100%	105%	97%	186%	106%	103%	100%	91%	32%
R2000	100%	105%	110%	193%	183%	103%	102%	95%	47%
Market on Close %									
NYSE	3.2%	2.7%	11.6%	4.6%	50.9%	8.6%	7.7%	3.6%	1.3%
NASDAQ	3.9%	3.1%	10.5%	6.0%	102.8%	10.0%	8.2%	4.5%	1.7%
SP500	3.3%	2.9%	10.2%	4.4%	20.8%	7.8%	7.8%	3.7%	0.9%
R2000	4.0%	3.1%	11.6%	6.1%	110.7%	10.3%	8.4%	4.5%	2.1%
Market on Open %									
NYSE	0.9%	0.8%	18.3%	1.9%	0.9%	0.8%	0.7%	1.0%	0.8%
NASDAQ	1.2%	1.1%	27.5%	2.3%	1.2%	1.0%	0.9%	1.4%	1.3%
SP500	0.7%	0.7%	15.6%	1.3%	0.7%	0.7%	0.6%	0.8%	0.6%
R2000	1.3%	1.1%	29.8%	2.5%	1.3%	1.1%	0.9%	1.4%	1.4%

Source: Kissell & Lie (2011).

The analysis evaluated trading patterns during four different periods: full day, intraday, opening auction, and closing auction. These periods are defined as:

Full Day: the full trading day from 9:30 a.m. to 4:00 p.m. and includes opening and closing auction volume.

Intraday: the full trading day from 9:30 a.m. to 4:00 p.m. and excludes opening and closing auction volumes. This consists of the volume and trading patterns that are actionably by traders and algorithms.

Opening Auction: volume that trades in the opening auction volume on the day.

Closing Auction: volume that trades in the closing auction volume on the day.

Share volumes in each of these periods were compared to the average volumes that trade in these periods on normal trading days in order to gain insight into the changing trading behavior by special event day.

The main findings from the paper are:

- Small cap stocks trade a higher percentage in the opening and closing auctions than large cap stocks.
- NYSE large cap stocks trade a higher percentage of the day's volume in auctions than NASDAQ large cap stocks.
- NASDAQ small cap stocks trade a higher percentage of the day's volume in auctions than NYSE small cap stocks.
- *FOMC days.* There was 4% more trading volume on the day but 5% more volume during the intraday period. Both opening auction volume and closing auction volume were lower than on normal trading days. On FOMC days traders concentrated trading around the time of the FOMC meeting and new announcement.
- *Triple Witching.* Large caps stocks traded 20% more volume and small cap stocks traded 45% more volume than on normal days. However, the change in intraday volume (total volume not including the opening and closing auctions) was not nearly as dramatic. Large cap volume was 3% lower than normal and small cap volume was only 10% more than normal. The opening auction volume was 15% more for large cap and 29% more for small cap, and the closing auction volume was 10% more for large cap and 12% more for small cap stocks.
- *Company Earnings.* Volumes on company earnings days were on average almost twice as high as normal days and this was consistent across all categories of the trading day. For the full day, large cap volume was 84% higher and small cap volume was 92% higher. Intraday volumes were 86% higher for large caps and 92% for small caps. The auction volumes were also about twice as high as on normal days. But in times where the earnings announcement was during the trade day the closing auction volume was considerably higher. In times where the earnings announcement was the prior evening or before market hours the opening auction volume was much higher.
- *Index Change.* A large difference across market cap volumes for large and small cap stocks on days of index changes. Large cap traded 24% more volume but small cap volume was 185% more than on normal day's. Large caps intraday volume was 6% more than normal and small cap was 83% more. The large majority of the increase in volume, as expected, was traded in the closing auction. Large cap stocks traded 20.8% of a normal day's volume in the closing auction compared to 3.3% on a normal day, and small cap stocks traded 110.7% of a normal

days volume in the closing auction compared to 4.0% on a normal day. There was no change in the volumes traded in the opening auction on index change days.

- *Month End.* Total volumes were 7% higher for large cap stocks and 9% for small cap stocks. Intraday volumes were only 3% higher for both large and small caps compared to normal days. Closing auction volumes were 8.6% of a normal day for large caps and 10.0% of a normal day for small caps. Opening auction volume was slightly less than what is normally experienced.
- *Quarter End.* Surprisingly, closing auction volumes, while higher than a normal day's, was actually lower than month end volumes. Large cap volumes were 5% higher and small cap volumes were 6% higher than normal days. There were no differences in intraday volumes compared to normal trading days. The big difference was similar to month end. Large caps traded 7.8% of a normal day's volume and small caps traded 8.4% of a normal day's volume in the closing auction.
- *Before/After Holidays.* Volumes were lower as expected. Large caps traded 8% lower and small caps traded 4% lower than on normal days. Intraday volumes changes were pretty similar. Intraday large cap was 9% and intraday small cap was 5% lower than on normal days.
- *Early Close.* Large cap volumes were −67% of normal days and small cap volumes were −52% of normal days. Closing auction volumes were also much lower, large cap was 0.9% of a normal day and small cap was 2.1% of a normal day. Opening auction volumes, however, were consistent with normal days.

Each of these categories has different repercussions for trading, traders, and algorithms and insight into each can result in dramatically improved trading performance. Conducting on-going market microstructure research for these special event days provides analysts and developers with essential insight into how best to transact stocks on these days. This could lead to improved trading performance and higher portfolio returns (Table 2.10).

FLASH CRASH

The flash crash of May 6, 2010 was a very significant financial event for many reasons. On the morning of May 6, 2010 the market opened slightly down from the previous close. The financial markets were worried about the debt crisis in Greece—potential defaults—which was dominating the news. At about 2:45 p.m. the flash crash hit. The SPX index lost 102 points

(−8.7%), the DJIA lost 1010 points (−9.3%), the VIX volatility index surged 40% from 24 to 41, some stocks and ETFs declined 60%, 70%, 80% and more—all the way down to pennies. The entire crash lasted about 20 minutes and almost fully rebounded by the end of the day.

At the end of the day the SPX was only down 3.24%, the DJIA was only down 3.20%, and the VIX dropped to 33. While there is much criticism about the structure of the US markets and potential for these large drops, critics have not given nearly enough credit to the overall system and its ability to self-correct. Consider this: The market dropped almost 10% in about a 20 minute period, but then was resilient enough to recover before the end of the day.

I recall the events of May 6 pretty well. I was on a client visit watching the financial news in a hotel lobby at the time of the crash. The network was showing the events in Greece and discussing the potential for a Greece default. Then the crash began, the television showed the declines in the S&P and Dow Jones, but it was too early for any type of discussion or commentary on the situation.

I called the trading desk within minutes and asked what was going on. They were also unaware of the crash in progress, but aware of the decline in prices. I heard people on the trading floor asking if the data was bad, there must have been some lost data connection. No one knew the reason or the extent of what was happening—it was so quick. I heard others in the background stating that Greece was rioting and a default was imminent. When I arrived at the client meeting the financial media showed that the S&P and Dow Jones were down only slightly, but nothing to signal a large crash. I started believing my colleagues, and thought that the decline must have been bad data and not a crash. But then the discussion quickly turned to what happened. It was a "Flash Crash" although the term wasn't coined until afterwards.

The initial theories were that it was a "fat" finger error, where a single trader entered an order with too many zeros, or a futures trader trading too large a position in too short of a period of time, or something to do with high frequency traders. Questions were being asked if the high frequency traders rigged the market for gain, if they did not provide enough liquidity to the market (when in fact they were not under any obligation to provide liquidity), and whether they have unfair advantage over regular investors because of faster access to the market and more efficiently written programming code.

Regardless of the true reason, many market participants turned on the high frequency traders. It was their fault. What I thought was humorous at the time was that no one was able to isolate what it was that the high frequency traders were doing at the time that was not everyday business. But there was a large market belief that it was the high frequency traders' fault. They served as a nice scapegoat. Especially since it has been reported that over the previous few years they all made boat loads of profits; so they must be bad.

The SEC issued a report in September 2010 discussing the flash crash. While the report put forth many theories and hypotheses for its cause, it did not provide ample statistical evidence supporting the theories behind the cause. Many of these potential reasons stated as potential causes of the crash also happened in the past and without any issues at all, so there must have been, as many have stated, a perfect storm. But what really was that perfect storm?

There was also a lot of discussion surrounding the events and people stating that the markets are fragile and the current structure is flawed. Many have stated that regulators need to make some significant and drastic changes to the structure. Some have stated that spreads need to increase to nickels while others stated that spreads need to decrease to sub-penny.

The important take-away points from the flash crash for algorithmic trading:

- Algorithmic and electronic trading is exponentially quicker than humans transacting orders manually. The nice thing about algorithms is that they will do exactly what you program them to do, and the not-so-nice thing about algorithms is that they will do exactly what you program them to do. So if you do not program every potential event into its logic there will be times when there are unforeseen market events and this could lead to dire results.
- Specialists and market makers always played a very important role in financial markets regarding price discovery. While this process failed during the program crash of 1987, I do not recall any other event since then leading up to the flash crash where there was such a large market decline. Specialists and market makers have been halting orders when appropriate and spending time to understand what is going on. We are now in a system where there is no single specialist and/or market maker controlling stocks and halting trading when necessary.
- Developers need to incorporate safeguards into algorithms to protect investors from the ominous days. However, these safeguards cannot be too strict because they may exclude investors from transacting in a rapidly moving market where there are no issues.

■ Market crashes happen. There has been a long history of crashes and bubbles from tulip bulbs, the crash of 1929, the program crash, technology bubble, housing, flash crash, to name just a few. These events have happened in the past and will happen again. As financial professionals, we need not only to be prepared when they happen, but knowledgeable enough to be able to react in the proper way when surprise events occur.

Empirical Evidence from the Flash Crash

Investigation into actual trading data on flash crash day confirmed atypical trading behavior in most stocks. However, what struck as much different from other crashes was the market's ability to rebound and self-correct. Our investigation focused on traded volume and price movement over the full day, as well as a review of the intraday trading patterns.

Our main findings are (Table 2.11):

■ Volumes were about twice as high as normal. Large cap stocks traded 200% of average, small cap stocks traded 183% of average. These volumes were significantly higher than normal, especially for a non-special event day.

Table 2.11 May 6, 2010—Flash Crash

				Percentage Change			Z-Score		
Category	30ADV	06/05/2010 Volume	Avg Ratio	Close-Close	Open-Close	High-Low	Close-Close	Open-Close	High-Low
Market Cap									
LC	7,088,778	13,551,862	2.00	−3.2%	−2.7%	−10.7%	−1.8	−1.5	−6.1
SC	502,188	950,161	1.83	−3.8%	−3.4%	−11.8%	−1.4	−1.2	−4.4
Exchange									
Listed	2,875,366	5,515,632	1.95	−3.7%	−3.2%	−11.7%	−1.6	−1.3	−5.2
NASDAQ	1,089,932	2,051,336	1.79	−3.7%	−3.3%	−11.5%	−1.4	−1.2	−4.4
Listed									
LC	6,773,377	12,869,070	1.99	−3.1%	−2.7%	−10.5%	−1.8	−1.5	−6.1
SC	666,493	1,348,684	1.92	−4.1%	−3.5%	−12.4%	−1.5	−1.2	−4.6
NASDAQ									
LC	8,297,812	16,169,230	2.02	−3.6%	−3.0%	−11.4%	−1.9	−1.6	−6.1
SC	395,032	690,254	1.77	−3.7%	−3.3%	−11.5%	−1.3	−1.2	−4.3

- Large cap stocks finished the day down 3.2% from the prior night's close and down 2.7% from the opening. This is equivalent to a 1.8 (close-to-close) and 1.5 (open-to-close) sigma event and is not something considered out of the norm in itself. The sigma-event measure is a standard z-score measurement and is computed by dividing the price movement by the daily volatility as a means to normalize across stocks. Z-scores or sigma events between $+/-1$ are considered normal movement. Z-scores between $+/-2$, while high, are still considered relatively normal since 95% of the observations should be between a z-score of $+/-2$. Theoretically, z-Scores should only be >3 or < -3 0.27% of the time.
- Small cap stocks finished the day down 3.8% from the prior night's close and down 3.4% from the opening. This is equivalent to a 1.4 (close-to-close) and 1.2 (open-to-close) sigma event. Again, this is not something unexpected or something that should be considered out of the ordinary.
- Analysis of the high-low range, however, showed that the day was much out of the ordinary. The high-low percentage range measured as the change from the high of the day to the low of the day was -10.7% for large cap stocks and -11.8% for small cap stocks. This is equivalent to a z-score of 6.1 for large caps and 4.4 for small caps. To put this into perspective, we should only see price movement of this magnitude 1 in 7.5 million years for large cap and 1 in 738 years for small cap.
- Even though prices returns have much fatter tails than the normal distribution, this analysis shows that the intraday price movement was much different than ordinary. And then couple this price movement with an almost equal and opposite reversal and we can start to see the significance of this flash crash event.

The next phase of the flash crash analysis focused on evaluating the intraday behavior for the SPY exchange traded fund (ETF), as a proxy for the SP500 index and a sample of ten stocks. The analysis really shed light onto what happened on flash crash day.

The SPY volume on flash crash day was 3.3 times its 30 day average. This was much higher than the average for all stocks which was about 2 times its average. The close-to-close price change was -3.4% which was consistent with other stocks, but due to much lower volatility this movement was equivalent to a 3.5 sigma event ($Z = -3.5$). The high-low price change for SPY was -10.8% which was also consistent with the average stock high-low range (-10.7% for large cap stocks).

Figure 2.8 illustrates the intraday prices and volumes for SPY. Price remained relatively flat throughout the day. Then beginning at

Figure 2.8 SPY Intraday Trading Behavior.

approximately 1 p.m. prices started to decline slowly. Around 2:40 p.m. there was a very sharp decline in prices that lasted for about 20 minutes, followed by a price reversal. SPY did experience another slight decrease towards the close but this was due to offsetting positions that were acquired during the day. Volume in SPY followed almost an identical pattern. Volumes were pretty constant until about 2:00 p.m. when they started to increase. Then there was the sharp increase in volume traded at the time of the flash crash right around 2:40 p.m. Volumes leveled off slightly after the reversal but still remained much higher than normal into the close.

Our sample of ten stocks has some similar patterns to SPY (Table 2.12). The average volume ratio was 3.6 compared to 3.3 (SPY). The close-to-close price change was −3.4% (same as SPY) and the open-close change was −4.3% compared to −2.9% for SPY. The most dramatic difference, however, was the magnitude of the high-low price change for these stocks. The average percentage high-low price change was −35.3% which was 3.3 times higher than the high-low percentage range for SPY.

The intraday price movement (normalized starting at 100) for these names is shown in Figure 2.9. Notice that most of the stocks had price movement that was relatively constant up until about 2:40 p.m. Some stocks opened much lower or declined over the first half-hour but these also remained pretty constant through the flash crash. Each experienced a very sudden and sharp decline which was quickly followed by a reversal. Note how quickly prices declined for all of these stocks only to be followed by another quick and sharp reversal. The close-to-close price change of −3.4% was equivalent to a −2.1 sigma event. On a 250 day trading year we should expect declines of

Table 2.12 Sample of Actual Volumes and Price Movement (Flash Crash)

Symbol	Volume Ratio	Percentage Change			Z-Score		
		Close-Close	Open-Close	High-Low	Close-Close	Open-Close	High-Low
IPXL	2.6	−4.4%	−4.4%	−62.1%	−2.2	−2.2	−30.8
SAM	5.6	−0.7%	−6.1%	−52.5%	−0.4	−3.6	−31.0
PG	2.6	−2.3%	−1.9%	−37.2%	−2.9	−2.4	−46.9
GTY	2.9	−3.9%	−3.5%	−35.9%	−2.5	−2.2	−22.6
MAIN	6.2	−4.7%	−4.3%	−33.0%	−2.8	−2.5	−19.7
CNP	3.1	−3.5%	−3.5%	−31.4%	−2.7	−2.8	−24.4
ESRX	2.7	−2.8%	−2.9%	−27.6%	−2.0	−2.0	−19.2
GGC	3.1	−2.4%	−5.7%	−27.5%	−0.8	−1.9	−9.4
CMO	2.7	−4.9%	−4.6%	−25.3%	−2.7	−2.6	−14.0
FSYS	4.3	−4.3%	−5.8%	−20.4%	−1.6	−2.2	−7.6
Avg	**3.6**	**−3.4%**	−4.3%	**−35.3%**	**−2.1**	**−2.4**	**−22.6**
SPY	3.3	−3.4%	−2.9%	−10.8%	−3.5	−3.0	−11.3

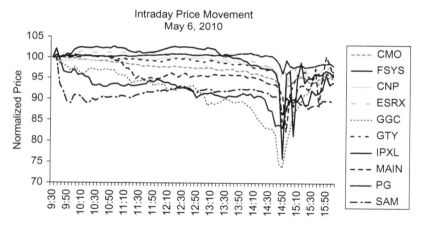

Intraday Price Movement
May 6, 2010

Legend: CMO, FSYS, CNP, ESRX, GGC, GTY, IPXL, MAIN, PG, SAM

■ **Figure 2.9** Intraday Price Movement (Sample of Stock).

this magnitude to occur about 5 times a year. So this type of movement is not that out of the ordinary or unexpected. However, the average high-low price range was −35.3% and is equivalent to a −22.6 sigma event! We have heard of 6 sigma before, well this is almost 4 times higher than a 6 sigma event! This magnitude decline in prices followed by a price reversal of almost the same magnitude is highly unlikely and unexpected. Volumes

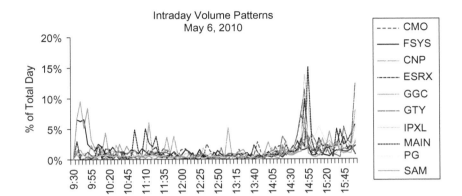

■ Figure 2.10 Intraday Volume Patterns (Sample of Stock).

were slightly higher at the open for some stocks—these were the stocks that experienced the price declines early in the trading day. The stocks experienced some slight spikes in volumes during the day leading up to the flash crash but then all stocks experienced dramatic increases that far exceeded historical trading patterns. Furthermore, the reduction in trading volume tapered off slightly after the rebound but was still dramatically higher than normal days and there were further spikes at the close (Figure 2.10).

What Should Regulators do to SafeGuard Investors from Potential Future Flash Crashes?

In order to safeguard investors (both retail and professional) from potential future crashes and other types of avoidable market disruptions, there are several steps regulators could take to improve market quality. These steps are:

- Regulations are needed to safeguard investors from any type of market disruptions: price movement, volume spikes, volatility measures, and information content.
- Market-wide trading halts are needed in order to allow all market participants to catch their breath and process information and ensure a level playing field.
- Circuit breakers based on price movement. This is needed at the index and stock level.
- Circuit breakers based on volume deviation from the norm. This is needed at the stock level.
- Accommodations can be made to these triggers and circuit breakers based on special event days (as described above).

Comparison with Previous Crashes

Investors and regulators need to keep in mind that our financial markets are very resilient. On the day of the flash crash markets recovered within a 20–30 minute time period. The mispricing of securities self-corrected itself much quicker than in previous crashes which took several years for full recovery.

To show this, we compared the Dow Jones Industrial Average (DJIA) for the May 6, 2010 flash crash to the crash of October 1929 and the 1987 program trading crash. First, in October 1929 the DJIA declined 13.7% on October 28, 1929 and declined another 12.5% on the 29th. Over this two day period the index was down 32.9% from its high to its low. This decline was further followed by an economic depression. The program trading crash took place on Monday October 19, 1987, and the DJIA declined 25.6% that day. The index was also down 4.7% the previous Friday. The recovery period following the 1987 program crash was several years but was not nearly as long as the 1929 crash.

Now compare these crashes to the May 6, 2010 flash crash. On the day, the DJIA declined 9.7% from its high to its low. This decline was quite rapid and took place almost entirely over a 30 minute period. On the day, the DJIA was down 3.2%, not nearly the same magnitude as the previous crashes. Furthermore, the DJIA recovered immediately following the flash crash. Figure 2.11 shows the movement of the DJIA for each of these crashes starting from thirty days before the crash to one year after the crash (250 days). The index values were normalized to a value of 100 at

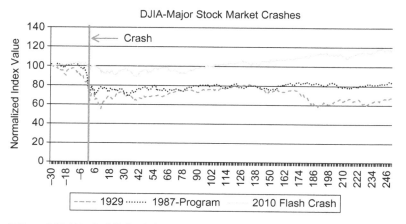

■ **Figure 2.11** Major Stock Market Crashes.

time t − 30. One year following the 1929 crash the DJIA was still down 30%. One year after the 1987 program trading crash markets were still down 20%. But one year following the flash crash markets were actually up 15%.

CONCLUSION

In this chapter we have provided an overview of market microstructure research. We have provided some of the more important industry findings and leading edge papers that provided the basis for many of today's execution and black box trading algorithms. Today's investors find themselves in a much different environment than the two exchange structure that was mutually exclusive for NYSE and OTC stocks. Traders today are faced with thirteen different displayed venues and from thirty to forty different dark pools and crossing networks, and different pricing models for each such as the maker-taker model, the inverse taker-maker model, as well as traditional straight commission. The chapter concluded with an overview of the current state of market microstructure and trading research that is being performed on a daily basis. These findings are being used by developers to improve and better calibrate trading algorithms, by traders to make improved trading decisions, and by portfolio managers to improve stock selection and portfolio construction. A thorough understanding of current market microstructure topics, market dynamics, and the current state of research models is essential for any investor seeking to achieve best execution.

Algorithmic Transaction Cost Analysis

INTRODUCTION

Transaction cost analysis (TCA) has regained a new found interest in the financial community as a result of the proliferation of algorithmic trading. Portfolio managers and traders are using TCA to evaluate performance of brokers and their algorithms. Furthermore, TCA is used by portfolio managers to improve performance as part of their stock selection and portfolio construction process.

Currently, there are many investors who utilize TCA to select their trading algorithms and make informed trading decisions. Those investors who are not yet utilizing TCA as a decision-making tool are missing valuable opportunities to improve portfolio performance and increase returns.

TCA has evolved significantly over the last several years, though it is still commonly conceptualized as a vague and unstructured concept. The accompanying literature and research still remains muddled due to misrepresentation by many brokers, vendors, and industry participants. We set out to shed new light below.

In order to fully assist investors' algorithmic transaction cost performance, we have developed a framework that consists of pre-, intra-, and post-trade analysis. Our framework is based on an unbundling scheme where costs are classified by ten components and categorized by where they occur during implementation. This scheme is based on the work of Perold (1988) and Wagner and Edwards (1993), and has been described in Journal of Trading, "The Expanded Implementation Shortfall: Understanding Transaction Cost Components," Kissell (2006), and in *Optimal Trading Strategies* (2003). Madhavan (2000, 2002) provides a detailed investigation of financial literature discussing transaction cost components and is considered by many as the gold standard of TCA literature review.

The Science of Algorithmic Trading and Portfolio Management. DOI: http://dx.doi.org/10.1016/B978-0-12-401689-7.00003-9

What Are Transaction Costs?

In economics, transaction costs are the fees paid by buyers, but not received by sellers, and/or the fees paid by sellers, but not received by buyers. In finance, transaction costs refers to the premium above the current market price required to attract additional sellers into the market, and the discount below the current market price required to attract additional buyers into the market. Transaction costs are described by Ronald Coarse (1937) in "The Nature of the Firm" as an unavoidable cost of doing business. He was subsequently awarded the Economics Nobel Prize in 1991 for his leading edge work.

What Is Best Execution?

The perception that best execution is an elusive concept has become severely overplayed in the industry. In reality, "best execution" is a very simple and direct concept:

Best execution (as stated in *Optimal Trading Strategies*) is the process of determining the strategy that provides the highest likelihood of achieving the investment objective of the fund. The strategy consists of managing transaction costs during all phases of the investment cycle, and determining when it is appropriate to take advantage of ever-changing market conditions.

Wayne Wagner described best execution in even simpler terms:

It is the process of maximizing the investment idea.

Best execution does not depend on how close the execution price occurs to an arbitrary benchmark price (such as the open, close, high, low, VWAP, etc.). Rather, it does depend on the investor's ability to make proper trading decisions by incorporating all market uncertainties and the current market conditions. The ultimate goal of best execution is to ensure that the trading decisions are consistent with the overall investment objectives of the fund. (See Kissell and Malamut (2007) for a discussion on ensuring consistency between investing and trading consistency.)

To determine whether or not best execution has been met requires the performance evaluation to be made based on the "information set" that was available at the beginning of trading combined with the investment objective of the fund. If either the information set or the underlying investment objective is not known or is not available it is simply not possible to determine if best execution was achieved—regardless of how close the transaction prices were to any benchmark price.

What Is the Goal of Implementation?

Implementation is the process of determining suitable appropriate trading strategies and adaptation tactics that will result in best execution. Unfortunately, it is not possible for investors to pre-evaluate and determine the best way to execute a position under all possible scenarios, but investors can develop rules and guidelines to make these tasks quicker, easier, and more efficient during trading.

In Wayne Wagner's terminology,

> *Implementation is the Journey to Best Execution.*

UNBUNDLED TRANSACTION COST COMPONENTS

We have identified ten distinct transaction cost components: commissions, taxes, fees, rebates, spreads, delay cost, price appreciation, market impact, timing risk, and opportunity cost. These are described below following the definitions in Kissell (2003, 2006).

1. Commission

Commission is payment made to broker-dealers for executing trades and corresponding services such as order routing and risk management. Commissions are commonly expressed on a per share basis (e.g., cents per share) or based on total transaction value (e.g., some basis point of transaction value). Commission charges may vary by:

i. Broker, fund (based on trading volume), or by trading type (cash, program, algorithms, or DMA).
ii. Trading difficulty, where easier trades receive a lower rate and the more difficult trades a higher rate. In the current trading arena commissions are highest for cash trading followed by programs, algorithms, and DMA.

2. Fees

Fees charged during execution of the order include ticket charges assessed by floor brokers, exchange fees, clearing and settlement costs, and SEC transaction fees. Very often brokers bundle these fees into the total commissions charge.

3. Taxes

Taxes are a levy assessed based on realized earnings. Tax rates will vary by investment and type of earning. For example, capital gains, long-term earnings, dividends, and short-term profits can all be taxed at different percentages.

4. Rebates

The rebate component is a new transaction cost component that is the byproduct of the new market environment (see Chapter 2). Trading venues charge a usage fee using a straight commission fee structure, a maker-taker model, or a taker-maker (inverted) model. In a straight commission model, both parties are charged a fee for usage of the system. In the maker-taker model, the investor who posts liquidity is provided with a rebate and the investor who takes liquidity is charged a fee. In an inverted or taker-maker model, the investor posting liquidity is charged a fee and the investor who takes liquidity is provided with a rebate. In both cases the fee charged will be higher than the rebate provided to ensure that the trading venue will earn a profit. Brokers may or may not pass this component onto their clients. In the cases when it does not pass through the component the broker will pay the fee or collect the rebate for their own profit pool. The commission rate charged to investors in these cases is likely to already have this fee and/or rebate embedded in its amount.

Since the fee amount or rebate collected is based on the trading venue and whether the algorithm posts or takes liquidity, the selection of trading venue and smart router order logic could be influenced based on the net incremental cost or rebate for the broker rather than the investor. Many questions arise (and rightly so) as to whether or not the broker is really placing orders correctly based on the needs of their investor or are looking to capture and profit from the rebates themselves. Analysts are highly encouraged to inquire about and challenge the logic of rebate-fee payment streams generated by various types of trading algorithms and smart routers in order to confirm the logic is in their best interest.

5. Spreads

The spread is the difference between best offer (ask) and best bid price. It is intended to compensate market makers for the risks associated with acquiring and holding an inventory while waiting to offset the position in the market. This cost component is also intended to compensate for the risk potential of adverse selection or transactions with an informed investor (i.e., acquirement of toxic order flow). Spreads represent the round-trip cost of transacting for small orders (e.g., 100 share lots) but do not accurately represent the round-trip cost of transacting blocks (e.g., 10,000 + shares).

6. Delay Cost

Delay cost represents the loss in investment value between the time the manager makes the investment decision and the time the order is released to the market. Managers who buy rising stocks and sell falling stocks will incur a delay cost. Delay cost could occur for many reasons.

First, delay cost may arise because traders hesitate in releasing the orders to the market. Second, cost may occur due to uncertainty surrounding who are the most "capable" brokers for the particular order or trade list. Some brokers are more capable at transacting certain names or more capable in certain market conditions. Third, traders may decide to hold off the transaction because they believe better prices may occur. However, if the market moves away, e.g., an adverse momentum, then the delay cost can be quite large. Fourth, traders may unintentionally convey information to the market about their trading intentions and order size (information leakage). Fifth, overnight price change movement may occur. For example, stock price often changes from the close to the open. Investors cannot participate in this price change, so the difference results in a sunk cost or savings depending on whether the change is favorable. Investors who are properly managing all phases of the investment cycle can minimize (if not avoid completely) all delay cost components except for the overnight price movement.

7. Price Appreciation

Price appreciation represents how the stock price would evolve in a market without any uncertainty (natural price movement). Price appreciation is also referred to as price trend, drift, momentum, or alpha. It represents the cost (savings) associated with buying stock in a rising (falling) market or selling (buying) stock in a falling (rising) market. Many bond pricing models assume that the value of the bond will appreciate based on the bond's interest rate and time to maturity.

8. Market Impact

Market impact represents the movement in the price of the stock caused by a particular trade or order. It is one of the more costly transaction cost components and always results in adverse price movement and a drag on performance. Market impact will occur due to the liquidity demand (temporary) of the investor and the information content (permanent) of the trade. The liquidity demand cost component refers to the situation where the investors wishing to buy or sell stock in the market have insufficient counterparties to complete the order. In these situations, investors will have to provide premiums above the current price for buy orders or discount their price for sell orders to attract additional counterparties to complete the transaction. The information content of the trade consists of inadvertently providing the market with signals to indicate the investor's buy/sell intentions, which in turn the market often interprets the stock as under-or overvalued, respectively.

Mathematically, market impact is the difference between the price trajectory of the stock with the order and what the price trajectory would have been had the order not been released to the market. Unfortunately, we are not able to simultaneously observe both price trajectories and measure market impact with any exactness. As a result, market impact has been described as the "Heisenberg uncertainty principle of trading." This concept is further described and illustrated in Chapter 4, Market Impact Models.

9. Timing Risk

Timing risk refers to the uncertainty surrounding the estimated transaction cost. It consists of three components: price volatility, liquidity risk, and parameter estimation error. Price volatility causes the underlying stock price to be either higher or lower than estimated due to market movement and noise. Liquidity risk drives market impact cost due to fluctuations in the number of counterparties in the market. Liquidity risk is dependent upon volumes, intraday trading patterns, as well as the aggregate buying and selling pressure of all market participants. Estimation error is the standard error (uncertainty) surrounding the market impact parameters.

10. Opportunity Cost

Opportunity cost is a measure of the forgone profit or avoided loss of not being able to transact the entire order (e.g., having unexecuted shares at the end of the trading period). The main reasons that opportunity cost may occur are adverse price movement and insufficient liquidity. First, if managers buy stocks that are rising, they may cancel the unexecuted shares of the order as the price becomes too expensive, resulting in a missed profit. Second, if managers cannot complete the order due to insufficient market liquidity (e.g., lack of counterparty participation) the manager would again miss out on a profit opportunity for those unexecuted shares due to favorable price movement.

TRANSACTION COST CLASSIFICATION

Transaction costs can be classified into investment related, trading related, and opportunity cost components shown above.

Investment-Related Costs are the costs that arise during the investment decision phase of the investment cycle. They occur from the time of the investment decision to the time the order is released to the market. These costs often arise due to lack of communication between the portfolio manager and trader in deciding the proper implementation objective (strategy), or due to a delay in selecting the appropriate broker or algorithm. The longer it takes for the manager and trader to resolve these issues, the higher potential for adverse price movement and higher investment cost. Traders often spend valuable time investigating how trade lists should be implemented and what broker or trading venue to use. The easiest way to reduce investment-related transaction cost is the use of proper pre-trade analysis, alternative strategy evaluations, and algorithm selections in order for the manager and traders to work closely together to determine the strategy most consistent with the investment objective of the fund.

Trading-Related Costs. Trading-related transaction costs comprise the largest subset of transaction costs. They consist of all costs that occur during actual implementation of the order. While these costs cannot be eliminated, they can be properly managed based on the needs of the fund. The largest trading-related transaction costs are market impact and timing risk. However, these two components are conflicting terms and often referred to as the "trader's dilemma," as traders need to balance this trade-off based on the risk appetite of the firm. Market impact is highest utilizing an aggressive trading strategy and lowest utilizing a passive strategy. Timing risk, on the other hand, is highest with a passive strategy and lowest with an aggressive strategy. Market impact and timing risk are two conflicting terms.

Opportunity Cost. Opportunity cost, as stated above, represents the foregone profit or loss resulting from not being able to fully execute the order within the allotted period of time. It is measured as the number of unexecuted shares multiplied by the price change during which the order was in the market. Opportunity cost will arise either because the trader was unwilling to transact shares at the existing market prices (e.g., prices were too high) or because there was insufficient market liquidity (e.g., not enough sellers for a buy order or buyers for a sell order) or both. The best way to reduce opportunity cost is for managers and traders to work together to determine the number of shares that can be absorbed by the market within the manager's specified price range. If it is predetermined that the market is not able to absorb all shares of the order within the specified prices, the manager can modify the order to a size that can be easily transacted at their price points.

TRANSACTION COST CATEGORIZATION

Financial transaction costs are comprised of fixed and variable components and are either visible or hidden (non-transparent).

Fixed cost components are those costs that are not dependent upon the implementation strategy and cannot be managed or reduced during implementation. Variable cost components, on the other hand, vary during implementation of the investment decision and are a function of the underlying implementation strategy. Variable cost components make up the majority of total transaction costs. Money managers, traders, and brokers can add considerable value to the implementation process simply by controlling these variable components in a manner consistent with the overall investment objective of the fund.

Visible or transparent costs are those costs whose fee structure is known in advance. For example, visible costs may be stated as a percentage of traded value, as a $/share cost applied to total volume traded, or even as some percentage of realized trading profit. Visible cost components are primarily attributable to commissions, fees, spreads, and taxes. Hidden or non-transparent transaction costs are those costs whose fee structure is unknown. For example, the exact cost for a large block order will not be known until after the transaction has been completed (if executed via agency) or until after the bid has been requested (if principal bid). The cost structures for these hidden components are typically estimated using statistical models. For example, market impact costs are often estimated via non-linear regression estimation.

Non-transparent transaction costs comprise the greatest portion of total transaction cost and provide the greatest potential for performance enhancement. Traders and/or algorithms need to be especially conscious of these components in order to add value to the implementation process. If they are not properly controlled they can cause superior investment opportunities to become only marginally profitable and/or profitable opportunities to turn bad. Table 3.1 illustrates our Unbundled Transaction Costs categories. Table 3.2 illustrates our Transaction Cost classification.

TRANSACTION COST ANALYSIS

Transaction cost analysis (TCA) is the investor's tool to achieve best execution. It consists of pre-trade, intraday, and post-trade analysis.

Pre-trade analysis occurs prior to the commencement of trading. It consists of forecasting price appreciation, market impact and timing risk for the

Table 3.1 Unbundled Transaction Costs

	Fixed	**Variable**
Visible:	Commission	Spreads
	Fees	Taxes
	Rebates	
		Delay Cost
Hidden:	n/a	Price Appreciation
		Market Impact
		Timing Risk
		Opportunity

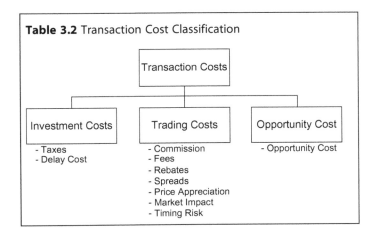

Table 3.2 Transaction Cost Classification

specified strategy, evaluating alternative strategies and algorithms, and selecting the strategy or algorithm that is most consistent with the overall investment objective of the fund.

Intraday analysis is intended to ensure that the revised execution strategies will continuously be aligned with the high level trading decisions. It consists of specifying how these strategies are to adapt to the endlessly changing market conditions (e.g., price movement and liquidity conditions). The only certainty in trading is that actual conditions will differ from expected. Participants need to understand when it is appropriate to change their strategy and take advantage of these changing market conditions.

Both pre-trade and intraday analysis consist of making and revising execution strategies (in real-time) to ensure trading goals are consistent with overall investment objectives. Best execution is determined more on decisions made pre-trade than post-trade. Most analysts are very good Monday morning quarterbacks. However, investors need a quality coach who can make and execute decisions under pressure with unknown conditions.

Post-trade analysis, on the other hand, does not consist of making any type of trading decision (either pre-trade or intraday). Post-trade analysis is used to determine whether the pre-trade models give accurate and reasonable expectations, and whether pre-trade and intraday decisions are consistent with the overall investment objectives of the fund. In other words, it is the report card of execution performance.

Post-trade analysis consists of two parts: measuring costs and evaluating performance. All too often, however, there is confusion regarding the meaning of these parts. For example, comparison of the execution price to the VWAP price over the day is not a trading cost—it is a proxy for performance. Comparison to the day's closing price is not a cost—it is a proxy for tracking error. And, comparison of execution price to the opening price on the day or the market price at time of order entry is a cost to the fund and does not give insight into the performance of the trade.

Post-trade analysis needs to provide a measurement of cost, and evaluation of performance at the broker, trader, and algorithm level. When appropriate, the post-trade report should provide universe comparisons, categorization breakdowns (large/small orders, adverse/favorable price movement, high/low volatility, market up/down, etc.) and trend analysis.

Measuring/Forecasting

A cost measure is an "ex-post" or "after the fact" measure, and is determined via a statistical model. It is always a single value and can be either positive (less favorable) or negative (savings). It is computed directly from price data. A cost forecast, on the other hand, occurs "ex-ante" or "prior to trading." It is an estimated value comprised of a distribution with an expected mean (cost) and standard deviation (timing risk).

The average or mean trading cost component is comprised of market impact and price appreciation. The forecasted market impact estimate will always be positive and indicate less favorable transaction prices. The price appreciation component, on the other hand, could be zero (e.g., no expectation of price movement), positive, indicating adverse price movement and less favorable expected transaction prices, or negative, indicating favorable price momentum and better transaction prices. The trading cost standard error term is comprised of price volatility, liquidity risk, and parameter estimation error from the market impact model.

Cost versus Profit and Loss

There is not much consistency in the industry regarding the terminology or sign to use when measuring and forecasting costs. Many participants state cost as a positive value while others state cost as a negative value. For example, some participants refer to a positive cost of $+30$ bp as underperformance and a negative cost of -30 bp as outperformance (savings). Others treat this metric in the opposite way with the $+30$ bp indicating better transaction prices and -30 bp indicating worse transaction prices.

To avoid potential confusion, our "Cost" and "Profit and Loss" terminology throughout the text will be as follows:

> A "**Cost**" metric will always use a positive value to indicate underperformance and a negative value to indicate better performance. For example, a cost of 30 bp indicates less favorable execution than the benchmark and -30 bp cost indicates better performance than the benchmark.
>
> A "**Profit and Loss**" or "**PnL**" metric will always use a negative value to indicate underperformance and a positive value to indicate better performance. For example, a PnL of -5 bp indicates less favorable execution than the benchmark and a PnL of $+5$ bp indicates better performance compared to the benchmark.

IMPLEMENTATION SHORTFALL

Implementation shortfall (IS) is a measure that represents the total cost of executing the investment idea. It was introduced by Perold (1988) and is calculated as the difference between the paper return of a portfolio where all shares are assumed to have transacted at the manager's decision price and the actual return of the portfolio using actual transaction prices and shares executed. It is often described as the missed profiting opportunity as well as the friction associated with executing the trade. Many industry participants refer to implementation shortfall as slippage or simply portfolio cost.

Mathematically, implementation shortfall is written as:

$$IS = Paper\ Return - Actual\ Return \tag{3.1}$$

Paper return is the difference between the ending portfolio value and its starting value evaluated at the manager's decision price. This is:

$$Paper\ Return = S \cdot P_n - S \cdot P_d \tag{3.2}$$

Here S represents the total number of shares to trade, P_d is the manager's decision price, and P_n is the price at the end of period n. $S \cdot P_d$ represents the starting portfolio value and $S \cdot P_n$ represents the ending portfolio value. Notice that the formulation of the paper return does not include any transaction costs such as commissions, ticket charges, etc. The paper return is meant to capture the full potential of the manager's stock picking ability. For example, suppose a manager decides to purchase 5000 shares of a stock trading at $10 and by the end of the day the stock is trading at $11. The value of the portfolio at the time of the investment decision was $50,000 and the value of the portfolio at the end of the day was $55,000. Therefore, the paper return of this investment idea is $5000.

Actual portfolio return is the difference between the actual ending portfolio value and the value that was required to acquire the portfolio minus all fees corresponding to the transaction. Mathematically, this is:

$$Actual\ Portfolio\ Return = \left(\sum s_j\right) \cdot P_n - \sum s_j p_j - fees \tag{3.3}$$

where,

$\left(\sum s_j\right)$ represents the total number of shares in the portfolio

$\left(\sum s_j\right) \cdot P_n$ is the ending portfolio value

$\sum s_j p_j$ is the price paid to acquire the portfolio

and fees represent the fixed fees required to facilitate the trade such as commission, taxes, clearing and settlement charges, ticket charges, rebates, etc. s_j and p_j represent the shares and price corresponding to the jth transaction.

For example, suppose a manager decides to purchase 5000 shares of stock trading at $10. However, due to market impact, price appreciation, etc., the average transaction price of the order was $10.50 indicating that the manager invested $52,500 into the portfolio. If the stock price at the end of the day is $11 the portfolio value is then worth $55,000. If the total fees were $100, then the actual portfolio return is $55,000 − $52,500 − $100 = $2400.

Implementation shortfall is then computed as the difference between paper return and portfolio return as follows:

$$IS = \underbrace{S \cdot P_n - S \cdot P_d}_{\text{Paper Return}} - \underbrace{\left(\sum s_j\right) \cdot P_n - \sum s_j p_j - fees}_{\text{Actual Portfolio Return}} \qquad (3.4)$$

In our example above, the implementation shortfall for the order is:

$$IS = \$5000 - \$2400 = \$2600$$

The implementation shortfall metric is a very important portfolio manager and trader decision making metric. It is used to select stock picking ability, measure trading costs, and as we show below, measure broker and algorithmic performance.

Implementation shortfall can be described in terms of the following three examples:

1. Complete Execution
2. Opportunity Cost (Andre Perold)
3. Expanded Implementation Shortfall (Wayne Wagner)

Complete Execution

Complete execution refers to the situation where the entire order is transacted in the market. That is $\sum s_j = S$. Suppose a manager decides to purchase S shares of stock that is currently trading at P_d and at the end of the trading horizon the price is P_n. Then implementation shortfall is computed following the above calculation as follows:

$$IS = (S \cdot P_n - S \cdot P_d) - \left(\left(\sum s_j\right) \cdot P_n - \sum s_j p_j - fees\right)$$

Since $\sum s_j = S$ this equation reduces to:

$$IS = \sum s_j p_j - S \cdot P_d + fees$$

This could also be written in terms of the average execution price P_{avg} for all shares as follows:

$$IS = S \cdot P_{avg} - S \cdot P_d + fees = S \cdot (P_{avg} - P_d) + fees$$

since $\sum s_j p_j = S \cdot P_{avg}$. Notice that when all shares are executed the implementation shortfall measure does not depend on the future stock price P_n at all.

Example: A manager decided to purchase 5000 shares when the stock was at $10. All 5000 shares were transacted in the market, but at an average transaction price of $10.50. If the commission fee was $100 then implementation shortfall of the order is:

$$IS = 5000 \cdot (\$10.50 - \$10.00) + \$100 = \$2600$$

Opportunity Cost (Andre Perold)

The opportunity cost example refers to a situation where the manager does not transact the entire order. This could be due to prices becoming too expensive or simply a lack of market liquidity. Either way, it is essential that we account for all unexecuted shares in the implementation shortfall calculation. This process is a follows:

First, compute the paper portfolio return:

$$Paper\ Return = S \cdot P_n - S \cdot P_d$$

Next, compute the actual portfolio return for those shares that were executed:

$$Actual\ Return = \left(\sum s_j\right) P_n - \sum s_j p_j + fees$$

Then, the implementation shortfall is written as:

$$IS = (S \cdot P_n - S \cdot P_d) - \left(\left(\sum s_j\right) P_n - \sum s_j p_j + fees\right)$$

Let us now expand on this formulation. Share quantity S can be rewritten in terms of executed shares $\sum s_j$ and unexecuted shares $(S - \sum s_j)$ as follows:

$$S = \underbrace{\sum s_j}_{Executed} + \underbrace{\left(S - \sum s_j\right)}_{Unexecuted}$$

If we substitute the share quantity expression above into the previous IS formulation we have:

$$IS = \left(\sum s_j + \left(S - \sum s_j\right) \cdot P_n - \sum s_j + \left(S - \sum s_j\right) \cdot P_d\right)$$
$$- \left(\left(\sum s_j\right) P_n - \sum s_j p_j + fees\right)$$

This equation can be written as:

$$IS = \underbrace{\sum s_j p_j - \sum s_j P_d}_{Execution\ Cost} + \underbrace{\left(S - \sum s_j\right) \cdot (P_n - P_d)}_{Opportunity\ Cost} + fees$$

This is the implementation shortfall formulation of Perold (1988) and differentiates between execution cost and opportunity cost. The execution cost component represents the cost that is incurred in the market during trading. Opportunity cost represents the missed profiting opportunity by not being able to transact all shares at the decision price.

Example: A manager decides to purchase 5000 shares of a stock at $10 but the manager is only able to execute 4000 shares at an average price of $10.50. The stock price at the end of trading is $11.00. And the commission cost is $80, which is reasonable since only 4000 shares traded in this example compared to 5000 shares in the above example. Then implementation shortfall including opportunity cost is:

$$IS = \left(\sum s_j p_j - \sum s_j P_d \right) + \left(S - \sum s_j \right) \cdot (P_n - P_d) + \text{fees}$$

It is important to note that in a situation where there are unexecuted shares then the IS formulation does depend upon the ending period stock price P_n but in a situation where all shares do execute then the IS formulation does not depend upon the ending period price P_n.

Furthermore, in situations where we have the average execution price of the order, IS further simplifies to:

$$IS = \sum s_j \cdot \left(P_{avg} - P_d \right) + \left(S - \sum s_j \right) \cdot (P_n - P_d) + \text{fees}$$

In our example we have:

$$IS = 4000 \cdot (\$10.50 - \$10.00) + 1000 \cdot (\$11.00 - \$10.00) + \$80$$
$$= \$2000 + \$1000 + \$80 = \$3080$$

The breakdown of costs following Perold is: execution cost = $2000, opportunity cost = $1000, and fixed fee = $80.

Expanded Implementation Shortfall (Wayne Wagner)

Our third example shows how to decompose implementation shortfall based on where the costs occur in the investment cycle. It starts with opportunity cost, and further segments the cost into a delay component which represents the missed opportunity of being unable to release the order into the market at the time of the investment decision. "Expanded Implementation Shortfall" is based on the work of Wayne Wagner and is often described as Wagner's Implementation Shortfall. This measurement provides managers with valuable insight into "who" is responsible for which costs. It helps us understand whether the incremental cost was due to a delay in releasing the order to the market or due to inferior performance by the trader or by the algorithm. Knowing who is responsible for cost will help investors improve the process of lowering transaction costs going forward. Wagner's expanded implementation shortfall categorizes cost into delay, trading, and opportunity related cost. Perold's original formulation did not separate delay and trading related costs when they occurred during the implementation phase. Wagner's formulation of implementation shortfall is what makes it possible to measure performance across traders, brokers, and algorithms.

The derivation of the expanded implementation shortfall is as follows.

First, define two time horizons: investment and trading. The investment horizon is the time from the investment decision t_d to beginning of trading t_0. The trading horizon is the time from beginning of trading t_0 to the end of trading t_n. The corresponding prices at these time intervals are P_d, which is the decision price, P_0 which is the price at beginning of trading, also known as the arrival price, and P_n, which is the price at the end of trading. All prices are taken as the mid-point of the bid-ask spread if during market hours or the last traded price or official close if after hours.

Next, rewrite the price change over these two intervals as follows:

$$(P_n - P_d) = (P_n - P_0) + (P_0 - P_d)$$

Now substitute this price into Perold's implementation shortfall:

$$IS = \left(\sum s_j p_j - \sum s_j P_d\right) + \left(S - \sum s_j\right) \cdot (P_n - P_d) + fees$$

This is:

$$IS = \left(\sum s_j p_j - \sum s_j P_d\right) + \left(S - \sum s_j\right) \cdot ((P_n - P_0) + (P_0 - P_d)) + fees$$

This expression can then be written based on our investment and trading horizons and is known as the Expanded Implementation Shortfall or Wagner's Implementation Shortfall. This is as follows:

$$Expanded\ IS = \underbrace{S(P_0 - P_d)}_{Delay\ Related} + \underbrace{\sum s_j p_j - \left(\sum s_j\right) P_0}_{Trading\ Related} + \underbrace{\left(S - \sum s_j\right)(P_n - P_0)}_{Opportunity\ Cost} + fees$$

This could also be written in terms of the average transaction price P_{avg} as follows:

$$Expanded\ IS = \underbrace{S(P_0 - P_d)}_{Delay\ Related} + \underbrace{\left(\sum s_j\right)(P_{avg} - P_0)}_{Trading\ Related} + \underbrace{\left(S - \sum s_j\right)(P_n - P_0)}_{Opportunity\ Cost} + fees$$

This is the expanded implementation shortfall metric proposed by Wayne Wagner that makes a distinction between the investment and trading horizons. It was first identified in Wagner (1975) and later explained in Wagner (1991) and Wagner and Edwards (1993). The delay related component has also been referred to as the investment related cost. The delay cost component could be caused by the portfolio manager, buy-side trader, or broker-dealer. For example, see Almgren and Chriss (2000), Kissell and Glantz (2003), or Rakhlin and Sofianos (2006).

Example: A manager decides to purchase 5000 shares of a stock at $10. By the time the order is finally released to the market the stock price has increased to $10.25. If the manager is only able to execute 4000 shares at an average price of $10.50 and the stock price at the end of trading is $11.00 what is the expanded implementation shortfall cost by components? Assume total commission cost is $80.

The calculation of the expanded implementation shortfall is:

$$Expanded\ IS = \underbrace{5000 \cdot (\$10.25 - \$10.00)}_{Delay\ Related} + \underbrace{4000 \cdot (\$10.50 - \$10.25)}_{Trading\ Related}$$

$$+ \underbrace{1000 \cdot (\$11.00 - \$10.25)}_{Opportunity\ Cost} + \$80 = \$3080$$

The delay related component is: $1250
The trading related component is: $1000
The opportunity cost component is: $750
Fixed fee amount is: $80
Total expanded implementation shortfall = $3080.

Notice that Wagner's expanded IS cost is the same value as Perold's IS. However, the opportunity cost in this example is $750 compared to $1000 previously. The reason for this difference is that the expanded IS measures opportunity cost from the time the order was released to the market as opposed to the time of the manager's decision. In actuality, the delay related cost component above can be further segmented into a trading related delay cost and an opportunity related delay cost. This is shown as follows:

$$Delay\ Cost = S \cdot (P_0 - P_d) = \underbrace{\left(S - \sum s_j\right) \cdot (P_0 - P_d)}_{Opportunity\ Related\ Delay} + \underbrace{\left(\sum s_j\right)(P_0 - P_d)}_{Trading\ Related\ Delay}$$

Analysts may wish to include all unexecuted shares in the opportunity cost component as a full measure of missed profitability.

It is important to point out that in many cases the analysts will not have the exact decision price of the manager since portfolio managers tend to keep their decision prices and reasons for the investment to themselves. However, analysts know the time the order was released to the market. Hence, the expanded implementation shortfall would follow our formulation above where we only analyze costs during market activity, that is, from t_0 to t_n. This is:

$$Market\ Activity\ IS = \underbrace{\left(\sum s_j\right)\left(P_{avg} - P_0\right)}_{Trading\ Related} + \underbrace{\left(S - \sum s_j\right)(P_n - P_0)}_{Opportunity\ Cost} + fees$$

Implementation Shortfall Formulation

The different formulations of implementation shortfall discussed above are:

$$IS = S \cdot \left(P_{avg} - P_d\right) + fees \tag{3.5}$$

$$Perold\ IS = \sum s_j \cdot \left(P_{avg} - P_d\right) + \left(S - \sum s_j\right) \cdot (P_n - P_d) + fees \tag{3.6}$$

$$Wagner\ IS = S(P_0 - P_d) + \left(\sum s_j\right)\left(P_{avg} - P_0\right) + \left(S - \sum s_j\right)(P_n - P_0) + fees \tag{3.7}$$

$$Mkt\ Act.\ IS = \left(\sum s_j\right)\left(P_{avg} - P_0\right) + \left(S - \sum s_j\right)(P_n - P_0) + fees \tag{3.8}$$

Trading Cost/Arrival Cost

The trading cost component is measured as the difference between the average execution price and the price of the stock at the time the order was entered into the market (arrival price). It is the most important metric to evaluate broker, venue, trader, or algorithmic performance, because it quantifies the cost that is directly attributable to trading and these specific parties. It follows directly from the trading related cost component from the expanded implementation shortfall. The investment related and opportunity cost components are more attributable to investment managers than to the trading party.

The trading cost or arrival cost component is:

$$Arrival\ Cost_\$ = \sum s_j p_j - \left(\sum s_j\right) P_0 \tag{3.9}$$

$$S, s_j > 0 \ for\ buys$$

$$S, s_j < 0 \ for\ sells$$

In basis points this expression is:

$$Arrival\ Cost_{bp} = \frac{\sum s_j p_j - \left(\sum s_j\right) P_0}{\left(\sum s_j\right) P_0} \cdot 10^4_{bp} \tag{3.10}$$

In general, arrival costs can be simplified as follows:

$$Arrival\ Cost_{bp} = Side \cdot \frac{P_{avg} - P_0}{P_0} \cdot 10^4_{bp} \tag{3.11}$$

where,

$$Side = \begin{cases} 1 & if\ Buy \\ -1 & if\ Sell \end{cases}$$

EVALUATING PERFORMANCE

In this section we describe various techniques to evaluate performance (note: we will use the profit and loss (PnL) terminology). These methods can be used to evaluate and compare trade quality for a single stock or basket of trades, as well as performance across traders, brokers, or algorithms. It can also serve as the basis for universe comparisons. In the following section we provide non-parametric statistical techniques that are being used to compare algorithmic performance.

Techniques that will be discussed in this section include: market or index adjusted cost, benchmark comparisons, various volume weighted average price (VWAP), participation weighted average price (PWP), relative performance measure (RPM), and z-score statistical measures.

Trading Price Performance

Trading price performance or simply trading PnL is identical to the trading cost component above and is measured as the difference between the average execution price and the price of the stock at the time the order was entered into the market (arrival price). A positive value indicates more favorable transaction prices and a negative value indicates less favorable transaction prices. Trading PnL is a measure of the cost during trading and reports whether the investor did better or worse than the arrival price. For example, a trading PnL of -10 bp indicates the fund underperformed the arrival price benchmark by 10 bp. The formulation for trading PnL multiplies the arrival cost calculation above by minus 1. This is:

$$Trading\ PnL_{bp} = -1 \cdot Side \cdot \frac{P_{avg} - P_0}{P_0} \cdot 10^4_{bp} \tag{3.12}$$

Benchmark Price Performance

Benchmark price performance measures are the simplest of the TCA performance evaluation techniques. These are intended to compare specific measures such as net difference and tracking error, or to distinguish between temporary and permanent impact. Some of the more commonly used benchmark prices include:

- Open—as a proxy for arrival price.
- Close—insight into end-of-day tracking error and is more commonly used by index funds that use the closing price in valuation of the fund.
- Next Day Open—as a way to distinguish between temporary and permanent market impact.
- Next Day Close or Future Day Close—also to distinguish between temporary and permanent impact.

The benchmark PnL calculation is:

$$Trading\ PnL_{bp} = -1 \cdot Side \cdot \frac{P_{avg} - P_B}{P_B} \cdot 10^4_{bp} \tag{3.13}$$

where $P_B = benchmark\ price$.

VWAP Benchmark

The VWAP benchmark is used as a proxy for fair market price. It helps investors determine if their execution prices were in line and consistent with fair market prices.

The calculation is:

$$VWAP\ PnL_{bp} = -1 \cdot Side \cdot \frac{P_{avg} - VWAP}{VWAP} \cdot 10^4_{bp} \tag{3.14}$$

where VWAP is the volume weighted average price over the trading period. A positive value indicates better performance and a negative value indicates underperformance.

Interval VWAP comparison serves as a good measure of execution quality and does a nice job of accounting for actual market conditions, trading activity, and market movement. The interval VWAP, however, does suffer from three issues. First, the larger the order the closer the results will be to the VWAP price, as the order price will become the VWAP price. Second, actual performance can become skewed if there are large block trades that occur at extreme prices (highs or lows) in crossing venues, especially in cases where investors have limited opportunity to participate with those trades. Third, the VWAP measure does not allow easy comparison across stocks or across the same stock on different days. For example, it is not possible to determine if missing VWAP by 3 bps in one stock is better performance than missing VWAP by 10 bps in another stock. If the first stock has very low volatility and the second stock has very high volatility, missing VWAP by 10 bps in the second name may in fact be better performance than missing VWAP by 3 bps in the first name.

There are three different VWAP performance metrics used: full day, interval, and VWAP to end of day.

Full Day VWAP: Used for investors who traded over the entire trading day from open to close. There is currently no "official" VWAP price on the day but many different providers, such as Bloomberg, Reuters, etc., do offer one. These vendors determine exactly what trades will be included in the VWAP calculations but they may not use all the market trades. For example, some providers may filter trades that were delayed or negotiated because they do not feel these prices are indicative of what all market participants had fair access to.

Interval VWAP: Used as a proxy for the fair market price during the time the investor was in the market trading. The interval VWAP is a specific VWAP price for the investor over their specific trading horizon and needs to be computed from tic data. This is in comparison to a full day VWAP price that is published by many vendors.

VWAP to End of Day: Used to evaluate those orders that were completed before the end of the day. In these cases, the broker or trader made a conscious decision to finish the trade before the end of the day. This VWAP to End of Day provides some insight into what the fair market price was including even after the order was completed. It helps determine if the decision to finish the order early was appropriate. This is a very useful metric to evaluate over time to determine if the trader or broker is skilled at market timing. But it does require a sufficient number of observations and a large tic data set.

It is worth noting that some B/Ds and vendors refer to the VWAP comparison as a cost rather than a gain/loss or performance indication. For those parties, a positive value indicates a higher cost (thus underperformance) and a negative value indicates a lower cost (thus better performance) and is the complete opposite of the meaning in the formula above. Unfortunately, representation of costs, P/L, or G/L as a metric is not consistent across industry participants and investors need to be aware of these differences.

Participation Weighted Price (PWP) Benchmark

Participation weighted price (PWP) is a variation of the VWAP analysis. It is intended to provide a comparison of the average execution price to the likely realized price had they participated with a specified percentage of volume during the duration of the order.

For example, if the PWP benchmark is a 20% POV rate and the investor transacted 100,000 shares in the market starting at 10 a.m. the PWP-20% benchmark price is computed as the volume weighted average price of the first 500,000 shares that traded in the market starting at 10 a.m. (the arrival time of the order). It is easy to see that if the investor transacted at a 20% POV rate their order would have been completed once 500,000 shares traded in the market since $0.20*500,000 = 100,000$ shares. The number of shares in a PWP analysis is equal to the number of shares traded divided by the specified POV rate.

The PWP PnL metric is computed as follows:

$$PWP\ Shares = \frac{Shares\ Traded}{POV\ Rate} \tag{3.15}$$

$$PWP\ Price = volume\ weighted\ price\ of\ the\ first\ PWP\ shares\ starting$$
$$at\ the\ arrival\ time\ t_0$$

$$PWP\ PnL_{bp} = -1 \cdot \frac{P_{avg} - PWP\ Price}{PWP\ Price} \cdot 10^4_{bp} \tag{3.16}$$

The PWP benchmark also has some inherent limitations similar to the VWAP metric. First, while PWP does provide insight into fair and reasonable prices during a specified time horizon it does not allow easy comparison across stocks or across days due to different stock price volatility and daily price movement. Furthermore, investors could potentially manipulate the PWP by trading at a more aggressive rate to push the price up for buy orders or down for sell orders, and give the market the impression that they still have more to trade. Since temporary impact does not dissipate instantaneously, the PWP price computed over a slightly longer horizon could remain artificially high (buy orders) or artificially low (sell orders) due to temporary impact cost. Participants may hold prices at these artificially higher or lower levels waiting for the non-existent orders to arrive. The end result is a PWP price that is more advantageous to the investor than what would have occurred in the market if the order had actually traded over that horizon.

Relative Performance Measure (RPM)

The relative performance measure (RPM) is a percentile ranking of trading activity. It provides an indication of the percentage of total activity that the investor outperformed in the market. For a buy order, it represents the percentage of market activity that transacted at a higher price and for a sell order it represents the percentage of market activity that transacted at a lower price. The RPM is modeled after the percentile ranking used in standardized academic tests and provides a descriptive statistic that is more consistent and robust than other measures.

The RPM was originally presented in *Optimal Trading Strategies* (2003) and Kissell (2007) and was based on a volume and trade metric. That original formulation, however, had at times small sample size and large trade percentage limitations bias. For example, the original formulation considered all of the investor's trades at the average transaction price as outperformance. Therefore, in situations where the investor transacted a large size at a single price all the shares were considered as outperformance and the end result would overstate the actual performance. Leslie Boni (2009) further elaborates on this point in her article "Grading Broker Algorithms," Journal of Trading, Fall 2009, and provides some important insight and improvements.

To help address these limitations, we revised the RPM formulation as follows:

The RPM is computed based on trading volume as follows:

$$RPM = \frac{1}{2} \cdot \left(\left(\% \text{ of volume traded at a price less favorable or equal to } P_{avg} \right) \right.$$
$$\left. + \left(1 - \% \text{ of volume traded at a price less favorable or equal to } P_{avg} \right) \right)$$
(3.17)

This metric can also be formulated for buy and sell orders as follows:

$$RPM_{Buy} = \frac{1}{2} \cdot \left(\frac{Total\,Volume + Volume\,at\,Price > P_{avg} - Volume\,at\,Price < P_{avg}}{Total\,Volume} \right)$$

(3.18)

$$RPM_{Sell} = \frac{1}{2} \cdot \left(\frac{Total\,Volume + Volume\,at\,Price < P_{avg} - Volume\,at\,Price > P_{avg}}{Total\,Volume} \right)$$

(3.19)

This formulation of RPM is now the average of the percentage of volume that traded at our execution price or better and 1 minus the average of the percentage of volume that traded at our price or worse. Thus, in effect, it treats half of the investor's orders as better performance and half the order as worse performance. As stated, the original formulation treated all of the investor's shares as better performance and inflated the measure.

The RPM is in many effects a preferred measure to the VWAP metric because it can be used to compare performance across stocks, days, and volatility conditions. And it is not influenced to the same extent as VWAP when large blocks trade at extreme prices.

The RPM will converge to 50% as the investor accounts for all market volume in the stock on the day similar to how the VWAP converges to the average execution price for large orders.

Brokers achieving fair and reasonable prices on behalf of their investors should achieve an RPM score around 50%. RPM scores consistently greater than 50% are an indication of superior performance and scores consistently less than 50% are an indication of inferior performance. The RPM measure can also be mapped to a qualitative score, for example:

RPM	Quality
0–20%	Poor
20–40%	Fair
40–60%	Average
60–80%	Good
80–100%	Excellent

Pre-Trade Benchmark

The pre-trade benchmark is used to evaluate trading performance from the perspective of what was expected to have occurred. Investors compute the difference between actual and estimated to determine whether performance was reasonable based on how close they came to the expectation.

Actual results that are much better than estimated could be an indication of skilled and quality execution, whereas actual results that are much worse than estimated could be an indication of inferior execution quality.

The difference between actual and estimated, however, could also be due to actual market conditions during trading that are beyond the control of the trader—such as sudden price momentum, or increased or decreased liquidity conditions. (These are addressed below through the use of the z-score and market adjusted cost analysis.)

The pre-trade performance benchmark is computed as follows:

$$Pre\text{-}Trade\ Difference = Estimated\ Arrival\ Cost - Actual\ Arrival\ Cost \quad (3.20)$$

A positive value indicates better performance and a negative value indicates worse performance.

Since actual market conditions could have a huge influence on actual costs, some investors have started analyzing the pre-trade difference by computing the estimated market impact cost for the exact market conditions—an ex-post market impact metric. While this type of measure gives reasonable insight in times of higher and lower volumes, on its own it does not give an adequate adjustment for price trend. Thus investors also factor out price trend via a market adjusted performance measure.

Index Adjusted Performance Metric

A market adjusted or index adjusted performance measure is intended to account for price movement in the stock due to the market, sector, or industry movement. This is computed using the stock's sensitivity to the underlying index and the actual movement of that index as a proxy for the natural price appreciation of the stock (e.g., how the stock price would have changed if the order was not released to the market).

First compute the index movement over the time trading horizon:

$$Index\ Cost_{bp} = \frac{Index\ VWAP - Index\ Arrival\ Cost}{Index\ Arrival\ Cost} \cdot 10^4_{bp} \quad (3.21)$$

Index arrival is the value of the index at the time the order was released to the market. Index VWAP is the volume weighted average price for the index over the trading horizon. What is the index volume weighted price over a period? Luckily there are many ETFs that serve as proxies for various underlying indexes such as the market (e.g., SPY), or sectors, etc., and thus provide easy availability to data to compute volume weighted average index prices.

If the investor's trade schedule sequence followed a different weighting scheme than volume weighting, such as front-or back-loaded weightings, it would be prudent for investors to compute the index cost in each period. In times where the index VWAP is not available, it can be approximated as $Index\ VWAP = 1/2 \cdot R_m$, where R_m is the total return in basis points of the underlying index over the period. The $\frac{1}{2}$ is the adjustment factor to account for continuous trading Kissell (2008).

The index adjusted cost is then:

$$Index\ Adjusted\ Cost_{bp} = Arrival\ Cost_{bp} - \hat{b}_{KI} \cdot Index\ Cost_{bp} \qquad (3.22)$$

$\hat{b}_{KI}$ is the stock k's sensitivity to the index. It is determined via linear regression in the same manner we calculate beta to the market index. Notice all we have done is subtract out the movement in the stock price that we would have expected to occur based only on the index movement. The index cost is not adjusted for the side of the trade.

Z-Score Evaluation Metric

The z-score evaluation metric provides a risk adjusted performance score by normalizing the difference between estimated and actual by the timing risk of the execution. This provides a normalized score that can be compared across different stocks and across days. (A z-score measure is also used to measure the accuracy of pre-trade models and to determine if these models are providing reasonable insight to potential outcomes cost.)

A simple statistical z-score is calculated as follows:

$$Z = \frac{Actual - Expected}{Standard\ Deviation}$$

For transaction cost analysis, we compute the normalized transaction cost as follows:

$$Z = \frac{Pre\text{-}Trade\ Cost\ Estimate - Arrival\ Cost}{Pre\text{-}Trade\ Timing\ Risk} \qquad (3.23)$$

Notice that this representation is opposite the statistical z-score measure ($z = (x-u)/sigma$). In our representation a positive z-score implies performance better than the estimate and a negative value implies performance worse than the estimate. Dividing by the timing risk of the trade normalizes for overall uncertainty due to price volatility and liquidity risk. This ensures that the sign of our performance metrics are consistent—positive indicates better performance and negative indicates lower quality performance.

If the pre-trade estimates are accurate, then the z-score statistic should be a random variable with mean zero and variance equal to one. That is, $Z \sim (0, 1)$. There are various statistical tests that can be used to test this joint hypothesis.

There are several points worth mentioning with regards to trading cost comparison. First, the test needs to be carried out for various order sizes (e.g., large, small, and mid-size orders). It is possible for a model to overestimate costs for large orders and underestimate costs for small orders (or vice versa) and still result in $Z \sim (0, 1)$ on average. Second, the test needs to be carried out for various strategies. Investors need to have a degree of confidence regarding the accuracy of cost estimates for all of the broker strategies. Third, it is essential that the pre-trade cost estimate be based on the number of shares traded and not the full order. Otherwise, the pre-trade cost estimate will likely overstate the cost of the trade and the broker being measured will consistency outperform the benchmark giving the appearance of superior performance and broker ability. In times where the order was not completely executed, the pre-trade cost estimates need to be adjusted to reflect the actual number of shares traded. Finally, analysts need to evaluate a large enough sample size in order to achieve statistical confidence surrounding the results, as well as conduct cross-sectional analysis in order to uncover any potential bias based on size, volatility, market capitalization, and market movement (e.g., up days and down days).

It is also important to note that many investors are using their own pre-trade estimates when computing the z-score measure. There is a widespread resistance to using a broker's derived pre-trade estimate to evaluate their own performance. As one manager stated, everyone looks great when we compare their performance to their cost estimate. But things start to fall into place when we use our own pre-trade estimate. Pre-trade cost comparison needs to be performed using a standard pre-trade model to avoid any bias that may occur with using the provider's own performance evaluation model.

Market Cost Adjusted Z-Score

It is possible to compute a z-score for the market adjusted cost as a means of normalizing performance and comparing across various sizes, strategies, and time periods similar to how it is used with the trading cost metric. But in this case, the denominator of the z-score is not the timing risk of the trade since timing risk accounts in part for the uncertainty in total price movement (adjusted for the trade schedule). The divisor in this case has to be the tracking error of the stock to the underlying index (adjusted for the trading strategy). Here the tracking error is identical to the standard deviation of the regression equation:

$$Index\ Adj\ Cost = Arrival\ Cost - \hat{b}_{kI} \cdot Index\ Cost + \varepsilon$$

Where the adjusted tracking error to the index is $\sqrt{\sigma_\varepsilon^2}$

Here we subtract only estimated market impact cost (not total estimated cost) for the market adjusted cost since we already adjusted for price appreciation using the stock's underlying beta and index as its proxy.

$$Mkt\ Adj\ Z\text{-}Score = \frac{Pre\text{-}Trade\ Estimate - Mkt\ Adj\ Cost}{Adj\ Tracking\ Error\ to\ the\ Index} \qquad (3.24)$$

Adaptation Tactic

Investors also need to evaluate any adaptation decisions employed during trading to determine if traders correctly specify these tactics and to ensure consistency with the investment objectives. For example, many times investors instruct brokers to spread the trades over the course of the day to minimize market impact cost, but if favorable trading opportunities exist then trading should accelerate to take advantage of the opportunity. Additionally, some instructions are to execute over a predefined period of time, such as the next two hours, but with some freedom. In these situations, brokers have the opportunity to finish earlier if favorable conditions exist, or extend the trading period if they believe the better opportunities will occur later in the day.

The main goal of evaluating adaptation tactics is to determine if the adaptation decision (e.g., deviation from initial strategy) was appropriate given the actual market conditions (prices and liquidity). That is, how good a job does the broker do in anticipating intraday trading patterns and favorable trading opportunities.

The easiest way to evaluate adaptation performance is to perform the interval VWAP and interval RPM analyses (see above) over the time period specified by the investor (e.g., a full day or for the specified two hour period) instead of the trading horizon of the trade. This will allow us to determine if the broker actually realized better prices by deviating from the initially prescribed schedule and will help distinguish between skill and luck.

As with all statistical analyses, it is important to have a statistically significant sample size and also perform cross-sectional studies where data points are grouped by size, side, volatility, market capitalization, and market movement (e.g., up days and down days) in order to determine if there is any bias for certain conditions or trading characteristics (e.g., one broker or algorithm performs better for high volatility stocks, another broker or algorithm performs better in favorable trending markets, etc.).

COMPARING ALGORITHMS

One of the biggest obstacles in comparing algorithmic performance is that each algorithm trades in a different manner, under a different set of market conditions. For example, a VWAP algorithm trades in a passive manner with lower cost and more risk compared to an arrival price algorithm which will trade in a more aggressive manner and have higher cost but lower risk. Which is better?

Consider the results from two different algorithms. Algorithm A has lower costs on average than algorithm B. Can we conclude that A is better than B? What if the average cost from A and B are the same but the standard deviation is lower for A than for B. Can we now conclude that A is better than B? Finally, what if A has a lower average cost and also a lower standard deviation? Can we finally conclude that A is better than B? The answer might surprise some readers. In all cases the answer is no. There is simply not enough information to conclude that A is a better performing algorithm than B even when it has a lower cost and lower standard deviaion. We need to determine whether or not this is a statistical difference or due to chance.

One of the most fundamental goals of any statistical analysis is to determine if the differences in results are "true" differences in process or if they are likely only due to chance. To assist with the evaluation of algorithms we provide the following definition:

> *Performance from two algorithms is equivalent if the trading results are likely to have come from the same distribution of costs.*

There are two ways we can go about comparing algorithms: paired observations and independent samples.

A paired observation approach is a controlled experiment where orders are split into equal pairs and executed using different algorithms over the same time periods. This is appropriate for algorithms that use static trading parameters such as VWAP and percentage of volume (POV). These are strategies that will not compete with one another during trading and are likely to use the exact same strategy throughout the day. For example, trading 1,000,000 shares using a single broker's VWAP algorithm will have the same execution strategy as trading two 500,000 share orders with two different VWAP algorithms (provided that the algorithms are equivalent). Additionally, trading 1,000,000 shares with one broker's POV algorithm (e.g., POV = 20%) will have the same execution strategy as using two different broker POV algorithms at one-half the execution rate (e.g., POV = 10% each). A paired observation approach ensures that identical orders are executed under identical market conditions. Analysts can also choose between the arrival cost and VWAP benchmark as the performance metric. Our preference for the paired sample tests is to use the VWAP.

An independent sampling approach is used to compare orders that are executed over different periods of time using different algorithms. This test is appropriate for algorithms such as implementation shortfall that manage the trade-off between cost and risk and employ dynamic adaptation tactics. In these cases we do not want to split an order and trade in algorithms that adapt trading to real-time market conditions because we do not want these algorithms to compete with one another. For example, if a 1,000,000 shares order is split into two orders of 500,000 shares and given to two different brokers, these algorithms will compute expected impact cost based on their 500,000 shares not on the aggregate imbalance of 1,000,000 shares. This is likely to lead to less than favorable prices and higher than expected costs since the algorithms will likely transact at an inappropriately faster or slower rate. The algorithm may confuse the incremental market impact from the sister order with short-term price trend or increased volatility, and react in a manner inappropriate for the fund, resulting in higher prices. Our preference is to use the arrival cost as our performance metric in the independent sample tests.

A paired observation approach can use any of the static algorithms providing that the underlying trade schedule is the same across brokers and algorithms, e.g., VWAP and POV. An independent sampling approach needs to be used when we are evaluating performance of dynamic algorithms that adapt to changing market conditions.

Non-Parametric Tests

We provide the outline of six non-parametric tests that can be used to determine if two algorithms are equivalent. They are based on paired samples (Sign Test, Wilcoxon Signed Rank Test), independent samples (Median Test, Mann-Whitney U Test) and evaluation of the underlying data distributions (Chi-Square and Kolmogorov-Smirnov goodness of fit). Readers who are interested in a more thorough description of these tests as well as further theory are referred to Agresti (2002), De Groot (1986), Green (2000), and Mittelhammer, Judge and Miller (2000). Additionally, Journal of Trading's "Statistical Methods to Compare Algorithmic Performance" (2007) gives additional background and examples for the Mann-Whitney U test and the Wilcoxon signed rank test. We follow the mathematical approach presented in the JOT article below.

Each of these approaches consists of: (1) devising a hypothesis, (2) the calculation process to compute the test statistic, and (3) comparing that test statistic to a critical value.

Paired Samples

For paired samples the analysis will split the order into two equal pieces and trade each in a different algorithm over the same exact time horizon. It is important in these tests to only use algorithms that do not compete with one another such as VWAP, TWAP, or POV. A static trade schedule algorithm could also be used in these tests since the strategy is predefined and will not compete with another. The comparison metric used in these tests can be either arrival cost or VWAP performance.

Sign Test

The sign test is used to test the difference in sample medians. If there is a statistical difference between medians of the two paired samples we conclude that the algorithms are not equivalent.

Hypothesis:

H_0: Medians are the same ($p = 0.5$)
H_1: Medians are different ($p \neq 0.5$)

Calculation Process:

1. Record all paired observations.
   (X_i, Y_i) = paired performance observations for algorithms X and Y.
   Let $Z_i = X_i - Y_i$.
   k = number of times $Z_i > 0$.
   n = total number of pairs of observations.
2. T is the probability that $z \geq k$ using the binomial distribution

$$T = \sum_{j=k}^{n} \binom{n}{j} \cdot p^j \cdot (1-p_j)^{n-j} = \sum_{j=k}^{n} \binom{n}{j} \cdot (0.5)^j \cdot (0.5)^{n-j}$$

For a large sample the normal distribution can be used in place of the binomial distribution.

Comparison to Critical Value:

- α is the user specified confidence level, e.g., $\alpha = 0.05$.
- Reject the null hypothesis if $T \geq \alpha$ or $T \geq (1 - \alpha)$.

Wilcoxon Signed Rank Test

The Wilcoxon signed rank test determines whether there is a difference in the average ranks of the two algorithms using paired samples. This test can also be described as determining if the median difference between paired observations is zero. The testing approach is as follows:

Hypothesis:

H_0: Sample mean ranks are the same
H_1: Sample mean ranks are different

Calculation process:

1. Let (A_i, B_i) be the paired performance results.
   Let $D_i = A_i - B_i$ where $D_i > 0$ indicates algorithm A had better performance and $D_i < 0$ indicates algorithm B had better performance.
2. Sort the data based on the absolute values of differences $|D_1|, |D_2|, \cdots, |D_n|$ in ascending order.
3. Assign a rank r_i to each observation. The smallest absolute value difference is assigned a rank of 1, the second smallest absolute value difference is assigned a rank of 2, ..., and the largest absolute value difference is assigned a rank of n.
4. Assign a signed rank to each observation based on the rank and the original difference of the pair. That is:

$$S_i = \begin{cases} +r_i & if\ A_i - B_i > 0 \\ -r_i & if\ A_i - B_i < 0 \end{cases}$$

5. Let T_n be the sum of all ranks with a positive difference. This can be determined using an indicator function W_i defined as follows:

$$W_i = \begin{cases} 1 & if\ S_i > 0 \\ 0 & if\ S_i < 0 \end{cases}$$

$$T_n = \sum_{i=1}^{n} r_i \cdot W_i$$

Since the ranks r_i take on each value in the range $r_i = 1, 2, \ldots, n$ (once and only once) T_n can also be written in terms of its observation as follows:

$$T_n = \sum_{i=1}^{n} i \cdot W_i$$

❑ If the results are from the same distribution then the differences D_i should be symmetric about the point $\theta = 0 \rightarrow P(D_i \geq 0) = 1/2$ and $P(D_i \leq 0) = 1/2$.

❑ If there is some bias in performance then differences D_i will be symmetric about the biased value $\theta = \theta^* \rightarrow P(D_i \geq \theta^*) = 1/2$ and $P(D_i \leq \theta^*) = 1/2$.

❑ Most statistical texts describe the Wilcoxon signed ranks using a null hypothesis of $\theta^* = 0$ and alternative hypothesis of $\theta^* \neq 0$. This book customizes the hypothesis test for algorithmic comparison.

6. If performance across algorithms is equivalent then there is a 50% chance that $D_i > 0$ and a 50% chance that $D_i < 0$. The expected value and variance of our indicator function W is as follows:

$$E(W) = 1/2 \cdot 1 + 1/2 \cdot 0 = 1/2$$

$$V(W) = E(X^2) - [E(W)]^2 = 1/2 - \left(1/2\right)^2 = 1/4$$

7. This allows us to easily compute the expected value and variance of our summary statistic T_n. This is as follows:

$$E(T_n) = \sum_{i=1}^{n} i \cdot E(W_i) = \frac{1}{2} \cdot \sum_{i=1}^{n} i = \frac{n(n+1)}{4}$$

$$V(T_n) = \sum_{i=1}^{n} i^2 \cdot V(W_i) = \frac{1}{4} \cdot \sum_{i=1}^{n} i^2 = \frac{n(n+1)(2n+1)}{24}$$

As $n \rightarrow \infty$, T_n converges to a normal distribution and we can use the standard normal distribution to determine our critical value:

$$Z_n = \frac{T_n - E(T_n)}{\sqrt{V(T_n)}}$$

Comparison to Critical Value:

■ Reject the null hypothesis if $|Z_n| > C_{\alpha/2}$ where $C_{\alpha/2}$ is the critical value on the standard normal curve corresponding to the $1 - \alpha$ confidence level.

■ For a 95% confidence test (e.g., $\alpha = 0.05$) we reject the null hypothesis if $|Z_n| > 1.96$.

■ Above we are only testing if the distributions are different (therefore we use a two-tail test).

■ This hypothesis can also be constructed to determine if A has better (or worse) performance than B based on whether $D_i > 0$ or $D_i < 0$ and using a one-tail test and corresponding critical values.

Independent Samples

The independent samples can be computed over different periods, used for different stocks. The total number of observations from each algorithm can also differ. As stated above, it is extremely important for the analyst to randomly assign trades to the different algorithms, ensure similar trading characteristics (side, size, volatility, market cap) and market conditions over the trading period. Below are two non-parametric tests that can be used to compare algorithms using independent samples. It is best to compare like algorithms in these tests such as arrival price, IS, aggressive-in-the-money, etc. Since the orders are not split across the algorithms, they can be dynamic and will not compete with one another.

Mann-Whitney U Test

The Mann-Whitney U test compares whether there is any difference in performance from two different algorithms. It is best to compare "like" algorithms in this case (e.g., IS to IS, ultra-aggressive to ultra-aggressive, etc.). The arrival cost metric is the performance metric in this test.

Hypothesis:

H_0: Same performance
H_1: Different performance

Calculation Process:

1. Let m represent the number of orders transacted by broker A.
   Let n represent the number of orders transacted by broker B.
   Total number of orders $= m + n$.
2. Combine the samples into one group.
3. Order the combined data group from smallest to largest cost.
   For example, the smallest value receives a rank of 1, the second smallest value receives a rank of 2, ..., the largest value receives a rank of $m + n$.
   Identify each observation with an "A" if the observation was from algorithm A and "B" if it was from algorithm B.
4. The test statistic T is the sum of the ranks for all the observations from algorithm A.

This can be computed using help from an indicator function defined as follows:

$$W_i = \begin{cases} 1 & \text{if the observation was from algorithm A} \\ 0 & \text{if the observation was from algorithm B} \end{cases}$$

Then the sum of the ranks can be easily computed as follows:

$$T = \sum_{i=1}^{n} r_i \cdot W_i$$

- If the underlying algorithms are identical the actual results from each sample will be evenly distributed throughout the combined grouping. If one algorithm provides better performance results its sample should be concentrated around the lower cost rankings.
- In the situation where the null hypothesis is true the expected rank and variance of T are:

$$E(T) = \frac{m \cdot (m + n + 1)}{2}$$

$$V(T) = \frac{mn \cdot (m + n + 1)}{12}$$

As with the Wilcoxon signed rank test, it can be shown that as $n, m \to \infty$ the distribution of T converges to a normal distribution. This property allows us to test the hypothesis that there is no difference between broker VWAP algorithms using the standard normal distribution with the following test statistic:

$$Z = \frac{T - E(T)}{\sqrt{V(T)}}$$

Comparison to Critical Value:

- Reject the null hypothesis H_0 if $|Z| > C_{\alpha/2}$.
- $C_{\alpha/2}$ is the critical value on the standard normal curve corresponding to the $1 - \alpha$ confidence level.
- For example, for a 95% confidence test (i.e. $\alpha = 0.05$) we reject the null hypothesis if $|Z| > 1.96$. Notice here we are only testing if the distributions are different (therefore a two tail-test).
- The hypothesis can also be constructed to determine if A has better (or worse) performance than B by specifying a one-tail test. This requires different critical values.

Analysts need to categorize results based on price trends, capitalization, side, etc. in order to determine if one set of algorithms performs better or worse for certain market conditions or situations. Many times a grouping of results may not uncover any difference.

An extension of the Mann Whitney U test used to compare multiple algorithms simultaneously is the Kruskal-Wallis one way analysis of variance test. This test is beyond the scope of this reference book, but readers interested in the concept can reference Mansfield (1994) or Newmark (1988).

Median Test

The median test is used to determine whether or not the medians of two or more independent samples are equal. If the medians of the two samples are statistically different from one another then the algorithms are not equivalent. This test is as follows:

Hypothesis:

H_0: Same medians
H_1: Different medians

Calculation Process:

1. Use arrival cost as the performance measure.
2. Choose two algorithms that are similar (e.g., arrival, IS, etc.). This experiment can be repeated to compare different algorithms.
3. Use a large enough number of orders and data points in each algorithm so that each has a representative sample size. Make sure that the orders traded in each algorithm are similar: size, volatility, market cap, buy/sell, and in similar market conditions.
4. Let X = set of observations from algorithm A.
   Let Y = set of observations from algorithm B.
5. Determine the overall median across all the data points.
6. For each sample count the number of outcomes that are less than or equal ("LE") to the median and the number of outcomes that are greater than ("GT") the median. Use the table below to tally these results.

	Sample A	Sample B	Subtotal
LE overall median	a	b	(a + b)
GT overall median	c	d	(c + d)
Subtotal	(a + c)	(b + d)	(a + b + c + d) = n

7. Compute the expected frequency for each cell

$$ef_{ij} = \frac{\text{total observations in row i} + \text{total observations in column j}}{\text{overall total number of observations}}$$

8. Compute test statistic χ^2

$$\chi^2 = \sum \frac{(\text{number of observations} - ef)^2}{ef}$$

$$\chi^2 = \frac{(a - ef_{11})^2}{ef_{11}} + \frac{(b - ef_{12})^2}{ef_{12}} + \frac{(c - ef_{21})^2}{ef_{21}} + \frac{(d - ef_{22})^2}{ef_{22}}$$

David M. Lane (Rice University) provided an alternative calculation of the test statistic χ^2 that makes a correction for continuity. This calculation is:

$$\chi^2 = \frac{n\left(|ad - bc| - \frac{n}{2}\right)^2}{(a + b)(c + d)(a + c)(b + d)}$$

Comparison to Critical Value:

- $df = (\text{number of columns} - 1) \cdot (\text{number of rows} - 1) = 1$.
- Reject the null hypothesis if $\chi^2 \geq \chi^{2*}(df = 1, \ \alpha = 0.05) = 3.84$.

Distribution Analysis

Distribution analysis compares the entire set of performance data by determining if the set of outcomes could have been generated from the same data generating process ("DGP"). These tests could be based on either pair-samples or independent samples. Analysts need to categorize results based on price trends, capitalization, side, etc. in order to determine if one set of algorithms performs better or worse for certain market conditions or situations. Many times a grouping of results may not uncover any difference in process.

Chi-Square Goodness of Fit

The chi-square goodness of fit test is used to determine whether two data series could have been generated from the same underlying distributions. It utilizes the probability distribution function (pdf). If it is found that the observations could not have been generated from the same underlying distribution then we conclude that the algorithms are different.

Hypothesis:

H_0: Data generated from same distribution
H_1: Data generated from different distributions

Calculation Process:

1. Use the arrival cost as the performance measure.
2. Choose two algorithms that are similar (e.g., arrival, IS, etc.).
3. Trade a large enough number of orders in each algorithm in order to generate a representative sample size. Ensure that the orders traded in each algorithm have similar characteristics such as side, size, volatility, trade time, and market cap, and were traded in similar market conditions.
4. Let X = set of results from algorithm A.
   Let Y = set of results from algorithm B.
5. Categorize the data into groups of buckets.
   Combine the data into one series. Determine the bucket categories based on the combined data. We suggest using from ten to twenty categories based on number of total observations. The breakpoints for the category buckets can be determined based on the standard deviation of the combined data or based on a percentile ranking of the combined data. For example, if using the standard deviation method use categories such as $< -3\sigma$, -3σ to -2.5σ, ..., 2.5σ to 3σ, $3\sigma +$. If using the percentile ranking method order all data points from lowest to highest and compute the cumulative frequency from 1/n to 100% (where n is the combined number of data point). Select break points based on the values that would occur at 10%, 20%, ..., 100% if ten groups, or 5%, 10%, ..., 95%, 100% if twenty buckets. Count the number of data observations from each algorithm that fall into these bucket categories.
6. Compute the test statistic χ^2

$$\chi^2 = \sum_{k=1}^{m} \frac{(\text{observed sample X in bucket k} - \text{observed sample Y in bucket k})^2}{\text{observed sample Y in bucket k}}$$

m = number of buckets.

Comparison to Critical Value:

■ Reject the null hypothesis if $\chi^2 \geq \chi^{2*}(df = m - 1, \alpha = 0.05)$

Kolmogorov-Smirnov Goodness of Fit

The Kolmogorov-Smirnov goodness of fit test is used to determine whether two data series of algorithmic performance could have been generated from the same underlying distributions. It is based on the cumulative distribution function (cdf). If it is determined that the data samples could not have been generated from the generating process then we conclude that the algorithms are different.

Hypothesis:

> H_0: Data generated from same distribution
> H_1: Data generated from different distributions

Calculation Process:

1. Use the arrival cost as the performance measure.
2. Choose two algorithms that are similar (e.g., arrival, IS, etc.).
3. Trade a large enough number of orders in each algorithm in order to generate a representative sample size. Ensure that the orders traded in each algorithm have similar characteristics such as side, size, volatility, trade time, and market cap, and were traded in similar market conditions.
4. Let X = set of results from algorithm A−n observations in total.
   Let Y = set of results from algorithm B−m observations in total.
5. Construct empirical frequency distributions for each data series by ranking the data from smallest to lowest. Let $F_A(x)$ be the cumulative probability for data series A at value x and Let $F_B(x)$ be the cumulative probability for data series B at value x. That is, these functions represent the number of data observations in each respective data series that are less than or equal to the value x.
6. Compute the maximum difference between these cumulative functions over all values. That is:

$$D_n = \left(\frac{mn}{m+n}\right)^{1/2} \max_x \left|F_A(x) - F_B(x)\right|$$

Mathematicians will often write this expression as:

$$D_n = \left(\frac{mn}{m+n}\right)^{1/2} \sup_x \left|F_A(x) - F_B(x)\right|$$

Comparison to Critical Value:

- The critical value is based on the Kolmogorov distribution.
- The critical value based on $\alpha = 0.05$ is 0.04301.
- Reject the null hypothesis if $D_n \geq 0.04301$.

EXPERIMENTAL DESIGN

There are five concerns that need to be addressed when performing the statistical analyses described above. These are: (1) Proper Statistical Test; (2) Small Sample Size; (3) Data Ties; (4) Categorization of Data; and (5) Balanced Sample Set.

Proper Statistical Tests

In statistical testing, the preferred process is a controlled experiment so that the analyst can observe the outcomes from two separate processes under identical market conditions (e.g., Wilcoxon signed rank test). While this is an appropriate technique for static strategies such as VWAP and POV algorithms, it is not an appropriate technique for those algorithms with dynamic trading rates and/or those that employ real-time adaptation tactics. Employing a controlled experiment for dynamic algorithms will likely cause the algorithms to compete with one another and will lead to decreased performance. For dynamic algorithms (e.g., implementation shortfall and ultra-aggressive algorithms) it is recommended that investors utilize the two sample non-pair approach and the Wilcoxon-Mann-Whitney ranks test.

In theory, it is appropriate to compare algorithms with static strategies (e.g., VWAP and POV) with the Wilcoxon-Mann-Whitney ranks test. However, doing so causes more difficulty with regards to robust categorization and balanced data requirements. It is recommended that algorithms with static parameters be compared via the Wilcoxon signed rank test approach.

Small Sample Size

In each of these statistical techniques it is important to have a sufficiently large data sample in order to use the normal approximation for hypothesis testing. In cases where the sample sizes are small (e.g., n and/or m small) the normal distribution may not be a reasonable approximation methodology and analysts are advised to consult statistical tables for the exact distributions of T_n and T. We recommend using at least $n > 100$ and $m > 100$ for statistically significant results.

Data Ties

It is assumed above the results are samples from a continuous distribution (i.e., statistically there will never be identical outcomes). Due to finite precision limitations, analysts may come across duplicate results, inhibiting a unique ranking scheme. In these duplicate situations, it is recommended that the data point be included in the analysis twice. In the case that algorithm "A" is the better result for one data point and algorithm "B" is the better result for the second data point, a unique ranking scheme will exist. If the tail areas of the results are relatively the same this approach should not affect the results. If the tail areas are different this may be a good indication that the data is too unreliable and further analysis is required. Analysts with strong statistical training may choose alternative ranking schemes in times of identical results.

Proper Categorization

When analyzing algorithmic performance it is important to categorize trades by side (buy/sell/short), size, market conditions (such as up and down days), company capitalization (large, mid, and small cap), etc. Categorization allow analysts to determine if one algorithm works statistically better or worse in certain situations. For example, if VWAP algorithm "A" makes market bets by front-loading executions and VWAP algorithm "B" makes market bets by back-loading, "A" will outperform "B" for buys on days with a positive drift and for sells on days with a negative drift. Conversely, algorithm "B" will outperform "A" for buys on days with a negative drift and for sells on days with a positive drift. A statistical test that combines executions from a large array of market conditions may miss this difference in performance especially if we are comparing averages or medians. It is essential that analysts perform robust statistical hypothesis testing for all performance testing techniques.

Balanced Data Sets

It is imperative that analysts utilize a random selection process for submitting orders to algorithms and ensure that the data sets are balanced across the specified categorization criteria, e.g., size, side, capitalization, market movement, etc. This basically states that the percentage breakdown in the categorization groups described above will be similar. Otherwise, the statistical results may fall victim to Simpson's Paradox (e.g., dangers that arise from drawing conclusions from aggregate samples).

FINAL NOTE ON POST-TRADE ANALYSIS

One final note on post-trade analysis is the following. Consider the possibility that performance is equivalent across all families of algorithms. For example, there is no difference across VWAP algorithms, IS algorithms, ultra-aggressive algorithms, etc. Subsequently, two important issues arise. First, can brokers still add value to the trading process? Second, is there any need for third party post-trade services? The answer to both these questions is yes.

Brokers can still add value to the process by providing appropriate pre-trade analysis to ensure proper selection of algorithms and algorithmic parameters. Furthermore, brokers can partner with investors to customize algorithms to ensure consistency across the investment and trading decisions. For example, see Kissell and Malamut (2006) and Engle and Ferstenberg (2006). Most importantly, however, broker competition propels innovation and advancement that continue to benefit investors.

Third party consultants also serve as an essential service to the industry. Not only can they be used by the buy-side to outsource numerical analysis, but more importantly, these consultants have access to a larger universe of trades for various investment styles and algorithms, both robust and balanced, and are thus positioned to provide proper insight into performance and trends. Comparatively, brokers typically only have access to trades using their algorithms and investors only have access to their trades. Access aside, the statistical testing procedure of these consultants cannot remain a black box; transparency is crucial in order for the industry to extract value from their services. Transaction cost analysis remains an essential ingredient to achieve best execution. When administered properly, improved stock selection and reduced costs have proven to boost portfolio performance. As such, advancement of TCA models is an essential catalyst to further develop the algorithmic trading and market efficiency space.

Market Impact Models

INTRODUCTION

This chapter provides an overview of market impact models with emphasis on the "Almgren & Chriss" and the "I-Star" approaches. The Almgren and Chriss (AC) model, introduced by Robert Almgren and Neil Chriss (1997), is a path-dependent approach that estimates costs for an entire order based on the sequence of trades. This is referred to as a bottom-up approach because the cost for the entire order is determined from the actual sequence of trades.

The I-Star model, introduced by Robert Kissell and Roberto Malamut (1998), is a top-down cost allocation approach. First, we calculate the cost of the entire order, and then allocate to trade periods based on the actual trade schedule (trade trajectory). The preferred I-Star formulation is a power function incorporating imbalance (size), volatility, liquidity, and intraday trading patterns.

Alternative market impact modeling approaches have also appeared in the academic literature. For example, Wagner (1991); Kissell and Glantz (2003); Chan and Lakonishok (1997); Keim and Madhavan (1997); Barra (1997); Bertismas and Lo (1998); Breen, Hodrick, and Korajczyk (2002); Lillo, Farmer, and Mantegna (2003); and Gatheral (2010, 2012).

DEFINITION

Market impact is the change in price caused by a particular trade or order. It is one of the more costly transaction cost components and always causes adverse price movement. Market impact is often the main reason managers lag behind their peers. Market impact costs will occur for two reasons: liquidity needs and urgency demands (temporary), and information content (permanent).

Temporary Impact represents the liquidity and urgency cost component. This is the price premium buyers have to provide the market to attract additional sellers and the price discount sellers need to provide

The Science of Algorithmic Trading and Portfolio Management. DOI: http://dx.doi.org/10.1016/B978-0-12-401689-7.00004-0

to attract additional buyers. This cost component can be effectively managed during implementation of an investment decision.

Permanent Impact represents the information cost component. The information content, whether real or perceived, causes market participants to adjust their prices to a new perceived fair value. The rationale is that informed investors typically buy undervalued stock and sell overvalued stock. As participants observe buy orders their perception (at least to some extent) is that the stock is undervalued and they will adjust their offer prices upwards. As participants observe sell orders their perception (again, at least to some extent) is that the stock is overvalued and they will adjust bid prices downward. It is an unavoidable trading cost.

Mathematically, we define market impact as the difference between the actual price trajectory after the order is released to the market and the price trajectory that would have occurred if the order were never released. Regrettably, we cannot observe both price trajectories simultaneously and it is not possible to construct a controlled experiment to measure both trajectories at the same time. As a result, market impact is often referred to as the *Heisenberg uncertainty principle of finance*.

Example 1: Temporary Market Impact

A trader receives a buy order for 50,000 shares of RLK. Market quotes show 1000 shares at $50; 2000 shares at $50.25; 3000 shares at $50.50; and 4000 shares at $50.75. The trader can only execute 1000 shares at the best available price and another 9000 shares at the higher prices for an average price of $50.50. But this only represents 10,000 shares of the original 50,000 share order. In order to attract additional seller liquidity into the market, the trader must offer the market an incremental premium above $50.75. The liquidity and urgency need of this trade caused the trader to incur impact. Another option available to the traders is to wait for additional sellers to arrive at the current market prices. If this occurs the trader will be able to transact at a better price, but if prices move higher due to general market movement while the trader is waiting for sellers to arrive the price could become even higher and the cost more expensive. Waiting for additional counterparties to arrive is always associated with market risk.

Example 2: Permanent Market Impact

A trader receives a buy order for 250,000 shares of RLK currently trading at $50. However, inadvertently this information is released to the market signaling that the stock is undervalued. Thus, investors who currently own stock will be unwilling to sell shares at the undervalued price of $50 and will adjust their price upwards to reflect the new information requiring the buyer to pay, say, an additional $0.10/share higher or $50.10 total. This is an example of the information content cost and permanent market impact.

GRAPHICAL ILLUSTRATIONS OF MARKET IMPACT

This section provides graphical illustrations of market impact from different perspectives.

Illustration 1—Price Trajectory

Madhavan (2000, 2002) presents a lucid graphical description of a sell order's temporary and permanent market impact cost. We use the same technique to graphically illustrate these concepts placing a buy order and sell order.

Figure 4.1a illustrates market impact for a buy order. Following a $30.00 opening trade, the stock fluctuates between $29.99 and $30.01, the result

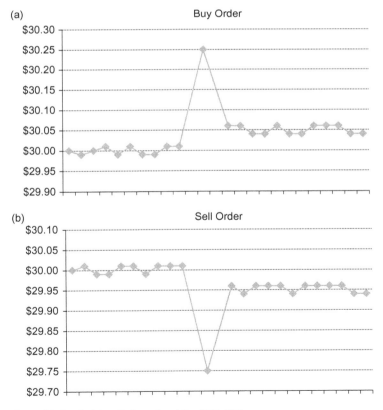

■ **Figure 4.1** Price Evolution. *Madhavan Formulation (2000, 2002).*

of the bid-ask bounce. At this time, an investor enters the market and submits a sizable buy order that immediately pushes the price to $30.25. The premium above current market price serves to attract additional sellers to complete the entire position. Price reversion immediately follows this transaction; but the price reverts to $30.05 not the original price of $30.00. Market participants inferred this trade as information based, due likely to the stock being undervalued. As a result, participants looking to sell additional shares were no longer willing to sell shares at $30.00 but would be willing to offer the shares at $30.05—what they perceive to be the new fair value. The investor incurred $0.25 of total market impact with $0.20 temporary impact (liquidity needs) and $0.05 permanent impact (information content).

Figure 4.1b illustrates the same concept but for a sell order. The stock begins trading at $30.00 and fluctuates between $29.99 and $30.01 due to the bid-ask bounce. An investor with a large order enters the market and immediate pushes the price down to $29.75. The investor needed to discount the price to attract additional buyers. After the transaction, we once again observe price reversion but the price only returns to $29.95, not $30.00. Market participants believed the price was overvalued causing them to re-establish a fair value price. Total cost to the investor is $0.25 with $0.20 being temporary (liquidity needs) and $0.05 being permanent (information content).

Illustration 2—Supply-Demand Equilibrium

We present a second illustration of the market impact concept through traditional economic supply-demand curves (Figure 4.2a–d). We use these curves to show the effect of a buy order on the stock price. Figure 4.2a depicts a system currently in equilibrium with q* shares transacting at price of p*. Figure 4.2b shows the effect of a new buyer entering the market on the equilibrium of the system. Assume the new buyer wishes to transact an incremental Δq shares. This results in a shift in the demand curve from D to D' to reflect the increased demand $q_1 = q^* + \Delta q$. It appears that the new equilibrium price for q_1 shares is p_1 but this is incorrect. Immediately after the new buyer enters the market, the group of existing sellers is likely to believe the demand was driven because the market price was undervalued and they will raise their selling prices. This results in an upwardshift in the supply curve from S to S' and causes the price to increase to p_2 from p_1 (Figure 4.2c) for q_1 shares. The impact from the incremental demand of Δq is p_2-p^*. After the trade, participants re-evaluate the price due to the information content of the trade. Their belief is likely to be that the incremental demand was due to the price

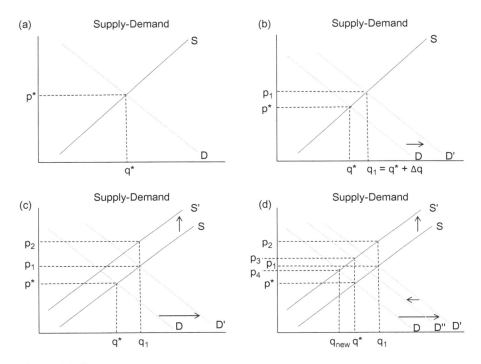

■ Figure 4.2 Supply-Demand Equilibrium.

being undervalued. Sellers will thus increase their asking price in the presence of the newly discovered information causing buyers to pay a higher price.

After shares transact, we face some uncertainty—what happens next?

After the trade the demand curve will shift back to its previous level. But will the supply curve remain at S′ or will it return to S? Will equilibrium quantity return to its original pre-incremental investor level $q_{new} = q^*$ or will equilibrium quantity decrease $q_{new} < q^*$ due to higher market prices? This is shown in Figure 4.2d.

One scenario assumes reduced market volume following the incremental demand. Since price increased due to the trade's information content (shift in the supply curve from S to S′) fewer buyers are willing to transact at higher prices matching the original equilibrium quantity q^*. For example, value managers buy stock only if they are within a specified price range because these prices can generate a higher profit for the manager. Therefore, once the market adjusts its pricing to reflect higher prices managers no longer will purchase those shares because they are outside the specified price range. The result: lower trading volume. The new equilibrium point will be the intersection of the original demand curve and the new supply curve S', in agreement with a new equilibrium price of p_4 and a new equilibrium quantity of q_{new}. Here we expect a post-trade price increase ($p_4 > q^*$) and a post-trade volume decrease ($q_{new} < q^*$). A breakdown of the market impact cost in this scenario is total impact $= p_2 - p^*$, with temporary impact $= p_2 - p_4$ and permanent impact $= p_4 - p^*$.

In a second scenario the original number of buyers may continue to purchase the same number of shares even at higher prices. For example, index managers hold certain stocks and quantities in their portfolio regardless of their market prices because they need to mimic the underlying index. Therefore, after the increment shares Δq are transacted the number of buyers returns to pre-trade levels. Since they are willing to transact at higher prices the demand curve returns to a higher level D''. The new equilibrium point: the point of intersection between S' and D''. Demand is identical to the pre-trade level q^* while the price will be p_3 (higher than the original equilibrium level of p^* and also higher than p_4 (from the first scenario) where we assumed fewer buyers post-trade due to the higher prices). A breakdown of the market impact cost in this scenario is total impact $= p_2 - p^*$, with temporary impact $= p_2 - p_3$ and permanent impact $= p_3 - p^*$. In both scenarios, the total impact of the trade is identical except for new post-trade equilibrium points. This results in computations for permanent and temporary impact along with different expectations for forward looking market volumes.

New equilibrium demand level and price uncertainty are major reasons behind the difficulty in distinguishing between temporary and permanent market impact cost. Regrettably, this is rarely addressed in the financial literature.

The question remains: Does excess market volume lead to more or less volume going forward? We often find that excess market volume corresponds with excess volume in the short-term. However, the higher volume is generally attributed to news—for example, earnings, major macroeconomic events or new announcements, corporate actions, and so on. Higher volume can also tie to investors implementing substantial orders executed over multiple days. We have uncovered evidence of volume returning to its original state as well as volume levels returning to lower levels. No statistical evidence exists suggesting that levels would remain at a higher state. In the rare cases where volume stood higher than pre-trade levels, the reasoning was: (1) the stock joined an index, (2) merger or acquisition, or (3) new product introduction. The best explanation we can offer concerning volume level expectations following a large trade is it will depend on investor mix prior to the large trade/order execution and the underlying reason of the order transaction. Additional research is needed in this area.

Illustration 3—Temporary Impact Decay Function

Temporary market impact is short-lived in the market. But how long exactly does it take for the price to move from the higher levels to the new equilibrium levels? This is referred to as temporary impact decay or dissipation of temporary impact. Figure 4.3a illustrates an example where the price is $30 but a large buy order pushes the price to $30.25. After time, this price reverts to a new equilibrium price of $30.05. But the price does not revert in a single jump but rather over time. Figure 4.3b depicts dissipation of temporary impact for three different decay rates: fast, medium, and slow. This figure shows that analysts need to not only understand the effect of impact but also the speed at which temporary

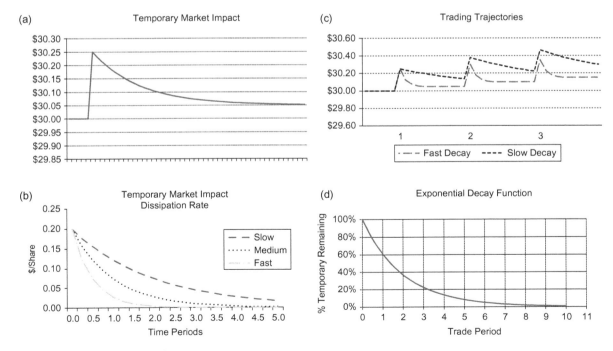

■ **Figure 4.3** Decay of Temporary Market Impact.

impact decays. Investors must understand differences in dissipation rate when they structure trade schedules otherwise they may pay prices much higher than expected.

Figure 4.3c illustrates prices that could occur under a fast and a slow decay rate under the same sequence of trades. In this example, the starting price is $30.00 and each trade's market impact is $0.25, with $0.20 temporary and $0.05 permanent. The first trade price under both trajectories is $30.25. The second trade price is comprised of temporary and permanent impact for the second trade, plus permanent impact of the first trade plus the remaining temporary impact of the first trade. A fast decay rate will cause investors to incur a lower amount of temporary impact than a slow decay rate. In the example, the second trade price is $30.30 for the fast decay rate and $30.38 for the slow decay rate. The slower decay rate causes the investor to incur $0.08/share more than the fast decay rate due to the amount of temporary impact still present in the market from the first trade. The spacing between these trades was just long enough for the temporary impact to dissipate fully for the fast decay function but not nearly enough for the slow decay function. The third trade price is equal to the permanent impact from the first two trades, plus the temporary impact amount of the first trade still, plus the temporary impact of the second trade price (if temporary impact still exists at the time of these trades). Under the fast decay schedule the third trade price is $30.35, which is comprised of permanent impact from all three trades ($0.05/share each) and temporary impact from the third trade only ($0.20/share) since the temporary impact from the preceding two trades has already been fully dissipated in the market. Under the slow decay rate the third trade price is $30.46, which is comprised of permanent impact from all three trades (0.05/share each—permanent impact is a cumulative effect) plus $0.033/share of remaining temporary impact from the first trade, plus $0.081/share of remaining temporary impact from the second trade, plus $0.20/share of temporary impact from the third trade. We see that the average execution price is $30.30 for the fast decay rate and $30.37 for the slow decay rate. Investors must understand the temporary impact function's decay rate to avoid the cumulative effect of temporary impact. The prices under the different decay schedules are shown in Table 4.1.

Table 4.1 Temporary Impact Rate of Decay

Temporary Impact	Trade 1	Trade 2	Trade 3	Avg Price
Fast Decay	$30.25	$30.30	$30.35	$30.30
Slow Decay	$30.25	$30.38	$30.46	$30.37

Figure 4.3d depicts how temporary decay can be determined from an exponential decay function. First, it is important to note the required properties of a decay function include: decreasing over time; always non-negative; approaches zero asymptotically, otherwise the decay function may include some permanent effect; and most important, provides an accurate representation of reality. Too often, we find quasi-quants selecting decay functions that possess only some of these required properties but when tested with real data the function does not provide an accurate description of the system. In many of these cases investors are better off using intuition than depending on insight from a faulty model.

A useful decay function that exhibits these properties and proves an accurate representation of reality is the exponential decay function. This function provides the percentage of temporary impact remaining over time (compared to $t = 0$) and is written as $d(t) = e^{-\gamma \cdot t}$. Here $\gamma > 0$ is the decay parameter that determines the rate of decay. Larger values of γ will decay at a faster rate than smaller values. From this expression the percentage of temporary impact that has already decayed at time t is: $1 - d(t)$.

An appealing property of the exponential decay function is that it decreases at a constant rate. In other words, the percentage reduction from one period to the next is the same. For example, with a parameter of $\gamma = 0.5$, the percentage of temporary impact remaining after the first period $d(1) = e^{-0.5 \cdot 1} = 0.6065$ and after two periods the percentage of temporary impact remaining is $d(21) = e^{-0.5 \cdot 2} = 0.3679$ and can also be written as $d(2) = d(2)^2$. The amount of temporary impact that has decayed after one period in this case is $1 - e^{-0.5 \cdot 1} = 0.3935$. After two periods the amount of temporary impact that has decayed is $1 - e^{-0.5 \cdot 2} = 0.632$. Figure 4.3d illustrates the quantity of temporary impact remaining for this function over several trade periods. Readers can verify that values in this figure match values computed above.

Example—Temporary Decay Formulation

The current price is \$30.00 and the temporary impact of each trade x_k is $f(x_k)$ (we exclude permanent impact here for simplicity). If the decay function parameter is γ the prices for our sequence of trades is:

$$P_0 = 30.00$$

The price of the first trade P_1 is the initial price plus the impact of the trade:

$$P_1 = P_0 + f(x_1)$$

The price of the second trade P_2 is the initial price plus the impact of the second trade plus the remaining impact from the first trade:

$$P_2 = P_0 + f(x_2) + f(x_1) \cdot e^{-\gamma \cdot 1}$$

The price of the third trade P_3 is the initial price plus the impact of the third trade plus all remaining temporary impact from all previous trades:

$$P_3 = P_0 + f(x_3) + f(x_2) \cdot e^{-\gamma \cdot 1} + f(x_1) \cdot e^{-\gamma \cdot 2}$$

Following, the price of the kth trade P_k is:

$$P_k = P_0 + f(x_k) + f(x_{k-1}) \cdot e^{-\gamma \cdot 1} + \cdots + f(x_{k-j}) \cdot e^{-\gamma \cdot j} + \cdots + f(x_1) \cdot e^{-\gamma \cdot (k-1)}$$

A general formulation of this expression is:

$$P_k = P_0 + \sum f(x_j) \cdot e^{-0.5 \cdot (k-j)}$$

Illustration 4—Various Market Impact Price Trajectories

Mathematically, market impact is the difference between the price trajectory of the stock with the order and the price trajectory that would have occurred had the order had not been released to or traded in the market. We are not able to observe both price paths simultaneously, only price evolution with the order or price evolution in the absence of the order. Scientists have not figured a way to construct a controlled experiment that will observe both situations simultaneously. Our failure to simultaneously observe both potential price trajectories' market impact has often been described as the Heisenberg uncertainty principle of finance.

Figure 4.4 illustrates four potential effects of market impact cost. Figure 4.4a shows the temporary impact effect of a trade. The buy order pushes the price up and then it reverts to its original path. Figure 4.4b depicts the permanent impact effect of a trade. The buy order pushes the price up. However, after the trade the price does not revert to its original path, but instead is parallel at a level higher than the original path. Figure 4.4c shows a combination of temporary and permanent impact. First, the order pushes the stock price up followed by some temporary reversion, but in this case the price trajectory remains just slightly higher than and parallel to the original trajectory. Figure 4.4d illustrates temporary impact disguised as permanent impact. In this example, the decay of market impact is extremely slow. It is so slow in fact that temporary impact has not

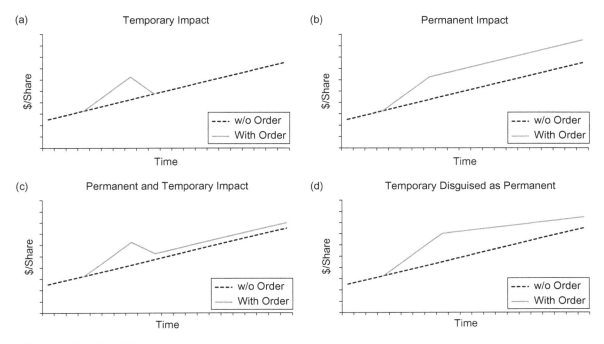

■ Figure 4.4 Market Impact Trajectories.

completely dissipated by the end of the trading horizon or trade day. Thus, the end of day price is composed of permanent impact with a large quantity of temporary impact still remaining. Uninformed analysis may mistakenly identify the full price dislocation as permanent impact. Incorrect identification can have a dire consequence on post-trade attribution and performance evaluation. Recognizing this prospect, many analysts have begun to employ future prices such as the next day's opening price or the closing price on the next trade day or two trade days hence to ensure temporary impact has fully dissipated.

DEVELOPING A MARKET IMPACT MODEL

To best understand the proposed market impact modeling approach it is helpful to review what has been uncovered in previous studies. First, cost is dependent on number of shares traded (e.g., trade size, total order size, or imbalance). This was demonstrated by Loebb (1983); Holtausen, Leftwich and Mayers (1987); Chan and Lakonishok (1993); Plexus Group (2000), etc. Second, costs vary by volatility and market capitalization, e.g., Stoll (1978); Amidhud and Mendelson (1980); Madhavan and Sofianos (1998);

Chan and Lakonishok (1995); Keim and Madhavan (1997); Breen, Hodrick and Korajczyk (2002). Third, price impact results from information leakage and liquidity needs. Fourth, market conditions affect the underlying costs, e.g., Beebower and Priest (1980); Wagner and Edwards (1993); Perold and Sirri (1993); Copeland and Galai (1983); Huang and Stoll (1994), etc. Finally, trading strategy (style) influences trading cost, e.g., Kyle (1985); Bertismas and Lo (1998); Grinold and Kahn (1999) and Almgren and Chriss (1999, 2000). Following these results, we are finally ready to define the essential properties for a market impact model.

Essential Properties of a Market Impact Model

Based on these research and empirical findings, we postulate the essential properties of a market impact model below. These expand on those published in *Optimal Trading Strategies* (2003) and *Algorithmic Trading Strategies* (2006).

P1. *Impact costs increase with size.* Larger orders will incur a higher impact cost than smaller orders in the same stock and with the same strategy.

P2. *Impact costs increase with volatility.* Higher volatility stocks incur higher impact costs for the same number of shares than for lower volatility stocks. Volatility serves as a proxy for price elasticity.

P3. *Impact cost and timing risk depend on trading strategy* (e.g., trade schedule, participation rate, etc.). Trading at a faster rate will incur higher impact cost but less market risk. Trading at a slower rate will incur less impact but more market risk. This is known as the trader's dilemma. Traders need to balance the trade-off between impact cost and risk.

P4. *Impact costs are dependent upon market conditions and trading patterns.* As the order is transacted with more volume the expected impact cost will be lower. As the order is transacted with less volume the expected impact cost will be higher.

P5. *Impact cost consists of a temporary and a permanent component.* Temporary impact is the cost due to liquidity needs and permanent impact is the cost due to the information content of the trade. They each have a different effect on the cost of the trade.

P6. *Market impact cost is inversely dependent upon market capitalization.* Large cap stocks have lower impact cost and small cap stocks have higher impact cost in general (holding all other factors constant). Some difference in cost across market capitalization categories, however, can be explained by volatility. For example, there are examples of small cap stocks with lower costs than some large cap stocks holding all other factors constant and there are examples of large cap stocks having higher costs than small cap stocks holding all other factors constant. This difference, however, can usually be explained through price volatility.

P7. *Trading costs increase with spreads.* Stocks with larger bid-ask spreads have higher trading costs than stocks with smaller spreads (all other factors held constant).

Additional factors that are found to explain differences in impact cost across stocks include:

P8. *Trading stability.* Differences in impact cost at the stock level are also dependent upon the stability of daily volumes and the intraday trading patterns (e.g., how the stock trades throughout the day and the quantity of block volume). Stocks with stable trading patterns (e.g., certainty surrounding day-to-day volumes, intraday volume profile, and quantity of block executions) are generally associated with lower impact cost than stocks exhibiting a high degree of instability (e.g., high uncertainty surrounding day-to-day volumes, large variations in intraday patterns, and choppy or sporadic block executions).

Since large cap stocks are associated with more stable trading patterns and small cap stocks generally relate to less stable trading patterns, market cap is a reasonable proxy for trading stability. However, at times mature small cap companies exhibits more stability than large caps, and vice versa.

P9. *Stock specific risk* (idiosyncratic risk). We found the error in trading cost (measured as the difference between the estimated cost and the actual cost) was correlated to the stock's idiosyncratic risk. This is an indication that price elasticity is dependent on stock volatility but there also appears to be a company specific component.

P10. *Spreads are a proxy for trading pattern uncertainty.* While spreads are treated as a separate transaction cost component, we have found that spreads are also correlated with company specific market impact cost. This finding, however, is more likely to be due to stocks specific trading stability than due to actual spreads (because spread costs were subtracted from the trading cost). Stocks with higher spreads were also usually those stocks with less stable intraday trading patterns. The higher spreads seemed to account for the instability in intraday trading patterns.

DERIVATION OF MODELS

We provide an in-depth discussion of two market impact modeling approaches: the Almgren & Chriss path dependent approach and the I-Star cost allocation approach.

Almgren & Chriss—Market Impact Model

The Almgren & Chriss (AC) market impact model is path dependent based on the actual sequence of trades and executions. Cost is computed as the difference between the actual transaction value of the sequence of trades and the transaction value that would have occurred had all the trades been executed at the arrival price. The AC model follows closely to the graphical representative shown in the price trajectory graphs of Madhavan (2000).

The cost function corresponding to the AC model is:

$$Cost = Side \cdot \left(\sum x_i P_0 - \sum x_i p_i \right) \tag{4.1}$$

where,

$$Side = \begin{cases} +1 & Buy\ Order \\ -1 & Sell\ Order \end{cases}$$

$x_i = shares\ traded\ in\ the\ i^{th}\ transaction$

$p_i = price\ of\ the\ i^{th}\ transaction$

$P_0 = arrival\ price$

$\sum x_i = total\ shares\ traded$

It is important to note here that this calculation only incorporates the trading related transaction cost component and not potential opportunity cost. For purposes of building a market impact model, one of the basic underlying assumptions is that all shares of the order X will be transacted, e.g., $\sum x_i = X$.

The Almgren & Chriss model computes market impact cost for each individual trade. The entire sequence of trades is then rolled up to determine total value traded and total trading cost. Because this approach is based on the sequence of trades the model is referred to as a path dependent approach. Additionally, because total cost is derived from trade level data it is also often referred to as a bottom-up approach.

The side indicator function above allows us to use a consistent expression for buy and sell orders. Many authors prefer to state the trading cost function separately for buys and sells as follows:

$$Cost = \begin{cases} \sum x_i P_0 - \sum x_i p_i & Buys \\ \sum x_i p_i - \sum x_i P_0 & Sells \end{cases}$$

In a later chapter (Chapter 8, Algorithmic Decision Making Framework) we expand on the order completion assumption and introduce ways investors can incorporate opportunity cost into the market impact model and decision making process. We show how investors can develop strategies to maximize the likelihood of executing an order within the desired price range (e.g., within their limit price) and hence minimize the probability of incurring opportunity cost due to adverse price movement.

The Almgren & Chriss model is comprised of three main components: temporary cost function, permanent cost function, and the market impact dissipation function. The temporary and permanent impact functions define how much the stock price will move based on the number of shares traded. The dissipation function defines how quickly the temporary price dislocation will converge or move back to its fair value (or in most situations, the new fair value which incorporates the permanent market impact cost).

Let us utilize a discrete time period random walk model. This process is described below.

Random Walk with Price Drift—Discrete Time Periods

Let the arrival price or starting price be P_0.

The price in the first period is equal to the starting price plus price drift in the first period plus noise (price volatility). That is, $P_1 = P_0 + \Delta P_1 + \varepsilon_1$.

Here ΔP_j represents the natural price movement of the stock in the jth period and is independent of the order (e.g., it would have occurred if the order was or was not transacted in the market), and ε_j is random noise (volatility) in the jth period.

The price in the second period is,

$$P_2 = P_1 + \Delta P_2 + \varepsilon_2$$

By substitution we have,

$$P_2 = P_1 + \Delta P_2 + \varepsilon_2 = (P_0 + \Delta P_1 + \varepsilon_1) + \Delta P_2 + \varepsilon_2 = P_0 + \Delta P_1 + \Delta P_2 + \varepsilon_1 + \varepsilon_2$$

This can also be written as:

$$P_2 = P_0 + \sum_{j=1}^{2} \Delta P_j + \sum_{j=1}^{2} \varepsilon_j$$

The discrete random walk model can then be generalized to determine the expected price P_k at any period of time k as follows:

$$P_k = P_0 + \sum_{j=1}^{k} \Delta P_j + \sum_{j=1}^{k} \varepsilon_j$$

In practice, we often make assumptions about the properties and distribution of the price drift ΔP_j and volatility ε_j terms such as a constant drift term or constant volatility.

In the case where there is no price drift term (e.g., no stock alpha over the period), the discrete random walk model simplifies to:

$$P_k = P_0 + \sum_{j=1}^{k} \varepsilon_j$$

Random Walk with Market Impact (No price drift)

Now let us consider the discrete random walk model without price drift but with impact cost.

Let, $P_0 =$ arrival price, $f(x_k) =$ temporary impact and $g(x_k) =$ permanent impact from x_k shares, and ε is random noise.

The first trade price is:

$$P_1 = P_0 + f(x_1) + g(x_1) + \varepsilon_1$$

The second trade price is the equal to the first trade price plus temporary and permanent impact caused by trading x_2 shares less the quantity of temporary impact from the first trade that has dissipated from the market price at the time of the second trade. This is:

$$P_2 = P_1 + f(x_2) + g(x_2) - \{f(x_1) \cdot (1 - e^{-\gamma \cdot 1})\} + \varepsilon_2$$

where, $f(x_1) \cdot \left(1 - e^{-\gamma \cdot 1}\right)$ represents the reduction of temporary market impact from the first trade.

Now, if we substitute our first trade price into the equation above we have:

$$P_2 = \{P_0 + f(x_1) + g(x_1) + \varepsilon_1\} + \{f(x_2) + g(x_2)\} - \{f(x_1) \cdot \left(1 - e^{-\gamma \cdot 1}\right)\} + \varepsilon_2$$

This reduces to:

$$P_2 = P_0 + \underbrace{\{f(x_2) + f(x_1) \cdot e^{-\gamma \cdot 1}\}}_{Cumulative\ Temporary} + \underbrace{\{g(x_1) + g(x_2)\}}_{Cumulative\ Permanent} + \underbrace{\{\varepsilon_1 + \varepsilon_2\}}_{Cumulative\ Noise}$$

where $f(x_1) \cdot e^{-\gamma \cdot 1}$ is the remaining temporary impact from the first trade.

Following, this formulation, the price in the third period is:

$$P_3 = P_0 + \underbrace{\{f(x_3) + f(x_2) \cdot e^{-\gamma \cdot 1} + f(x_1) \cdot e^{-\gamma \cdot 2}\}}_{Cumulative\ Temporary} + \underbrace{\{g(x_1) + g(x_2) + g(x_3)\}}_{Cumulative\ Permanent}$$

$$+ \underbrace{\{\varepsilon_1 + \varepsilon_2 + \varepsilon_3\}}_{Cumulative\ Noise}$$

After simplifying, we have:

$$P_3 = P_0 + \sum_{j=1}^{3} f(x_j) \cdot e^{-\gamma \cdot (3-j)} + \sum_{j=1}^{3} g(x_j) + \sum_{j=1}^{3} \varepsilon_j$$

In general, the price in period k is:

$$P_k = P_0 + \sum_{j=1}^{k} f(x_j) \cdot e^{-\gamma \cdot (k-j)} + \sum_{j=1}^{k} g(x_j) + \sum_{j=1}^{k} \varepsilon_j$$

With the addition of price drift ΔP into our formulation the equation becomes:

$$P_k = P_0 + \sum_{j=1}^{k} \Delta P_j + \sum_{j=1}^{k} f(x_j) \cdot e^{-\gamma \cdot (k-j)} + \sum_{j=1}^{k} g(x_j) + \sum_{j=1}^{k} \varepsilon_j$$

To estimate the AC model we need to first define our $f(x)$ and $g(x)$ impact functions and corresponding parameters, as well as the dissipation impact rate.

$$f(x) = side \cdot a_1 \cdot x^{a_2}$$

$$g(x) = side \cdot b_1 \cdot x^{b_2}$$

$$decay\ function = e^{-\gamma \cdot t}$$

Then, we have to estimate the following five parameters using actual trade data:

$$a_1, a_2, b_1, b_2, \gamma$$

In practice, it is often difficult to find statistically significant robust and stable parameters over time. Often parameters jump around from period to period and from stock to stock. Furthermore, these parameters frequently take on counterintuitive values such as the case if either $a_2 < 0$ or $b_2 < 0$ which would imply cheaper costs as we increase the quantity of shares traded. This would also create an arbitrage opportunity. For example, an investor would be able to purchase a large number of shares of stock and then sell smaller pieces of the order at higher prices. While this may be appropriate for a large bulk purchase at an outlet store it does not hold true for stock trading.

The Almgren & Chriss model is:

$$P_k = P_0 + \sum_{j=1}^{k} f(x_j) \cdot e^{-\gamma \cdot (k-j)} + \sum_{j=1}^{k} g(x_j) + \sum_{j=1}^{k} \varepsilon_j \tag{4.2}$$

I-STAR MARKET IMPACT MODEL

This section provides an overview of the I-Star market impact model. The model was originally developed by Kissell and Malamut (1998) and has been described in *Optimal Trading Strategies* (Kissell and Glantz, 2003), *A Practical Framework for Estimating Transaction Costs and Developing Optimal Trading Strategies to Achieve Best Execution*, (Kissell, Glantz, and Malamut (2004)), and *Algorithmic Trading Strategies* (Kissell, 2006). The model has greatly evolved since its inception in order to accommodate the rapidly changing market environment such as algorithmic trading, Reg-NMS, decimalization, dark pools, defragmentation, and a proliferation of trading venues, etc. A full derivation of the model is provided below with additional insight into where the model has evolved to incorporate industry and market microstructure evolution.

The I-Star impact model is:

$$I_{bp}^* = a_1 \cdot \left(\frac{Q}{ADV} \right)^{a_2} \cdot \sigma^{a_3} \tag{4.3}$$

$$MI_{bp} = b_1 \cdot I^* \cdot POV^{a_4} + (1 - b_1) \cdot I^* \tag{4.4}$$

$$TR = \sigma \cdot \sqrt{\frac{1}{250} \cdot \frac{1}{3} \cdot \frac{S}{ADV} \cdot \frac{1 - POV}{POV}} \cdot 10_{bp}^4 \tag{4.5}$$

MODEL FORMULATION

I-Star is a cost allocation approach where participants incur costs based on the size of their order and the overall participation with market volumes. The idea behind the model follows from economic supply-demand equilibrium starting at the total cost level[1]. The model is broken down into two components: Instantaneous Impact denoted as I-Star or I* and Market Impact which denoted as MI which represents impact cost due to the specified trading strategy. This impact function is broken down into a temporary and permanent term.

I-Star: Instantaneous Impact Equation

$$I_{bp}^* = a_1 \cdot \left(\frac{Q}{ADV}\right)^{a_2} \cdot \sigma^{a_3} \tag{4.6}$$

In trading, I-Star represents what we call theoretical instantaneous impact cost incurred by the investor if all shares were released to the market. This component can also be thought of as the total payment required to attract additional sellers or buyers to the marketplace. For example, the

---

[1]The reasoning behind this formulation and how it diverges from the Almgren & Chriss expression is simple. Prior to moving into the financial industry I was employed by R.J. Rudden Associates, Inc., a leading global consulting firm specializing in utility cost of service studies as part of rate cases. In these cases, utilities (both natural gas and electric companies) formulated studies to determine serving cost per customer class by mapping actual costs to usage point and allocating this quantity to each party based on usage percentage. This would ensure each customer paid only for services consumed based on the cost of providing that service. While many services were easy to compute usage by each customer, such as electric kwh or natural btu consumption, others were more difficult to compute due to services or mechanics that are shared across customers. For example, there are multiple parties sharing the same generators, overhead transmission lines, pipeline and natural gas storage facilities, as well as corporate functions such as strategy and administrative services. The basic concept of these studies was that we started with a total cost value that was known from accounting records, and then these costs were mapped and allocated to the appropriate customer based on usage and cost to provide the service. This was done to ensure a fair and equitable system across all customers so that no single customer class was being charged more than their fair usage. Those who played a large role (and unknowingly) in the development of an industry leading market impact model include: Rich Rudden, Steve Maron, John Little, Russ Feingold, Kevin Harper, and William Hederman. Thus fittingly, when I was presented with a project to compute and estimate market impact cost, the modeling approach I undertook followed this cost allocation methodology. The I-Star model follows directly from this system. Actual costs as mapped to their underlying components and allocated to point of usage. The methodology is described in this chapter and as we show has many appealing properties for execution strategies, algorithmic trading rules, as well as for portfolio optimization and basket trading strategies.

premium buyers must provide or discount sellers grant to complete the order within a specified timeframe.

In economics, I-Star represents the incremental cost incurred by demanders resulting from a supply-demand imbalance. We depicted this above via a graphical illustration. Following that example, our I-Star cost is determined directly from the imbalance Δq and the corresponding change in price Δp, that is, $I^* = \Delta q \cdot \Delta p = \Delta q \cdot (p_2 - p^*)$ (Figure 4.2d).

The variables of the instantaneous impact equation are:

Q = market imbalance (the difference between buying and selling pressure)
ADV = 30 day average daily volume (computed during exchange hours)
σ = 30 day volatility (day-to-day price change)
a_1, a_2, a_3 = model parameters (via non-linear regression analysis)

Market Impact Equation

$$MI_{bp} = \underbrace{b_1 \cdot I^* \cdot POV^{a_4}}_{Temporary\ Impact} + \underbrace{(1 - b_1) \cdot I^*}_{Permanent\ Impact} \qquad (4.7)$$

Market impact represents the cost that is expected to be borne by the trader based upon the underlying execution strategy, e.g., percentage of volume (POV), trade schedule, etc.

The variables of the model are:

I^* = instantaneous impact
POV = percentage of volume trading rate
b_1 = temporary impact parameters (via non-linear regression analysis)
a_4 = model parameters (via non-linear regression analysis)

Market impact further consists of the temporary and permanent cost component.

Derivation of the Model

Consider a situation where buyers have V shares to buy and sellers have V shares to sell—both within the same time period and urgency needs. In this situation we have an equilibrium condition where the shares to buy are equal to the shares to sell. Therefore, we expect there to be V shares transacted in the market without any extraordinary price movement (but there may be some price movement due to market, natural alpha, or noise).

Now suppose another participant (participant A) enters the market with an order to buy Q shares over the same time and with the same urgency needs. This creates a buy market imbalance equal to the Q shares. Notice that this is equivalent to the Δq shares from our supply-demand above. The new buy shares are $V + Q$ and the sell shares remain at V. In order for these additional Q shares to execute buyers will have to provide a premium to the market to attract the additional sellers.

Let's define this total premium as $I_{\$}^{*}$. Notice here that we are describing this process using dollar units. This process is the same whether dollars, dollars/share, or basis points are used. Our temporary impact parameter (b_1 from Equation 4.7) defines the breakdown between temporary and permanent cost. Total temporary cost is $b_1 \cdot I_{\$}^{*}$ and total permanent cost is $(1 - b_1) \cdot I_{\$}^{*}$.

In this formulation, it is not fair to assume that the entire temporary cost will be borne by participant A alone. The temporary cost will be rather shared (allocated) across all buyers. Think of this approach as an average costing methodology.

Since we now expect there to be $V + Q$ shares traded, that is, the original V shares plus the newly arrived Q shares, the portion of total temporary impact expected to be borne by investor A is calculated in proportion to their total trade volume. This is:

Cost Allocation Method

Temporary market impact cost is dependent upon the underlying trading rate. This rate is expressed in terms of percentage of volume or simply POV. It is:

$$POV = \frac{Q}{Q + V}$$

In this notation, Q is the net imbalance (absolute difference between buying and selling pressure), V is the expected volume excluding the order imbalance, and $Q + V$ is the total number of shares that is expected to trade in the market.

Therefore we have,

$$\text{Temporary Impact} = b_1 \cdot I_{\$}^{*} \cdot \frac{Q}{Q + V} \tag{4.8}$$

$$\text{Permanent Impact} = (1 - b_1) \cdot I_{\$}^{*} \tag{4.9}$$

Or alternatively,

$$MI = b_1 \cdot I_\$^* \cdot \frac{Q}{Q+V} + (1 - b_1) \cdot I_\$^* \tag{4.10}$$

We can see from Equation 4.10 that if participant A transacts more aggressively, say in a shorter period where only half of the expected market volume will transact, that is, $\frac{1}{2}V$, the temporary market impact cost allocated to participant A will now be:

$$\frac{Q}{Q + \frac{1}{2}V}$$

which is a higher percentage than previously.

If A trades over a longer period of time where $2V$ shares are expected to trade, market impact cost allocated to them will be:

$$\frac{Q}{Q + 2V}$$

which is a smaller percentage than previously.

This example helps illustrate that market impact cost is directly related to the urgency of the strategy. Quicker trading will incur higher costs on average than slower trading which will incur lower costs on average. Trading risk, on the other hand, will be lower for the more urgent orders and higher for the more passive orders, i.e. the trader's dilemma.

Due to the rapidly changing nature of the financial markets from regulatory change, structural changes, and investor confidence and perception of order flow information that often accompanies aggressive trading, many participants have begun to fit the market impact model using a more general form of the equation that incorporates an additional parameter a_4. This formulation is:

$$MI_{bp} = b_1 \cdot I^* \cdot \left(\frac{Q}{Q+V}\right)^{a_4} + (1 - b_1) \cdot I^* \tag{4.11}$$

Or in terms of POV we have:

$$MI_{bp} = b_1 \cdot I^* \cdot POV^{a_4} + (1 - b_1) \cdot I^* \tag{4.12}$$

The relationship between temporary impact and POV rate is shown in Figure 4.5. The percentage of temporary impact that will be allocated to the order is shown on the y-axis and the corresponding POV rate is shown on the x-axis. The figure shows the percentage allocated for various POV functions For example, when $a_4 = 1$ the relationship is linear and costs

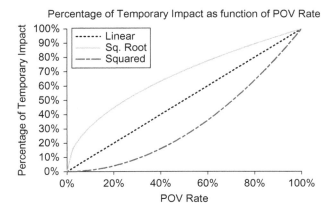

Figure 4.5 Percentage of Temporary Impact as a function of POV Rate.

change at the same rate. When $a_4 = \frac{1}{2}$ the relationship is a squareroot function. Costs increase much quicker for the lower POV rates and at a reduced rate for the higher POV rates. When $a_4 = 2$ the relationship is a squared function. Costs increase at a faster rate for the lower POV rates and at a slower rate for the higher POV rates. Depending upon the rate of change, the a_4 parameter, investors may structure their underlying trading schedule in different ways, either trading faster or slower than normally. Notice that at POV $= 100\%$ investors incur the entire temporary cost regardless of the function of the temporary impact rate.

Up to this point we have not yet defined the functional form of I^*. Our practical experience, empirical evidence, and data observations have found that trading cost is dependent upon size, volatility, and strategy. Thus, our functional form needs to include at least these variables. Some alternative or competing models have included market cap or other stock fundamental factors to differentiate between costs for different stocks even for the same relative size, e.g., 5% ADV. Our analysis has found that volatility provides a better fit than variables such as market cap, log of market cap, etc. At the very least, we need to ensure that the model adheres to the essential properties defined above.

I Formulation*

Our preferred functional form for I^* is the following power function:

$$I_{bp}^* = a_1 \cdot \left(\frac{Q}{ADV}\right)^{a_2} \cdot \sigma^{a_3} \qquad (4.13)$$

Notice the function form includes parameters a_1, a_2, a_3 so that we do not force any preconceived notions onto the model such as a square root function with size or a linear relationship with volatility. These parameter values are derived from our underlying data set.

At this point, we alert readers about setting fixed parameter values. For example, many industry participants set $a_2 = \frac{1}{2}$ which is the square root function—or more precisely, numerous industry participants assume that costs increase with the square root of size. Fixing this parameter value to $a_2 = \frac{1}{2}$ implies that investors will react to order and trade information in the same exact manner for all stocks and markets and in all time periods regardless of the underlying economic conditions. Which is simply not true. Investors react to trade information in different ways at different times. For example, recall the differences in market dynamics during the low volatility period of 2003–2004 and the high volatility period of Sept 2008 through Mar 2009.

Where did this $\frac{1}{2}$ or square root belief come from? We believe the $\frac{1}{2}$ power was set in place due to the way volatility scales with time or because of the $\frac{1}{2}$ parameter behind the optimal economic order quantity model. While in these cases there is a natural reason for the $\frac{1}{2}$ parameter, the same is not true when dealing with market impact cost modeling or price evolution and displacement based on order information.

There have been other functional forms of I^* that have been proposed. For example, *Optimal Trading Strategies* (2003) presents three forms of the I-Star model:

$$I_{bp}^* = a_1 \cdot \left(\frac{Q}{ADV}\right) + a_2 \cdot \sigma + a_3 \qquad (4.14)$$

$$I_{bp}^* = a_1 \cdot \left(\frac{Q}{ADV}\right) + a_2 \cdot \sigma^{a_3} + a_4 \qquad (4.15)$$

$$I_{bp}^* = a_1 \cdot \left(\frac{Q}{ADV}\right)^{a_2} \cdot \sigma^{a_3} \qquad (4.16)$$

We have performed significant testing on these models (as well as other function forms) using numerous data sets, time periods, and global regions, and found the power function formulation to be the most robust, stable, and accurate.

One important question that surfaces: Since the parameters a_1, a_2, a_3, b_1 are estimated across a data set of stocks and are identical for all stocks, how do we differentiate trading cost across different stocks and the same order size?

To address this let us revisit our graphical illustration of market impact using the supply-demand curves. When a new buyer enters the market the demand curve shifts out while the supply curve shifts up to account for order information. A new clearing price P_2 emerges at the intersection of these new curves, and is determined from the slope of the supply and demand curves. This slope happens to be the price elasticity of demand and supply. In actuality, it is often difficult to ascertain the correct price elasticity for a physical good, and even more difficult to ascertain the correct price elasticity for a financial instrument. However, the volatility term serves as an effective proxy for the financial instruments price elasticity term. Notice that volatility is present in each of the variations of I-Star above. Volatility is used in the model to assist us uncover how market impact will differ across stocks. This is explained as follows:

The instantaneous impact equation for a particular stock k is:

$$I_k^* = a_1 \cdot \left(\frac{Q_k}{ADV_k} \right)^{a_2} \cdot \sigma_k^{a_3} \tag{4.17}$$

Rewrite this expression as follows:

$$I_k^* = \underbrace{\{a_1 \cdot \sigma_k^{a_3}\}}_{Sensitivity} \cdot \underbrace{\left(\frac{Q_k}{ADV_k} \right)^{a_2}}_{Shape} \tag{4.18}$$

We now have a sensitivity expression $a_1 \cdot \sigma_k^{a_3}$ which is stock specific and a shape expression $\left(\frac{Q}{ADV} \right)^{a_2}$ which is a universal shape relationship across all stocks. If we have parameter $a_2 = 1$ then we have a linear function where its slope is $a_1 \cdot \sigma_k^{a_3}$. This is identical to the supply-demand representation we showed above. In our formulation, we allow for non-linear supply and demand curves where each stock has its own sensitivity but the shape of the curve is the same across all instruments (which has been found to be a reasonable relationship).

A natural question is why do we not estimate these parameters at the stock level? The answer—we do not have sufficient data to estimate these parameters for all stocks. If we look at market impact for a single stock, change is often dominated by market movement and noise making it very difficult to determine robust and stable parameters at the stock level. In the next chapter, we show challenges behind fitting a stock level model.

Comparison of Approaches

How do the Almgren & Chriss and I-Star models compare? Readers might be interested to know that both of these models will converge to the same trading trajectory for certain parameter values. This was shown by Roberto Malamut (Ph.D., Cornell University) via simulation techniques. In addition, even if the estimated parameters are close to true parameter values for both models the resulting trading trajectories from each model will be extremely close to one another, thus resulting in the same underlying trading strategy. We, however, have found an easier time finding a relationship between cost and order size using the I-Star impact model. But we encourage readers to experiment with both approaches to determine the modeling technique that works best for their needs. Calibration of the I-Star model and required data set is described next.

Underlying Data Set

The underlying data set needed to fit the I-Star model shown above includes: $Q = imbalance\ or\ order\ size$, $ADV = average\ daily\ volume$, $V^* = actual\ trading\ volume$, $\sigma = price\ volatility$, $POV = percentage\ of\ volume$, and $Cost = arrival\ cost$. These variables are described as follows.

Imbalance/Order Size

Recall, we based the I-Star model on total market imbalance (e.g., differences in buyer and seller transaction pressure). Regrettably, exact buy-sell imbalance at any point is neither known nor reported by any source so we have to infer this information from various data sets and techniques. The following methodologies have been used to infer order sizes and imbalances:

Lee and Ready Tick Rule—imbalance is defined as the difference between buy-initiated and sell-initiated trades in the trading interval. A positive imbalance signifies a buy-initiated order and a negative imbalance signifies a sell-initiated order. The Lee and Ready tick rule maps each trade to the market quote at that point in time. Trades at prices higher than the bid-ask spread mid-point are denoted as buy-initiated and trades at prices lower than the bid-ask spread mid-point are designated as sell initiated-trades. Trades exactly at the mid-point are designated based on the previous price change. If the previous change was an up tic, we designate the trade buy-initiated; down tic, a sell-initiated trade. The modified Lee and Ready tick rule assigns trades as buy- or sell-initiated based on price change. An up tic or zero-up tic is known as a buy-initiated trade and a down tic or zero-down tic represents a sell-initiated trade. The proliferation of dark pools prompted many practitioners to exclude trades that occurred inside the bid-ask spread from being designed as buy- or sell-initiated. The difference between buy-initiated and sell-initiated trades is denoted as the order imbalance, and the VWAP price over the period is used as a proxy for the average execution price for the order and is used to compute arrival cost.

Order Data—broker-dealers and vendors (including OMS and EMS companies) have large investor order databases. These actual order sizes and execution prices coupled with the stock specific information serve as input data for the model.

Customer Order Data—customers maintain their own inventory of trades and orders. Clients have access to the full order size and actual execution prices. These data points are used for input in to the model.

ADV = average daily trading volume. This metric is computed based on total market volume during exchange hours over a specified historical period such as a 20 or 30 day period. There has been much discussion regarding the appropriate time period to measure ADV and how the average should be calculated. In Chapter 7, Advanced Algorithmic Forecasting Techniques, we discuss many of the metrics being used and how an analyst can best determine the historical ADV measure for their needs.

V^* = actual market volume over the trading period. If the order was traded in the market over the period from 9:30 a.m. to 4:00 p.m. this measure is the actual volume on the day, but if the order was traded over the period form 10:00 a.m. to 2:00 p.m. then this statistic is measured as total market volume during the time the order was traded in the market, from 10:00 a.m. through 2:00 p.m.

σ = annualized volatility expressed as a decimal (e.g., 0.20 and not 20% or 20). It is computed as the standard deviation of log price returns (close-to-close) over the previous 20 or 30 days. In Chapter 6, Price Volatility, we discuss various different volatility forecasting methods used in the industry. The Journal of Trading article titled "Intraday Volatility Models: Methods to Improve Real-Time Forecasts" Kissell (2012) presents techniques on how analysts can develop real-time volatility forecasts to help improve trading decisions and algorithmic trading performance.

Percentage of Volume = computed as the market imbalance or customer order size divided by the actual market volume that traded in the market during the trading period. That is:

$$POV = \frac{Q}{V^*}$$

Arrival Cost = the difference between the execution price of the order and the arrival price of the order, e.g., the mid-point of the bid-ask spread when the order was released to the market. This measure is usually expressed in basis points. That is:

$$Cost = Side \cdot \frac{P_{avg} - P_0}{P_0} \cdot 10^4 \, bp$$

where,

P_{avg} = average execution price of the order

P_0 = arrival price

$$Side = \begin{cases} +1 & \text{if Buy} \\ -1 & \text{if Sell} \end{cases}$$

Pre-trade *of* pre-trades. Another technique that has become more popular and is being used by portfolio managers is known as the pre-trade of pre-trade approach. Investors use pre-trade estimates provided by multiple broker-dealers and/or vendors for various order sizes, strategies, and stocks. These data points are used as input to the I-Star model, the results of which form a general consensus model of the industry. The technique is further described in Chapter 5, Estimating I-Star Model Parameters, as well as "Creating Dynamic Pre-trade Models: Beyond the Black Box" (Journal of Trading, Fall 2011).

Imbalance size issues. Each of the previously discussed methodologies to derive our order imbalance size is accompanied by some inherent limitations. These include:

1. *Misidentification.* Imbalance is inferred from the trade and may misidentify buys as sells and vice versa.
2. *Survivorship Bias.* Investors often allow orders that are trading well (inexpensive) to continue to trade and cancel those orders that are underperforming (expensive)
3. *Small Orders.* Large concentrations of small orders cause results to be skewed to be more accurate for small trades and potentially less accurate for large trades.
4. *Incomplete Data Set.* B/Ds and vendors are often not familiar with investors and portfolio managers' intentions. They often observe day orders from the fund only (the fund may give a large multi-day order to different brokers each day in order to disguise their trading intentions).
5. *Over-fitting.* The universe of trades is executed in a very similar manner, making it difficult to perform what-if analysis and evaluate alternative trading strategies.

PARAMETER ESTIMATION TECHNIQUES

Estimating parameters of the I-Star model is not as direct as estimating the betas of a linear regression model. The I-Star model is non-linear and dependent on the product of parameters and these parameters are extremely sensitive in the marketplace. As a result it is difficult to find robust parameters that are stable over time.

For example, our model is:

$$I^* = a_1 \cdot \left(\frac{Q}{ADV} \right)^{a_2} \cdot \sigma^{a_3}$$

$$MI = b_1 \cdot I^* \cdot POV^{a_4} + (1 - b_1) \cdot I^*$$

After substitution we have:

$$MI = b_1 \cdot \left(a_1 \cdot \left(\frac{Q}{ADV} \right)^{a_2} \cdot \sigma^{a_3} \right) \cdot POV^{a_4} + (1 - b_1) \cdot \left(a_1 \cdot \left(\frac{Q}{ADV} \right)^{a_2} \cdot \sigma^{a_3} \right)$$

Below are three different techniques that have been used in the industry to estimate the parameters of this model. These techniques are:

1. Two-Step Process
2. Guesstimate Technique
3. Non-Linear Least Squares

Technique 1: Two-Step Process

The two step regression process involves fitting the parameters of the models in two stages. This technique was developed for the case where $a_4 = 1$ and there is a linear relationship between temporary impact and trading cost. This is as follows:

Step 1: Estimate Temporary Impact Parameter

Group the data into different POV rate categories such as 1% buckets. It is important to have a balanced data set so that the sizes and volatilities in each bucket are consistent across different POV rates. Otherwise, sample the data such that each POV bucket has identical order size and volatility characteristics.

One method to fit parameters is to group data into size and volatility buckets and compute each category's average market impact. This grouping is useful because the grouping eliminates noise (better shows a pattern) and provides a balanced data set (e.g., the underlying sample data is not skewed to certain sizes). When we group the data, we filter out small sample categories. For example, require at least 10 or 20 data points for each grouping bucket to be included in the estimation analysis.

Next, run a regression of the average cost of each POV category as a function of the POV rate. That is:

$$Avg\ Cost = \alpha_0 + \alpha_1 \cdot POV$$

Solve for α_0 and α_1 via linear regression analysis. Then estimate b_1 from the estimated alphas as follows:

$$\hat{b}_1 = \frac{\alpha_1}{\alpha_0 + \alpha_1}$$

Notice here that the intercept term α_0 corresponds to POV = 0. It further indicates the cost quantity that will be incurred regardless of the POV rate in the analysis[2].

Step 2: Estimate a_i Parameters

The process to estimate parameters a_1, a_2, a_3 follows from our estimated value for $\hat{b}_1$ above. We have:

$$MI = \hat{b}_1 \cdot I^* \cdot POV + (1 - \hat{b}_1) \cdot I^*$$

Next factor the equation as follows:

$$MI = I^* \cdot \left(\hat{b}_1 POV + \left(1 - \hat{b}_1 \right) \right)$$

Dividing both sides by $\hat{b}_1 POV + \left(1 - \hat{b}_1 \right)$ gives:

$$\frac{MI}{\left(\hat{b}_1 POV + \left(1 - \hat{b}_1 \right) \right)} = I^*$$

---

[2]During my economic consulting days with R.J. Rudden Associates, Inc., we used a similar analysis called a zero mains study. This study served to determine the breakdown between the fixed cost of installing a gas main and the variable cost of the gas main that is dependent upon the size of the main. The intercept term is equivalent to a gas main of zero inches and represents the fixed cost. The two-step regression analysis follows from this type of study.

Next, let $MI' = \dfrac{MI}{\left(\hat{b}_1 POV + \left(1 - \hat{b}_1\right)\right)}$ then we have:

$$MI' = I^*$$

Now rewrite I^* in its full functional form:

$$MI' = a_1 \cdot \left(\frac{Q}{ADV}\right)^{a_2} \cdot \sigma^{a_3}$$

The model parameters can now be estimated via non-linear regression techniques such as non-linear least squares or maximum likelihood estimates. It is important for analysts to evaluate heteroscedasticity and multicollinearity across the data.

After grouping the data, we often see the equation's LHS MI' will have all positive values (or only a few categories with negative values). In these cases we can perform a log transformation of the data. If all LHS data points are positive then we can transform all data points. If there are a few grouping categories with negative values we can transform all the positive LHS records and either eliminate the groupings with negative values or give these records a transformed LHS value of say -3 or -5 (the exact value will depend upon the positive LHS values and units). Another benefit of performing a log transformation is that this transformation will correct (approximately) for the heteroscedasticity—no other adjustment is required.

The log transformation of the equation is as follows:

$$ln(MI') = ln(\alpha_1) + \alpha_2 ln\left(\frac{Q}{ADV}\right) + \alpha_3 ln(\sigma)$$

This regression can now be solved via OLS—which is nice and direct and has easy statistics to interpret. The estimated I-Star parameters are then:

$$a_1 = exp(\alpha_1 + 0.5 \cdot SE), \ a_2 = \alpha_2, \text{ and } a_3 = \alpha_3$$

It is important to note that the two-step process can also be performed using the a_4 parameter and non-linear temporary market impact rate. In this case, we would estimate values for b_1 and a_4 in the first steps, and proceed in exactly the same manner in the second step starting with:

$$MI' = \frac{MI}{\left(\hat{b}_1 POV^{a_4} + \left(1 - \hat{b}_1\right)\right)}$$

Technique 2: Guesstimate Technique

The "guesstimate" technique is actually more complex than a simple guess at the solution. Rather this technique offers an iterative process where we make an educated guess for b_1 and then solve for a_1, a_2, a_3 similar to how we solve in the two-step process above after estimating b_1. For example, we have found from empirical data that for the most part we have $0.80 \leq b_1 \leq 1.00$. Thus, we can perform an iterative solution process where we let b_1 vary across these values. For example, allow b_1 to take on each of the values of $b_1 = 0.80, 0.81, \ldots, 1.00$. Then estimate the alpha parameters via non-linear regression analysis. The best fit solution, and thus, model parameter values, can be determined by the best fit non-linear R^2 statistic.

In the case where the temporary impact POV rate parameter $a_4 \neq 1$ we can devise an iterative approach to allow both b_1 and a_4 to vary over their feasible ranges and determine the best fit for all parameters: a_1, a_2, a_3, a_4, b_1. For example,

Let $b_1 = 0.80, 0.81, \ldots, 1.00$

Let $a_4 = 0.00, 0.05, \ldots, 2.00$

Analysts can choose the increment in these iterative loops to be smaller or larger as they feel necessary.

Technique 3: Non-Linear Optimization

Non-linear optimization is very sensitive to the actual convergence technique used, the starting solution, and of course, the level of correlation across the underlying input variables. The solution often has a difficult time distinguishing parameters and we are left with a good model fit but with estimated parameters based on the composite value $k_1 = b_1 \cdot a_1$. While the model may display an accurate fit, it would be difficult to distinguish between the effects of b_1 or a_1 individually. We could end up with a model that fits data well but which is limited in its ability to provide sensitivity and direction. Analysts choosing to solve the parameters of the model via non-linear regression of the full model need to thoroughly understand the repercussions of non-linear regression analysis as well as the sensitivity of the parameters, and potential solution ranges for the parameters. Non-linear regression estimation techniques are the main topic of Chapter 5.

Model Verification

We introduce methods to test and verify results by first forecasting market influences cost and timing risk using estimated parameters. Estimates are compared to actual costs in four different ways.

Model Verification 1: Graphical Illustration

Plot estimated and actual costs for various order sizes as a scatter graph (cost as y-axis and size as x-axis). Compute average cost for different order sizes to eliminate market noise, making sure to incorporate enough observations in each size category to eliminate the effect of market noise. Graphical illustration is the most helpful performance analysis for clients, although it is the least helpful from a statistical perspective.

Model Verification 2: Regression Analysis

Run a regression between actual and estimated costs using all data. If the forecasting model is accurate, then the regression results should show an intercept statistically equal to zero and a slope statistically equal to one. The R^2 may be lower due to noise but the t-stat and f-value should be very high implying a suitable model. This analysis will show visually whether or not the model is working well (e.g., all order sizes). Regression is the second most useful tool to help clients evaluate our model, and the second most effective statistical technique.

Model Verification 3: Z-Score Analysis

This technique allows us to jointly evaluate both the accuracy of the market impact and timing risk models. The test consists of computing a statistical z-score to determine the number of standard deviations the actual cost was from the estimated cost. The z-score is calculated as follows:

$$Z = \frac{Actual - Cost\ Estimated\ Market\ Impact}{Timing\ Risk}$$

If the model is accurate we should find the average z-score to be close to zero and the standard deviation (or variance) to be close to 1. That is, an accurate model will have:

$$Z \sim (0, 1)$$

It is important that we compute and evaluate the z-score for various order sizes and categories such as buys and sells, market cap, volatility, and so forth to ensure the model is robust or if deficiencies exist.

The distribution of the z-score and the chi-square goodness of fit of the data test will help evaluate the model statistically. This procedure has proven the most useful tool evaluating models from both a statistical basis and real-time TCA analysis (e.g., in the algorithms or from a reporting perspective).

Model Verification 4: Error Analysis

We analyze the error term (regression residual) to determine if we ignored factors driving trading cost. We compute the error term δ_i (difference between estimated and actual) as follows:

$$\delta_i = Estimated\ MI - Actual\ Cost$$

Then we regress δ_i on factors: market movement, side of order (buy vs. sell), sector, order size (to determine robustness of fit), market cap, and so forth. A statistically significant result would indicate the "factor" is a consistent contributor to trading cost.

Important Note: Analysts should perform data verification across all sample orders, grouping data into categories to determine bias. For example, you should perform data verification by order size categories, side of the order (buys and sells separate), by sector, volatility, and market movement (up days and down days). If bias is present, you need to discover where and why bias occurred and follow through to solutions.

Estimating I-Star Model Parameters

INTRODUCTION

In this chapter we introduce a framework enabling readers to build, test, and evaluate market impact models. The much celebrated "I-Star" model will be validated through this framework and techniques. Readers should experiment with alternative models and formulations of I-Star and determine the most suitable methodology given their trading and investing needs.

Our framework is based on the scientific method, a "process" quants use to achieve higher levels of knowledge. The scientific method works quite well—elementary school through Ph.D. dissertations. Yet, it appears, numerous Wall Street analysts seem to have abandoned the scientific in favor of models grounded in quark-gluon plasma ... —say another $3 billion dollar hedging loss!

The scientific method is an "experimentation process" whereby analysts ask and respond to questions objectively. It provides the tools needed to uncover the truth through rigorous experimentation and statistical testing.

Off the record comment: Managers, if your analysts are not following the steps provided here it might be time to replace your analysts. Analysts, if your managers are not asking questions relating to this process, or they are not properly scrutinizing results, it might be time to find a new job. Industry professionals, if your vendors, brokers, consultants, or advisors are not providing essential background material, statistical evidence, and model transparency it is time to find new partners.

Our objective in this chapter is twofold: (1) teach model building and parameter estimation, and (2) help analysts expand their knowledge of market impact models—significantly.

The Science of Algorithmic Trading and Portfolio Management. DOI: http://dx.doi.org/10.1016/B978-0-12-401689-7.00005-2

SCIENTIFIC METHOD

We employ the scientific method throughout this chapter to explore phenomena and observations, and pursue truth and the quest for greater knowledge. The steps are:

1. Ask a Question
2. Research the Problem
3. Construct a Hypothesis
4. Test the Hypothesis
5. Analyze the Data
6. Conclusions and Communication

Step 1: Ask a Question

A second grader may ask "Will my plant grow better using a white light bulb or a yellow light bulb?" A Ph.D. candidate may ask "What is the best monetary policy to stimulate GDP and hold back inflation in a new electronic economy with no country borders and minimal barriers to entry?" A Wall Street quant may simply ask "How much will my order move the market?"

Step 2: Research the Problem

The second step is to learn as much as possible about the problem. It is important to identify what worked and what failed in order to avoid reinventing the wheel and also to avoid potential dead-end approaches and other pitfalls. Ph.D. candidates will likely find that the literature review is one of the most important stages in the dissertation process. To paraphrase Bernard of Chartres and Isaac Newton, you will always reach higher heights when standing on the shoulders of giants.

Step 3: Construct the Hypothesis

The third step is to predict a solution to the problem. In scientific terminology, this is known as specifying the hypothesis. Our market impact model hypothesis includes formulating a mathematical model and incorporating those factors found to influence market price movement. It is important we develop a model that is easily measured and focused. Models that cannot be measured or fail to answer our questions are cast aside.

Step 4: Test the Hypothesis

Step four of the scientific method involves fair, objective, unbiased experiments. Here we perform hypothesis tests on parameters to ensure they are statistically significant and test the overall accuracy of the solution. In addition, we undertake sensitivity analysis on parameters to uncover any inherent limitations.

Step 5: Analyze the Data

We test our hypothesis to ensure the model's appropriateness and that it provides solid estimates of market impact. The experimental tests will either confirm our (model) formulation is appropriate and accurate, rule the model out, or suggest a revision or re-formulation of the model. Quite simply, does the model work? Does it require modification or should it be thrown out?

This will ensure your model offers a solid account of reality. This step also involves sensitivity analysis, evaluating errors, and performing what-if analysis surrounding extreme cases and possibilities (e.g., stress-testing the model). Here we want to learn just about everything we can—where it works well, and where its limitations may reside. We use a control data group for comparisons—that is, we perform an "out-of-sample" test utilizing a data set not included in the calibration phase. Control groups are used everywhere. In medicine—placebo drugs. Physics and engineering—controlled experiments both with and without the factors we are seeking to understand. Mathematicians and statisticians typically hold out, say one-third of the data sample to perform "out-of-sample" testing. If the model fails to predict outcomes accurately, revise the model formulation, your hypothesis, or return to the third step. You may determine in this phase that the model/hypothesis is an inappropriate solution, which is also a valuable piece of information. See *Simulating Neural Networks,* by Freeman (1994) for statistical out-of-sample testing procedures.

Step 6: Conclusion and Communication

Scientists communicate experiment results through wide ranging mediums.

Wall Street for the most, however, fails to share technological advances, particularly if a potential profit opportunity exists. The most accurate models are usually kept under lock and key to be paid for by their top investors or utilized by in-house trading groups.

Mathematical models, like I-Star, offer researchers both a workhorse model and a set of parameters to assist in the decision making process—stock selection, portfolio construction, optimization, trading algorithms, and black box modeling.

We now focus on applying the steps of the scientific method to estimate and test our market impact model.

SOLUTION TECHNIQUE
The Question

How much will my order cost me to trade?

In other words, how much premium is required to attract additional sellers so I can complete my order; how much do I need to discount my price to attract additional buyers so I can complete my order?

Research the Problem

The research step for our market impact modeling problem consists of both academic research (e.g., literature review), and observation and analysis of actual data.

First, let us start with the academic literature. Madhavan (2000, 2002), highly regarded as the ultimate starting point, provides a detailed review of relevant transaction cost analysis research and the market microstructure literature leading up to algorithmic trading. Almgren and Chriss (1997); Kissell and Glantz (2003); Kissell, Glantz and Malamut (2004); Wagner (1991); Gatheral (2010, 2012); and Domowitz and Yegerman (2006, 2011) provide us with a strong foundation and starting point for algorithmic trading findings.

Our review of the literature provided many key findings. Cost is dependent upon the number of shares traded (e.g., trade size, total order size, or imbalance). This has been shown by Loebb (1983); Holtausen, Leftwich and Mayers (1987); Chan and Lakonishok (1993); Plexus Group (2000) and others. Empirical evidence reveals that costs vary by volatility and market capitalization. For example, see Stoll (1978); Amidhud and Mendelson (1980); Madhavan and Sofianos (1998); Chan and Lakoniskhok (1995); Keim and Madhavan (1997); and Breen, Hodrick and Korajczyk (2002) to name a few.

Price impact results directly from trade and/or information leakage, as well as the liquidity needs of the investor or institutional fund. Market conditions over the trading period highly affect the underlying costs as well. See Beebower and Priest (1980); Wagner and Edwards (1993); Perold and Sirri (1993); Copeland and Galai (1983); and Stoll (1995). Additionally, there has been numerous evidence presented by Kyle (1985), Bertismas and Lo (1998); Grinold and Kahn (1999); and Almgren and

Chriss (1999, 2000) that finds that trading strategy (style) influences trading cost. Breen, Hodrick and Korajczyk (2002) provide a foundation for testing models which we utilize in the model error analysis section of the scientific method (Step 5: Analyze the Data).

From these publications a common underlying theme: price impact is caused by order size, trading strategy (e.g., level of transaction urgency, percentage of volume, participation rate, etc.), volatility, market capitalization, side (buy/sell)—as well as spreads and price.

Next we observe and analyze actual customer order data. Our data observation universe consists of actual executed trades during a three month period Jan. 2010 through Mar. 2010. While there are no specific steps to observe and analyze data, we recommend visualization by plotting data and simple analyses such as linear regression to uncover potential relationships.

As part of this step, we plotted the average trading cost (measured as the difference between the average execution price and mid-point of the spread at the time of the order arrival) as a function of several different variables including size, volatility, percentage of volume (POV) rate, and price. We segmented data into large cap and small cap stock (as per our literature research findings). Stocks with market capitalization of $2B or more were classified as large cap stocks, while stocks with market capitalization less than $2B were classified as small cap stocks. Traditionally, large cap stocks are categorized as stocks with a market cap greater than $5B, mid-cap stocks as stocks with a market cap between $2B and $5B, and small cap stocks as stocks with market caps less than $2B. We grouped mid-cap stocks in the large cap or small cap categories based on actual market cap. Readers are welcome to repeat steps in this chapter to determine if an additional category of stocks is needed.

There are several issues worth mentioning:

- First, share amounts traded were not necessarily the entire order. We were not able to observe the number of unexecuted shares or opportunity cost corresponding to the total order size. This may lead to survivorship bias where orders with favorable price momentum are completed more often than orders with adverse market movement resulting in lower than actual observed costs.
- Second, we were unable to observe actual specified trading strategy at the beginning of trading. We do not know whether traders or managers engaged in any opportunistic trading during the execution of the order. This would occur in situations whereby traders became more

aggressive in times of favorable price momentum and less aggressive in times of adverse price momentum and might give the impression that trading at faster rates could lead to lower costs than trading at a slower rate. We only had access to the actual trading conditions and market volumes over the trading horizon.

- Third, we did not discard data points or stocks. Readers and analysts should remember that our research step is a learning step. All data points should be included and observed to fully understand the underlying data set and system at hand—including outliers.
- Fourth, we require at least 25 data points for each bucket; less than 25 data points does not provide a reasonable point estimate of the trade cost for that particular interval. Unfortunately, the fourth requirement resulted in largely smaller trade sizes.
- Fifth, we did not know if these shares were part of a larger order where parts were executed the previous day(s) and/or portions were to be executed on subsequent day(s).
- Finally, the R^2 statistic reported is the R^2 for the grouped and averaged data set. In these cases, the measure is often inflated—however, it does provide insight into whether the dependent variables are related to the independent variable (e.g., trading cost). Here we are still in our learning mode—so the R^2 on grouped data indeed provides valuable insight.

Our graphical illustrations contained all data points. We computed buckets for the x-value and then computed the average cost for all data points that fell into that particular bucket. For example, we computed the average cost for all order sizes that were 5% ADV (rounded to nearest 1% ADV). The data points included in these intervals may have varying volatilities and POV rates as well as prices. All of these variables will have some effect on the actual cost of the trade. However, in this research and learn step, even this type of averaging approach will yield some insight into the underlying relationships between cost and variable.

Large Cap Stocks. Our large cap observations are shown in Figure 5.1. Findings mirror results in leading academic research. Costs were positively related to size, volatility, and POV rate, and negatively related to price. We ran a simple linear regression on the grouped data for each explanatory variable separately. The strongest relationship (based on R^2) for cost was size ($R^2 = 0.44$). The second strongest relationship was with volatility ($R^2 = 0.40$). This was followed by POV rate ($R^2 = 0.36$) and then price ($R^2 = 0.24$). In each of these cases, visual inspection shows that the relationship between cost and explanatory variable may be non-linear.

Small Cap Stocks. Our small cap observations are illustrated in Figure 5.2. Our findings also reflected the academic research with large

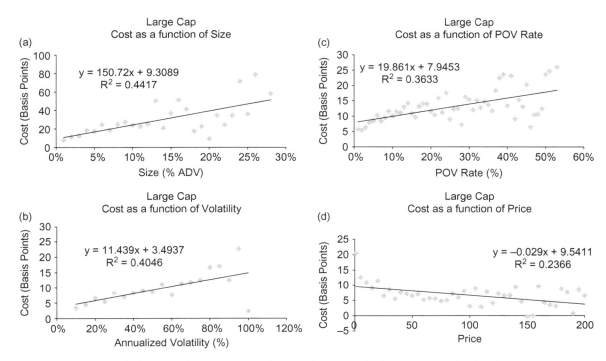

■ **Figure 5.1** Large Cap Stock Observations. (a) Large Cap Cost as a function of Size; (b) Large Cap Cost as a function of Volatility; (c) Large Cap Cost as a function of POV Rate; (d) Large Cap Cost as a function of Price.

cap stocks. Costs were positively related to size, volatility, and POV rate, and negatively related to price. The simple linear regression on the grouped data and each explanatory variable separately determined the strongest relationship for cost was with volatility ($R^2 = 0.71$). The second strongest relationship was with size ($R^2 = 0.40$). This was followed by POV rate ($R^2 = 0.39$) and then price ($R^2 = 0.12$). Similar to the large cap universe, our visual inspection of actual data shows that the relationship with small cap stocks and our variables also appears to be non-linear.

Market Cap. We analyzed the relationship between trading costs and natural log of market cap. This is shown in Figure 5.3. Here the relationship is negative—costs are lower for larger stocks as well as non-linear. Small cap stocks were more expensive to trade than large cap stocks. The relationship between trading cost and market cap was fairly strong with $R^2 = 0.70$.

Spreads. Figure 5.4 reveals costs as a function of spreads. Average spreads over the day were rounded to the nearest basis point (bp). The

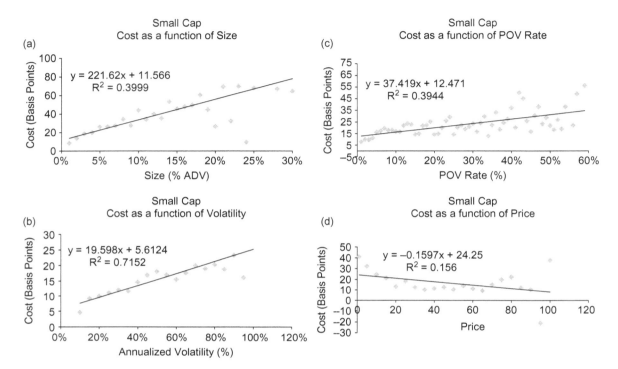

■ **Figure 5.2** Small Cap Stock Observations. (a) Small Cap Cost as a function of Size; (b) Small Cap Cost as a function of Volatility; (c) Small Cap Cost as a function of POV Rate; (d) Small Cap Cost as a function of Price.

relationship between trading cost and spreads is positive such that stocks with higher spreads show higher trading cost. The fit was $R^2 = 0.13$. Keep in mind—spreads are a separate transaction cost component.

Table 5.1 summarizes the results of our simple linear regression analysis of cost on each variable separately. We include the standard error and t-stat, along with coefficient and R^2 for each regression.

At this point, judicious analysts may prefer to perform additional analyses on the data to evaluate linear vs. non-linear relationship, as well as the correlation across explanatory factors and its effect on costs. For example, larger orders are usually executed with higher POV rates and smaller orders more often with lower POV rates. This unfortunately introduces a high degree of correlation. Furthermore, smaller cap stocks typically have higher volatility and larger cap stocks have lower volatility resulting in additional negatively correlated variables.

It is important to account for correlation across dependent variables when you estimate the model's parameters and test for statistical significance.

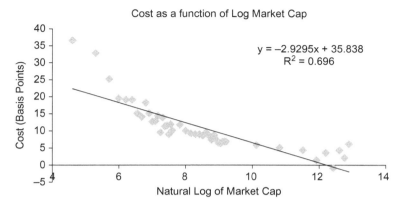

■ **Figure 5.3** Cost as a function of Log Market Cap.

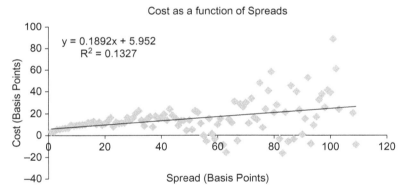

■ **Figure 5.4** Cost as a function of Spreads.

Our goal in the research step was simply to learn from the data and we found that our data set was consistent with the previous research findings and academic literature.

Construct the Hypothesis

Our hypothesis is that market impact cost follows a power function relationship with size, volatility, and strategy (POV rate), and is comprised of a temporary and a permanent impact component. The formulation is the now famous "I-Star" model:

$$I_{bp}^* = a_1 \cdot \left(\frac{S}{ADV} \right)^{a_2} \cdot \sigma^{a_3} \tag{5.1}$$

Table 5.1 Actual Trading Costs—Simple Linear Regression Results

Large Cap Stocks

	Size	Volatility	POV	Price
Est.	150.72	11.44	19.86	−0.03
SE	33.89	3.37	3.72	0.01
t-stat	4.45	3.40	5.34	−3.48
R^2	44%	40%	36%	24%

Small Cap Stocks

	Size	Volatility	POV	Price
Est.	221.62	19.60	37.42	−0.16
SE	53.24	3.00	6.25	0.09
t-stat	4.16	6.53	5.99	−1.87
R^2	40%	72%	39%	16%

Stock Characteristics

			LnMktCap	Price
Est.			−2.93	−876.64
SE			0.30	175.68
t-stat			−9.92	−4.99
R^2			70%	37%

Notes:
Simple linear regression results
Costs were analyzed compared to each variable separately
These results are not that of a multi-linear regression

$$MI_{bp} = b_1 \cdot I^* \cdot POV^{a_4} + (1 - b_1) \cdot I^* \tag{5.2}$$

where,

I = instantaneous impact cost expressed in basis points (bp)
MI = market impact cost expressed in basis points (bp)
S = shares to trade
ADV = average daily volume
POV = percentage of volume expressed as a decimal (e.g., 0.20)
σ = annualized volatility expressed as a decimal (e.g., 0.20)
a_1, a_2, a_3, a_4, b_1 = model parameters (estimated below)

Note that quite often we will rewrite our market impact model using the variable *Size* as follows:

$$I^*_{bp} = a_1 \cdot Size^{a_2} \cdot \sigma^{a_3} \tag{5.3}$$

where,

$$Size = \frac{Shares}{ADV}$$

Test the Hypothesis

The fourth step in the scientific method consists of testing our hypothesis. But before we start testing the hypothesis with actual data it is essential to have a complete understanding of the model and dependencies across variables and parameters. We need to understand what results are considered feasible values and potential solutions.

Let us start with the interconnection between the a_1 and a_2 parameters. Suppose we have an order of 10% ADV for a stock with volatility of 25%. If we have $I^* = 84$ bp and $\hat{a}_3 = 0.75$, then any combination of a_1 and a_2 that satisfies the following relationship are potential feasible parameter values:

$$I^*_{bp} = a_1 \cdot Size^{a_2} \cdot \sigma^{0.75}$$

or,

$$84 \text{ bp} = a_1 \cdot (0.10)^{a_2} \cdot (0.25)^{0.75}$$

Solving for a_2 in the above formula yields:

$$a_2 = ln\left(\frac{I^*}{a_1 \cdot \sigma^{a_3}}\right) \cdot \frac{1}{ln(Size)}$$

Using the data in our example we have:

$$a_2 = ln\left(\frac{84}{a_1 \cdot 0.25^{a_{0.75}}}\right) \cdot \frac{1}{ln(0.10)}$$

However, not all combinations of a_1 and a_2 are feasible solutions to the model. Having prior insight into what constitutes these feasible values will assist dramatically when estimating and testing the model parameters. For example, we know that $a_2 > 0$ otherwise costs would be decreasing with order size and it would be less expensive to transact larger orders than smaller orders and there would be no need to slice and order and trade over time (not to mention an arbitrage opportunity).

Figure 5.5a illustrates this point by showing the combinations of a_1 and a_2 that result in $I^* = 84$ for $100 < a_1 < 1250$. But notice that for values of $a_1 < 230$ the resulting a_2 value is negative thus violating one of our feasibility requirements.

Another relationship we have with the I-Star model is with parameter b_1, which is the temporary market impact parameter—in other words, the percentage of instantaneous impact (I^*) that is due to the liquidity needs of the investor and/or immediacy needs. Thus we have by definition $0 \le b_1 \le 1$.

Next, suppose that we have a market impact cost of MI $= 25.3$ bp for a stock with volatility $= 25\%$ and POV rate $= 30\%$ with known parameters $a_2 = 0.50$, $a_3 = 0.75$ and $a_4 = 0.50$.

In this example, we have a known value of MI but the value of I-Star is not known. Then as long as the combinations of a_1 and b_1 result in MI $= 84$ bp the equation is correct.

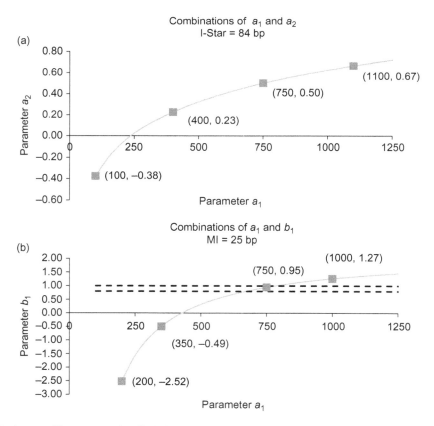

■ **Figure 5.5** (a) Combinations of Parameter: a_1 and a_2 (b) Combinations of Parameter: a_1 and b_1.

The set of potential feasible solutions are those where we have b_1 expressed in terms of a_1 as follows:

$$b_1 = \left(\frac{MI}{I^*} - 1 \right) \cdot \left(\frac{1}{POV^{a_4} - 1} \right)$$

Using the data in our example we have,

$$b_1 = \left(\frac{25.3}{I^*} - 1 \right) \cdot \left(\frac{1}{0.30^{0.5} - 1} \right)$$

Figure 5.5b depicts the combinations of b_1 and a_1 resulting in solutions with MI = 84 bp. But notice that there are several combinations of a_1 and b_1 that are not feasible solutions to the problem. Since we have a constraint on b_1 such that $0 \leq b_1 \leq 1$ we additionally have $425 < a_1 < 800$. Further, empirical evidence has found $b_1 > 0.70$, thus we have $625 < a_1 < 800$.

Performing these types of sensitivity analyses around the model parameters for those parameters with a known interval will greatly help analysts critique models and results.

Data Definitions

We are now ready to begin testing the I-Star model with actual data. As mentioned in Chapter 4, we will be calibrating our model using market tic data.

Using market data as opposed to actual customer order data will provide three major benefits. First, it provides us with a completely independent data set for estimation and allows us to use the customer order data set as our control group that will be used for comparison purposes. Second, using the market data universe will allow us to eliminate some inherent biases in the data due to potential opportunistic trading where investors trade faster and in larger quantities (shares are added to the order) in times of favorable price momentum, and slower and in smaller quantities (shares or cancelled/opportunity cost) in times of adverse price momentum. Third, this eliminates situations where we (i.e., the broker or vendor) are not provided with the complete order because it is traded over multiple days.

This section describes the data sets used to estimate the market impact parameters. The underlying data used for compiling the data statistics is actual tic data. This is also known as time and sales data, and includes the price of the trade, the number of shares transacted, and date and time of the trade. The data is available via the New York Stock Exchange (e.g., TAQ data for trade and quote data) for all securities traded in the US and/or from various third party data vendors.

Table 5.2 MI Data Sources

Factor	Data Source
Buy Volume	Tic Data
Sell Volume	Tic Data
Volume	Tic Data
Turnover	Tic Data
VWAP	Tic Data
First Price	Tic Data
Cost	Tic Data
Imbalance	Tic Data
ADV	End of Day
Volatility	End of Day
Size/Imbalance	Derived
POV	Derived

The data elements that need to be compiled and or recorded are shown in Table 5.2.

Universe of Stocks

The universe of stocks used for parameter estimation is the S&P1500. This provides a robust sample of 500 large cap stocks, 400 mid-cap stocks, and 600 small cap stocks.

Analysis Period

The time period for the analysis is the three month period Jan. 2010 to Mar. 2010 (and is the same period as is available for the control group).

Time Period

Data was compiled for three times periods. Full day 9:30 a.m. to 4:00 p.m. Morning 9:30 a.m. to 1:00 p.m. and Afternoon 1:00 p.m. to 4:00 p.m.

Number of Data Points

There were 1500 stocks, three periods per day, and about 65 trading days over the three months resulting in N = 292,500 data points.

Imbalance

Daily imbalance is estimated from actual tic data during exchange hours only (e.g., all trades between the hours of 9:30 a.m. and 4:00 p.m. or within our morning or afternoon periods). Data is first sorted in ascending order by time and trades are designated as buy-initiated or sell-initiated based on the modified Lee & Ready (1993) tic rule. Buy-initiated trades are those trades that occurred on an up tic or zero-up tic. Sell-initiated trades are those trades that occurred on a down tic or zero-down tic. Trades are not able to be designated as a buy-initiated or sell-initiated until after the first price change.

Imbalance is computed as the absolute difference between buy-initiated and sell-initiated volume for the particular period. The calculation is as follows:

$$Q = \left| \sum Buy\ Volume - \sum Sell\ Volume \right| \tag{5.4}$$

Side

The side of the imbalance is "buy" if there is more buy-initiated volume and "sell" if there is more sell-initiated volume. Mathematically, the side designation is:

$$Side = \begin{cases} +1 & if \sum Buy\ Volume > \sum Sell\ Volume \\ -1 & if \sum Sell\ Volume > \sum Buy\ Volume \end{cases} \tag{5.5}$$

Volume

Total market volume that traded over the same period used to calculate the imbalance.

$$V(t) = \sum_{i=1}^{t} v_i \tag{5.6}$$

Where t denotes the total number of trades during the period and v_i is the volume corresponding to the ith trade in the period.

Turnover

Turnover is the total dollar value traded during the trading period.

$$Turnover(t) = \sum_{i=1}^{t} p_i \cdot v_i \tag{5.7}$$

where p_i is the price of the ith trade.

VWAP

VWAP is the volume weighted average price during the trading period.

$$VWAP = \frac{\sum_{i=1}^{t} p_i \cdot v_i}{\sum_{i=1}^{t} v_i} \tag{5.8}$$

First Price

The mid-point of the bid-ask spread at the beginning of the trading interval. This is denoted as P_0.

Average Daily Volume

The average daily traded volume (ADV) in the stock over the previous T trading days. The value of T does vary by practitioner. For example, the more common historical periods are 10, 22, 30, and 66 days of data. Earlier we found that T = 30 days of data are a sufficient number of data points to measure the mean.

$$ADV = \frac{1}{T} \sum_{i=1}^{T} V_i(day) \tag{5.9}$$

where $V_i(day)$ is the total volume that traded on the ith historical day (e.g., i days ago).

Annualized Volatility

Annualized volatility is the standard deviation of the close-to-close logarithmic price change scaled for a full year using a factor of 250 days. Many practitioners use a 252 day scaling factor. However, for our purposes estimating market impact, the difference is negligible. Annualized volatility is included in the market impact model as a proxy for price volatility. For consistency, we use T = 30 days of data to compute our volatility estimate.

$$\sigma = \sqrt{\frac{250}{T-1} \sum_{i=2}^{T} (r_i - r_{avg})^2} \tag{5.10}$$

where, r_i is the log return on the ith historical day and r_{avg} is the average log return over the period. It is important to note that our annualized volatility is expressed as a decimal (e.g., $0.20 = 20\%$).

Size

The imbalance size expressed as a percentage of ADV. It is expressed as a decimal, that is, an imbalance size of 30% ADV is expressed as 0.30.

$$Size = \frac{Q}{ADV} \tag{5.11}$$

POV Rate

The percentage of volume rate (POV) is computed from imbalance and period volume. It is a proxy for trading strategy. It is important to note that percentage of volume is expressed as a decimal.

$$POV = \frac{Q}{V(t)} \tag{5.12}$$

Cost

Cost is defined as the difference between average execution price and the first price (expressed as a fraction of the initial price). It follows the definition of trading cost used in the implementation shortfall methodology (Perold, 1988). Here we compute cost as the logarithmic price change between average execution price and arrival price. We use the VWAP price over the interval as our proxy for average execution price. This calculation is as follows:

$$Cost = ln\left(\frac{VWAP}{P_0}\right) \cdot Side \cdot 10^4 bp \tag{5.13}$$

Additional Insight: we discuss the expanded implementation shortfall measure introduced by Wagner (1991) in Chapter 3, Transaction Cost Analysis. In Chapter 6, Price Volatility, we discuss techniques to estimate forward looking volatility. In Chapter 7, Advanced Algorithmic Forecasting Techniques, we discuss alternative methodologies for estimating daily volume. These include using various time periods, the mean vs. median, as well as advanced statistical measures such as autoregressive moving averages (ARMA).

Estimating Model Parameters

Estimation of the parameters for the complete I-Star model requires non-linear estimation techniques such as non-linear least squares, maximum likelihood, generalized method of moments, etc. In Chapter 4, Market Impact Models, we discuss three techniques including a two-step process, a guesstimate, and non-linear regression analysis.

In this section we use non-linear least squares regression techniques to estimate the parameters of our model.

$$I = a_1 \cdot Size^{a_2} \cdot \sigma^{a_3} \tag{5.14}$$

$$MI = b_1 \cdot I \cdot POV^{a_4} + (1 - b_1) \cdot I \tag{5.15}$$

These parameters are: $a_1, a_2, a_3, a_4,$ and b_1.

Outliers. To avoid potential issues resulting from outliers we filtered our data points based on daily stock volume and overall price movement. If a data point was outside of a specified range we excluded that data point. Filtering is commonly done on market impact data sets to avoid the effect of high price movement due to a force or market event that is not due to the buying or selling pressure of investors.

In our analysis we filtered the data to include only those data points with:

1. Daily Volume $\leq$ 3*ADV
2. $\frac{-4 \cdot \sigma}{\sqrt{250}} \leq$ Log Price Change (close-to-close) $\leq \frac{+4 \cdot \sigma}{\sqrt{250}}$

We decided to use four times the daily volatility to account for the potential incremental price movement due to the buying or selling pressure in an adverse momentum market. Analysts may choose to use different break points as well as filtering criteria.

Factor Independence. As is the case with any regression analysis we require explanatory factors to be independent. Unfortunately, our derivation process results in correlation across factors but this correlation is reduced by using multiple time horizons (full day, morning, and afternoon). Analysts can further reduce the correlation across factors through a sampling process of the data where we select a subset of data points such that the cross factor correlation is within a specified level (e.g., $-0.10 \leq rho \leq 0.10$). Our resulting data set had $N = 180,000$ points and the resulting correlation matrix is shown in Table 5.3.

Table 5.3 Factor Correlation Matrix

	Size	Volatility	POV
Size	1	−0.05	0.08
Volatility	−0.05	1	−0.03
POV	0.08	−0.03	1

Analysts can determine what is considered to be an acceptable level of cross factor correlation for their particular needs and determine the data sample set within these criteria through a random sampling process.

Heteroscedasticity. Analysis of the complete model above reveals potential heteroscedasticity of the error term. Each stock in the sample has different volatilities and POV rates (resulting in different trading times) and a different distribution of the error term. Kissell (2006) provides techniques to correct for heteroscedasticity in this model. One important note, however, is that after grouping the data into bins there is not much difference between the parameter estimation results without correcting for heteroscedasticity. We highly recommend analysts perform both analyses to understand the dynamics of this model before deciding if the heteroscedasticity step can be ignored.

Grouping Data. Prior to performing our non-linear regression we grouped our data into buckets in order to average away noise and to ensure a balanced data set. Data was bucketed into categories based on size, volatility, and POV rate according to the following criteria:

Size = 0.5%, 1%, 2%, . . ., 30%
Volatility = 10%, 20%, . . ., 80%
POV Rate = 1%, 5%, 10%, . . ., 65%

If we use too fine increments excessive groupings surface and we have found that a substantially large data grouping does not always uncover a statistical relationship between cost and our set of explanatory factors.

Next we averaged costs for each category above. We required at least 25 observations in order for the bucket to be included in the regression analysis. We required at least 25 data points in order to average away noise and determine the most likely cost given the category of size, volatility, and POV rate.

Sensitivity Analysis

We discussed briefly that we need to ensure a solution with a feasible set of parameter values. In other words, we are setting constraints on the model parameters. These feasible values are:

$$100 \leq a_1 \leq 1000$$
$$0.10 \leq a_2 \leq 1.0$$
$$0.10 \leq a_3 \leq 1.0$$
$$0.10 \leq a_4 \leq 1.0$$
$$0.70 \leq b_1 \leq 1$$

It is important to note here that the feasible range of model parameters is also dependent upon the current financial regime. For example, sensitivity and parameter values during the financial crisis of 2008–2009 could be much different than during a low volatility regime. Analysts need to continuously evaluate what constitutes a feasible range for the parameters.

The process we used to determine the sensitivity of model results to these parameters is as follows:

1. Start with parameter a_1.
2. Sets its value to $a_1 = 100$.
3. Solve the non-linear least squares model with $a_1 = 100$ and the above constraints on the other parameters.
4. Record the resulting parameter values and non-linear R^2 estimate e.g., $(a_1 = 100, a_2 = \hat{a}_2, a_3 = \hat{a}_3, a_4 = \hat{a}_4, b_1 = \hat{b}_1, NonR^2)$
5. Increase the value of a_1 (e.g., set $a_1 = 150$) and re-run the non-linear least squares regression, record the values, repeat until $a_1 = 1000$.
6. Repeat these steps for all the parameters., i.e., hold one parameter value constant and solve for the other four. Record results.
7. Plot and analyze the results.

We performed the above analysis for all feasible values of the parameters. For each parameter we plotted the specified parameter value and the non-linear R^2 from the best fit non-linear regression. For example, for $a_1 = 100$, the best fit non-linear R^2 was $R^2 = 0.23$. For $a_1 = 150$, the best fit non-linear R^2 was $R^2 = 0.38$, etc.

The results of our sensitivity analysis are shown in Figure 5.6a–e. Figure 5.6a shows the results for parameter a_1. The graph shows R^2 increasing from 0.28 (at $a_1 = 100$) to a maximum value of $R^2 = 0.41$ (at $a_1 = 700$), and then decreasing again to $R^2 = 0.38$. If we look at Figure 5.6a we find that the best fit R^2 value varies very little between the values $a_1 = 600$ to $a_1 = 800$. The best fit equation is pretty flat between these values. This type of result is not unique to the I-Star model; it is in fact pretty common across most non-linear equations and is the reason we have been stressing the need to perform a thorough sensitivity analysis on the data.

Figure 5.6b–e illustrates the sensitivity analysis for a_2 through b_1. Parameter a_2 has its best fit value at about $a_2 = 0.55$ and appears to have a range between $a_2 = 0.45$ to $a_2 = 0.65$. Parameter a_3 reaches its best fit at $a_3 = 0.75$ with a range of about $a_3 = 0.65$ to $a_3 = 0.80$. Parameter a_4 reaches its best fit at $a_4 = 0.45$ with a range of about $a_4 = 0.4$ to $a_4 = 1$. Parameter b_1 reaches its best fit at $b_1 = 0.92$ with a range of $b_1 = 0.87$ to

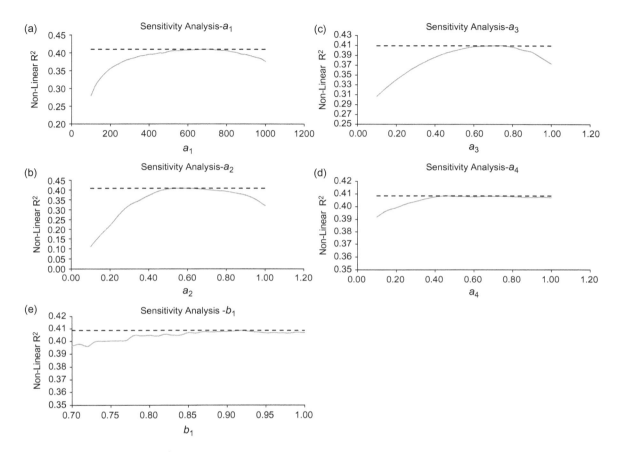

■ **Figure 5.6** Parameter Sensitivity Analysis.

$b_1 = 1.00$. It is important to mention that the model is not highly sensitive to parameters a_4 or b_1 and the best fit equations for these parameters vary very little within these ranges. For example, varying b_1 between 0.87 and 1.00 results in a non-R^2 of 0.4070 (min) to 0.4087 (max). Notice how flat this curve is even over the range 0.80 to 1.00. Thus it is no wonder why it has been so difficult in practice to uncover a difference between temporary and permanent market impact cost.

We learn a valuable lesson from the sensitivity analysis—it provides intuition surrounding feasible values of the parameters as well as how much we can expect those parameter values to vary. This is extremely useful in performing what if analysis and running alternative scenarios such as buy/sell, large cap/small cap, etc. (as we show below).

Table 5.4 Estimated MI Parameters					
Scenario	**a_1**	**a_2**	**a_3**	**a_4**	**b_1**
All Data	708	0.55	0.71	0.50	0.98
SE	100	0.03	0.02	0.05	0.04
R^2	0.42				

Next we performed our non-linear least squares regression for the full model without holding any parameter value fixed. We allowed the model to determine the set of parameters using the specified constraints to ensure feasible values. The results of the regression are shown in Table 5.4. The table includes parameter standard errors from the non-linear regression and has a non-linear $R^2 = 0.42$. These statistical results indicate a strong fit. Additionally, notice these results are all within the ranges we previously determined. This provides a higher degree of confidence in our results.

The best fit equation using these estimated parameters is:

$$I = 708 \cdot Size^{0.55} \cdot \sigma^{0.71} \tag{5.16}$$

$$MI = 0.98 \cdot I \cdot POV^{0.50} + (1 - 0.98) \cdot I \tag{5.17}$$

To further evaluate costs and determine differences across categories we further categorized the data into samples that consisted of large and small cap companies, buy and sell orders, and a breakdown by market cap and order size. In all cases, there was a high fit of the data. The non-linear R^2 ranged from 0.40 to 0.43. These results are shown in Table 5.5.

An analysis of the costs estimates over time did not find any differences across buy and sell orders when holding volatility and trading strategy constant. However, in practice, managers and traders often find sell orders to be more costly than buy orders for various reasons. First, buy orders are cancelled more often then sell orders. As the price moves too high the advantage and incremental alpha decreases and managers are better suited investing in an alternative stock. There are substitution stocks for buy orders but not for sell orders. Once a stock has fallen out of favor and the manager decides to remove the stock from the portfolio they will complete the order regardless of the price. Therefore, managers do not always realize the entire cost of the buy order because they rarely factor in opportunity cost. But the entire cost of the sell order is always realized. Second, managers typically sell stocks at a more aggressive rate

Table 5.5 Estimated MI Parameters

Scenario	a_1	a_2	a_3	a_4	b_1	R^2
All Data	708	0.55	0.71	0.50	0.98	0.42
Large Cap	687	0.70	0.72	0.35	0.98	0.43
Small Cap	702	0.47	0.69	0.60	0.97	0.43
Buy	786	0.58	0.74	0.60	0.90	0.43
Sell	643	0.44	0.67	0.60	0.98	0.43
Large—Buy	668	0.68	0.68	0.45	0.90	0.43
Large—Sell	540	0.52	0.64	0.45	1.00	0.41
Small—Buy	830	0.50	0.76	0.70	0.92	0.43
Small—Sell	516	0.71	0.69	0.10	0.90	0.40
Average:	675	0.57	0.70	0.48	0.95	0.42

than they buy stocks causing the cost to be higher due to the corresponding urgency level and not due to any systematic difference in order side. Third, when managers decide to sell stocks that have fallen out of favor it is often due to fundamentals and corresponds to increased volatility and decreased liquidity, further increasing the cost to trade. Managers often select stocks to hold in their portfolio under the most favorable of market conditions thus causing the buy orders to be less expensive than the sell orders. Again, this is due to a difference in company fundamental and less favorable explanatory factors. It is not due to any difference in cost due to the side of the order.

Analysis of costs by market cap, however, did find a difference in trading costs. Large cap stocks were less expensive to trade than small cap stocks. This difference was primarily due to small cap stocks having higher volatility and increased stock specific risk—both causing a higher price elasticity to order flow, i.e., increased market impact sensitivity. Additionally, large cap stocks usually have a larger amount of analyst coverage and therefore these stocks often have a lower quantity of information based trading and lower permanent impact. When the market observes increased trading activity in small cap stocks it appears that the belief is due to information based trading. This is also true with small cap index managers who do not try to hold the entire small cap universe but instead seek to minimize tracking error to the index by holding a smaller number of stocks from a universe that they believe will likely outperform the small cap index.

As stated previously, it is possible for the parameters of the model to vary but to still get the same cost estimates. Analysts interested in detailed differences across these categories can test the model using the parameters published in Table 5.5.

Cost Curves

Trading cost estimates can be computed for an array of order sizes and trading strategies expressed in terms of POV rate. For example, using parameters for the all data scenario (Equation 5.16 and Equation 5.17), trading an order that is 10% ADV for a stock with volatility = 25% utilizing a full day VWAP strategy is expected to cost 23.2 bp. Trading this order more aggressively, say with a POV rate of 20%, will cost 30.9 bp. Trading the same order more passively, say with a POV rate of 5%, will cost 17.2 bp.

Figure 5.7 graphically illustrates trading cost estimates for this stock (volatility = 25%) for various order sizes ranging from 1% ADV through 50% ADV, for four different trading strategies: VWAP, POV = 10%, POV = 20%, and POV = 40%. This figure also shows the model has the expected concave shape.

Table 5.6 provides the underlying cost curve data grids for this order. These cost curves provide the expected trading cost for various order sizes executed using various trading strategies (VWAP and POV rates) in tabular form. Cost curves (as will be discussed in later chapters) provide portfolio managers with essential data required for stock selection, portfolio construction, and optimization.

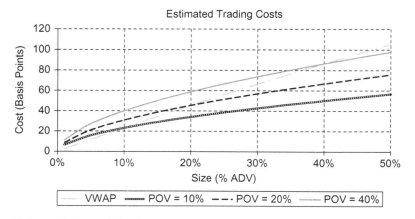

■ **Figure 5.7** Estimated Trading Costs.

Table 5.6 Estimated Market Impact Curves

Size (% ADV)	VWAP	POV = 5%	POV = 10%	POV = 15%	POV = 20%	POV = 25%	POV = 30%	POV = 35%	POV = 40%
					Trading Strategy				
1%	2.4	4.8	6.5	7.7	8.6	9.4	10.1	10.7	11.2
5%	11.7	11.7	15.8	18.8	21.1	23.0	24.6	26.0	27.3
10%	23.2	17.2	23.2	27.5	30.9	33.8	36.2	38.2	40.1
15%	34.5	21.6	29.1	34.5	38.8	42.3	45.3	47.9	50.2
20%	45.5	25.3	34.1	40.5	45.5	49.6	53.1	56.2	58.9
25%	56.1	28.6	38.6	45.8	51.5	56.1	60.1	63.6	66.6
30%	66.5	31.7	42.8	50.7	56.9	62.1	66.5	70.4	73.7
35%	76.6	34.5	46.6	55.2	62.0	67.7	72.5	76.6	80.3
40%	86.5	37.2	50.2	59.5	66.8	72.9	78.0	82.5	86.5
45%	96.1	39.7	53.6	63.5	71.3	77.8	83.3	88.1	92.3
50%	105.4	42.1	56.8	67.3	75.6	82.5	88.3	93.4	97.9

The data above is on the spreadsheet named Figure 5.7

Statistical Analysis

We are now up to step five of the scientific method where we analyze the data. In this step we compare the results of the model with the estimated parameter set to actual customer order data (the control group). We additionally perform an error analysis where we compare the estimated costs to the actual costs, and then perform a stock outlier analysis where we regress the model error on stock specific characteristics (such as market cap, spread, idiosyncratic risk, etc.).

Error Analysis

The first step in our error analysis was to compute the estimated market impact cost for each of the customer orders in the control group. But unlike the research step where we used all of the data points, here we filtered potential outliers in order to ensure we were analyzing the price impact due to the order's buying and selling pressure. We filtered data points from the control group identical to filtered data points derived from the market data cost section. We filtered days with volume greater than 3 times ADV and days with price movement greater than 4 times the stock's daily volatility. We conducted the filtering process because it is highly likely that on days with these types of volumes and/or price movement it was more likely to be due to market or stock specific new information rather than excessive buying or selling pressure.

Our error analysis consists of estimating the market impact cost for each of the remaining data points. We then grouped these data points into 1% ADV categories and graphically compared the results (Figure 5.8a). We observed a very strong and accurate fit for order size up to about 15% ADV and for sizes of 20% to 30% it appears that we are overestimating the actual cost. Figure 5.8b is an xy-scatter plot of estimated cost (y-axis) as a function of actual cost (x-axis). This graph shows that our market impact model is accurate for costs up to approximately 40 to 50 bp. Again, the model seems to overestimate costs for larger more expensive orders. Figure 5.8c plots the actual error measured as the difference between estimated cost and actual cost by order size. This figure gives the appearance that the model begins to overestimate costs at around 10%−15% ADV.

The difference emerging here between estimated cost and actual cost is not concerning for the larger more expensive orders due to survivorship bias and opportunistic trading. That is, investors are more likely to complete the larger size orders in times of favorable price momentum and a lower cost environment. Furthermore, investors are more likely to increase the original order in times of more favorable prices. In times of adverse price movement and a higher trading cost environment investors

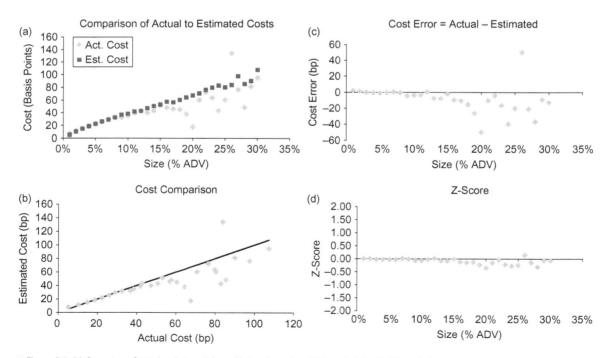

■ **Figure 5.8** (a) Comparison of Actual to Estimated Costs; (b) Cost Comparison; (c) Error Analysis; (d) Z-Score Analysis.

are more likely to not complete the order and cancel shares. This results in the actual measured costs being lower than they would have been had it not been for the price momentum and market conditions. Unfortunately we do not have a full audit trail in this case to be able to incorporate opportunistic trading and opportunity cost into our error analysis. But investors with a full audit trail will be equipped to properly incorporate opportunity cost as well as survivorship bias and opportunistic trading.

The next question is how far off the estimated cost is from the actual cost for the larger orders. Especially considering some orders are exposed to much greater market risk than others. To address this question we computed the z-score for each of the orders. That is,

$$Z = \frac{Actual\ Cost - Estimated\ MI}{Timing\ Risk} \tag{5.18}$$

The estimated timing risk is computed as:

$$TR = \sigma \cdot \sqrt{\frac{1}{3} \cdot \frac{1}{250} \cdot Size \cdot \frac{1 - POV}{POV}} \cdot 10^4 bp \tag{5.19}$$

Regardless of the distribution of the error, if the model is accurate the mean z-score will be zero and the variance will be one. That is,

$$Z \sim (0, 1)$$

We computed the average z-score in each size bucket. This is shown in Figure 5.8d. This analysis shows that the risk adjusted error is not as inaccurate as it first appears. The average z-score for all order sizes, while significantly different from zero, is still within $+/-1$ standard deviation. To be more exact, the z-score is $+/-0.25$ standard units for sizes up to 30% ADV. Thus, while the model is overestimating actual trading costs (likely due to opportunistic trading and survivorship bias), the risk adjusted error term is not considered grossly erroneous. The error is quite reasonable and thus not a large concern.

Stock Specific Error Analysis

The next step in our error analysis is to determine if there is anything specific to the stock or particular company that would help improve the accuracy of the model and reduce estimation error. For this step we follow the techniques presented by Breen, Hodrick and Korajczyk (2002).

Our error analysis was carried out by estimating the expected market impact cost using the parameters determined above and comparing these estimates to the actual order costs. The average error measured as the estimated cost

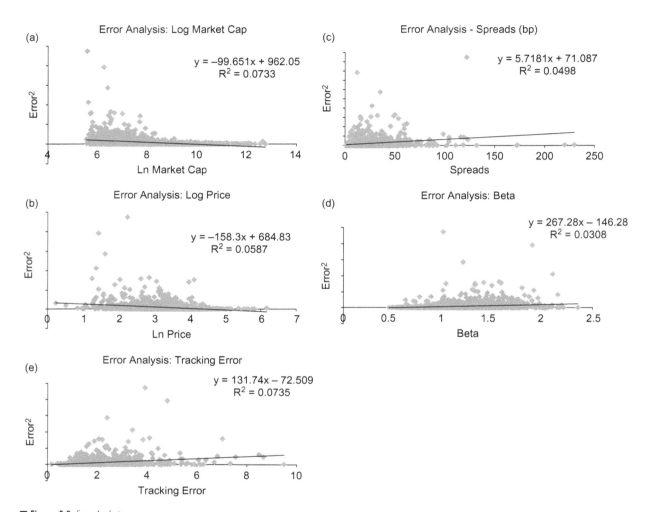

■ Figure 5.9 Error Analysis.

minus actual cost was determined for each stock and the average z-score was also computed for each stock. We then regressed the error squared and the z-score squared against stock specific variables including ln market cap, ln price, spreads, beta, and tracking error. We chose to regress the squared error and z-score metrics in order to determine which stock specific variables, if any, would assist us to understand and reduce the model error at the stocks level.

Figure 5.9a−e illustrates the regression results of the error squared as a function of the variables. As is consistent with previous research studies

Table 5.7 Stock Specific Error Analysis

	LnMktCap	LnPrice	Spreads	Beta	T.E.
Error²					
Est.	− 99.65	− 158.30	5.72	267.28	131.74
S.E.	8.75	15.64	0.62	37.03	11.54
t-stat	− 11.39	− 10.12	9.28	7.22	11.41
Z-Score²					
Est.	− 0.011	− 0.015	0.001	0.019	0.011
S.E.	0.002	0.003	0.000	0.008	0.002
t-stat	− 6.046	− 4.647	7.327	2.544	4.715

and academic reports, the error term is negatively related to market cap and price. Stocks with higher market capitalization and higher prices will have lower market impact cost. The error term is also positively related to spreads, beta, and tracking error (as a proxy for idiosyncratic risk). This is consistent with our expectations. Higher spreads are an indication of less stable trading patterns and more intraday risk. Higher beta is an indication of higher risk stocks and a higher level of price sensitivity. Higher tracking error or idiosyncratic risk is an indication of stock specific risk and potentially higher information content and permanent market impact cost.

This error analysis provides some valuable insight and potential variables for analysts to incorporate into the model to improve its accuracy. The results of the regression coefficients for the stock specific analysis for the error squared and the z-score squared are shown in Table 5.7.

Price Volatility

INTRODUCTION

In this chapter, we discuss price volatility and factor models and how they can be used to improve trading performance. We present various techniques that are used in the industry to forecast volatility as well as appropriate methods to calibrate these models.

Volatility is the uncertainty surrounding potential price movement, calculated as the standard deviation of price returns. It is a measure of the potential variation in price trend and not a measure of the actual price trend. For example, two stocks could have the exact same volatility but much different trends. If stock A has volatility of 10% and price trend of 20%, its one standard deviation return will be between 10 and 30%. If stock B also has volatility of 10% but price trend of 5%, its one standard deviation return will be between -5 and 15%. Stock with higher volatility will have larger swings than the stock with lower volatility, resulting in either higher or lower returns.

There are two volatility measures commonly used in the industry: realized and implied. Realized volatility is computed from historical prices and is often referred to as historical volatility. Realized volatility uses past history to predict the future. Implied volatility, on the other hand, is computed from the market's consensus of the fair value for a derivative instrument such as the S&P500 index option contract. Implied volatility is a "forward" looking or "future" expectation estimate.

> *Historical Volatility lets the data predict the future.*
> *Implied Volatility lets the market predict the future.*

We utilize volatility in many different ways. For example, traders use volatility to understand potential price movement over the trading day, as input into market impact models, to compute trading costs, and to select algorithms. Algorithms use volatility to determine when it is appropriate

The Science of Algorithmic Trading and Portfolio Management. DOI: http://dx.doi.org/10.1016/B978-0-12-401689-7.00006-4

to accelerate or decelerate trading rates in real-time. Portfolio managers use volatility to evaluate overall portfolio risk, as input into optimizers, for value-at-risk calculations, as part of the stock selection process, and to develop hedging strategies. Derivatives desks use volatility to price options and other structured products. In addition, plan sponsors use volatility to understand the potential that they will or will not be able to meet their long-term liabilities and financial obligations. Volatility is a very important financial statistic.

DEFINITIONS
Price Returns/Price Change

Price returns or price change can be computed using either the "percentage" returns or "log" returns formulation. But since returns have been found to be log-normally distributed it is appropriate to use the log returns calculation. We describe these two measures:

Percentage Price Return

$$r_t = \frac{p_t}{p_{t-1}} - 1 \tag{6.1}$$

Log Price Return

$$r_t = \ln\left(\frac{p_t}{p_{t-1}}\right) \tag{6.2}$$

Where $\ln(\cdot)$ represents the natural log function.

Average Return

This average period price return $\bar{r}$ is calculated differently for the "percentage" and "log" methods.

Average return—percentage method

The process for computing an n-period average return using the percentage change methodology is:

Start with:

$$p_n = p_0 \cdot (1 + \bar{r})^n$$

Then solving for $\bar{r}$ we have:

$$\bar{r} = \left(\frac{p_n}{p_0}\right)^{\frac{1}{n}} - 1 \tag{6.3}$$

Notice that the average return is not calculated as the simple average of all returns.

Average return—logarithmic change method

The average return using the log methodology, however, is determined directly from the simple average formula. This is:

$$\bar{r} = \frac{1}{n} \sum r_k \tag{6.4}$$

The average can also be computed directly from the starting and ending price as follows:

$$\bar{r} = \frac{1}{n} \cdot \ln\left(\frac{p_t}{p_0}\right) \tag{6.5}$$

How do percentage returns and log returns differ?

For small changes, there actually is not much difference between these two measures. Figure 6.1 shows a comparison percentage price change compared to log price change over the range of returns from −40% through +40%. As shown in this figure, there are very small differences in calculation process between −20 and +20%, and negligible differences between −10 and +10%. However, we start to see differences once we are outside of this range. For trading purposes where the daily price change is quite often less than ±10% either returns measure is reasonable. However, for a portfolio manager with a longer time horizon we will start to see some differences between these two measures.

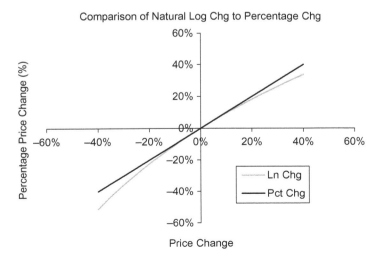

Figure 6.1 Comparison of Natural Log Price Change to Percentage Price Change.

Log returns can take on any value from $-\infty$ *to* $+\infty$ and the distribution of returns is symmetric. However, the percentage returns can only take on values between -100% and $+\infty$. The percentage return data values have a skewed distribution while the log returns distribution is symmetric.

The log returns method is a symmetric measure and provides mathematical simplification in many of our calculations. It is also more intuitive than the percentage change method. For example, if the price for a financial asset changes from 100 to 80 the percentage price change is -20%. However, if the price in the next period changes from 80 to 100 the percentage price change is $+25\%$. So this asset first lost 20% then increased 25%, but it is currently at its original starting value—it is not $+5\%$ overall. But using the log returns method we do not run into this inconsistency. If the stock price changes from 100 to 80 this corresponds to a log change of -22.31%. If the price in the next period increases from 80 to 100, this corresponds to a log change of $+22.31\%$. The sum of these returns states that the current price should be the same as the starting price since $+22.31\% -22.31\% = 0$ which is exactly what we have.

Price returns have been found to have a log-normal distribution. Thus using the log returns measure is more consistent with the underlying data. Going forward we will use the log returns methodology.

Volatility
Volatility is calculated as the standard deviation of price returns.

$$\sigma_i = \sqrt{\frac{1}{n-1}\sum_{k=1}^{n}(r_{ik}-\bar{r}_i)^2} \tag{6.6}$$

In the formulation above, there are n historical returns but we divide by $(n-1)$ to ensure an unbiased estimate. This formulation is also called the sample standard deviation.

Covariance
The covariance of returns for two stocks σ_{ij} is a measure of the co-movement of prices.

$$\sigma_{ij} = \frac{1}{n-2}\sum(r_{ik}-\bar{r}_i)(r_{jk}-\bar{r}_j) \tag{6.7}$$

Positive covariance means that the prices will move up and down together and negative covariance means that the prices will move in opposite directions.

Correlation

The correlation between two stocks ρ_{ij} is the covariance of the stocks divided by the volatility of each stock. This provides a correlation coefficient between one and minus one: $-1 \leq \rho_{ij} \leq 1$.

$$\rho_{ij} = \frac{\sigma_{ij}}{\sigma_i \cdot \sigma_j} \tag{6.8}$$

Stocks with a correlation of $\rho_{ij} = 1$ move perfectly with one another, stocks with a correlation of $\rho_{ij} = -1$ move perfectly in the opposite direction of one another, and stocks with a correlation of $\rho_{ij} = 0$ do not move together at all. Correlation provides a measure of the strength of co-movement between stocks.

Dispersion

The dispersion of returns is computed as the standard deviation of returns for a group of stocks. It is a cross-sectional measure of overall variability across stocks.

$$dispersion(r_p) = \sqrt{\frac{1}{n-1} \cdot \sum_{j=1}^{n} (\bar{r}_j - \bar{r}_p)^2} \tag{6.9}$$

where $\bar{r}_j$ is the average return for the stock and $\bar{r}_p$ is the average return across all stocks in the sample.

Dispersion is very useful to portfolio managers because it gives a measure of the directional movement of prices and how close they are moving in conjunction with one another. A small dispersion measure indicates that the stocks are moving up and down together. A large dispersion measure indicates that the stocks are not moving closely together.

Value-at-Risk

Value-at-risk (VaR) is a summary statistic that quantifies the potential loss of a portfolio. Many companies place limits on the total value-at-risk to protect investors from potential large losses. This potential loss corresponds to a specified probability α level or alternatively a $(1-\alpha)$ confidence.

If the expected return profile for a portfolio is $r \sim N(\bar{r}_p, \sigma_p^2)$. Then a $\alpha\%$ VaR estimate is the value return that occurs at the $1-\alpha$ probability level in the cumulative normal distribution. If $\alpha = 95\%$ this equation is:

$$0.05 = \int_{-\infty}^{r*} \frac{1}{\sqrt{2\pi\sigma^2}} \exp\left\{ -\frac{(r-\bar{r}_p)^2}{2\sigma_p} \right\} \tag{6.10}$$

Implied Volatility

Implied volatility is determined from the price of a call or put option. For example, the Black-Scholes option pricing model determines the price of a call option as follows:

$$C = S \cdot N(d_1) - X \cdot e^{-r_f T} \cdot N(d_2) \tag{6.11}$$

where,

$$d_1 = \frac{\ln(S/X) + (r_f + \sigma^2/2)T}{\sigma\sqrt{T}}$$

$$d_2 = d_1 - \sigma\sqrt{T}$$

C = call price
X = strike price
S = stock price
σ = stock volatility
$N(d)$ = probability that actual return will be less than d
r_f = risk free rate of return
T = future time period

The implied volatility is the value of the volatility in the above formula that will result in the current value of the call option. Since the call option price is determined by the market, we are able to back into the volatility terms that would provide this value, thus, the volatility is implied by the formulation. Implied volatility is most often solved via non-linear optimization techniques.

Beta

$$\beta_k = \frac{cov(r_k, r_m)}{var(r_m)} = \frac{\sigma_{ij}}{\sigma_m^2} \tag{6.12}$$

The beta of a stock represents the stock's sensitivity to a general market index. It is determined as the covariance of returns between the stock and the market divided by the variance of the index (volatility squared). The calculation is also the slope of the regression line of stock returns (y-axis) as a function of market returns (x-axis). Stocks with a positive beta, $\beta_k > 0$, move in the same direction as the market and stocks with a negative beta, $\beta_k < 0$, will move in the opposite direction of the market. Stocks with an absolute value of beta greater than one, $|\beta_k| > 1$, are more variable than the market and stocks with an absolute value of beta less than one, $|\beta_k| > 1$, are less variable than the market.

MARKET OBSERVATIONS—EMPIRICAL FINDINGS

In order that we better understand the meaning of these variability trends in the market, we evaluated volatility, correlation, and dispersion over the period January 1, 2000 through June 30, 2012. The universe used for our analysis was the SP500 index for large cap stocks and the R2000 index for small cap stocks. Stocks that were in each index on the last trading day in the month were included in calculations for that month.

Volatility: Volatility was computed as the 22-day standard deviation of log returns annualized using a scaling factor of $\sqrt{250}$. We computed volatility for all stocks in our universe on each day and reported the cross-sectional average as our daily data point. Figure 6.2a shows the average stock volatility for large cap stocks. At the beginning of 2000, volatility was fairly high. This was primarily due to the technology boom and increased trading volume. Volatility began the decade around 60% but decreased to about 20%. Volatility remained low until spring 2007 and then started to spike in August 2007 due to the quant crisis. But this was nothing compared to the financial crisis period of 2008−2009 where volatility spiked to over 100% and reached as high as 118%. Following the financial crisis volatility decreased to more reasonable levels of 20−30% and these lower levels persisted until fall 2011 due to the US debt ceiling crisis and the re-emergence of macroeconomic uncertainty in Europe. After the US addressed its debt issues volatility decreased shortly thereafter.

The average large cap volatility during our sample period was 36% and the average small cap volatility was 54%. Small cap volatility tends to be consistently higher than large cap volatility.

Correlation: Correlation was computed as the 22-day correlation measure across all pairs of stocks in each index. Figure 6.2b shows the average pair-wise correlation across all pairs of stock on each day for large cap stocks. For example, for a universe with 500 stocks there are 124,500 unique two pair combinations. Hence, we computed correlations across all 124,500 combinations of stock and show the average value in Figure 6.2b. For an index consisting of 2000 stocks there are just slightly less than 2 million unique combinations of stock (1,999,000 to be exact). Computing average pair-wise correlations is a very data intensive process. The graph of average large cap correlations over our period found relatively low correlation (less than 30%) at the beginning of the decade but then spiking at the time of the collapse bubble. This increase in correlation was due in part to the markets re-evaluation of stock prices to lower levels and a sell-off of stock from institutions that further pushed the prices in the

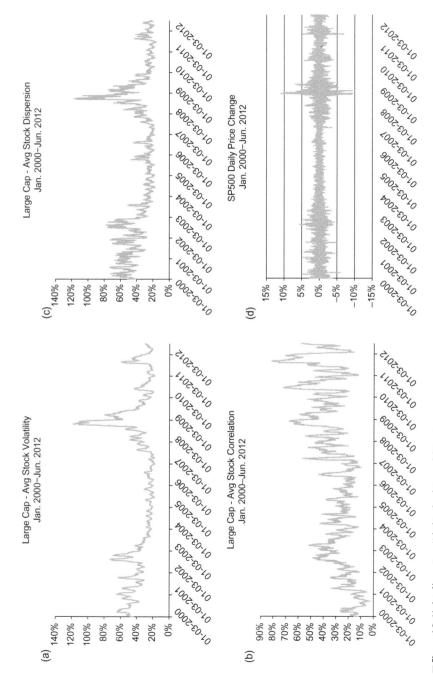

■ **Figure 6.2** Market Observations: Volatility, Correlation and Dispersion.

same direction. Since spring 2007 correlations have been on the rise and only recently seem to have leveled off, but at much higher levels than at the beginning of the decade. An interesting finding of our historical analysis is that the market experienced the highest level of correlation following the flash crash (but very short lived) and again during the US debt ceiling crisis. These correlation levels were even higher than the levels experienced during the financial crisis when markets experienced their highest levels of volatility.

The average large cap pair-wise correlation during our sample period was 32% and average small cap pair-wise correlation was 23%. Opposite to the volatility relationship mentioned above, small cap correlation tends to be lower than large cap correlation. This is primarily due to smaller companies having a larger amount of company risk. However, there are periods when correlation is greater across the small cap stocks.

Dispersion: We measured dispersion as the standard deviation of the 22-day stock return. For example, for the SP500 we computed the 22-day price return for each of the 500 stocks in the index. We then computed the standard deviation of these returns as our dispersion. Figure 6.2c illustrates large cap returns dispersion over our analysis period. As shown in the figure, dispersion was fairly constant at the beginning of the decade and then declined through August 2007 when it suddenly increased and spiked during the financial crisis. Dispersion decreased shortly thereafter but began increasing again during the US debt crisis of fall 2011. Comparison of dispersion to pair-wise correlation shows a slightly negative relationship. As correlation increases dispersion tends to fall and as correlation decreases dispersion tends to increase. This inverse relationship between dispersion and correlation is more dramatic for small cap stocks than for large cap stocks.

Over our sample period, the average large cap dispersion was 36% and the average small cap dispersion was 58%. Dispersion tends to be consistently larger for small caps than for large caps.

Volatility Clustering: Figure 6.2d illustrates the daily price change for the SP500 index. Notice the extent of volatility clustering. This means that large price swings tend to be followed by large price swings (either positive or negative) and small price swings tend to be followed by small price swings (either positive or negative). The beginning of the decade was associated with a higher volatility regime due to the technology bubble. This was followed by relatively lower volatility through the beginning of the quant breakdown starting in August 2007 followed by ultra-high volatility during the financial crisis of 2008−2009. The graph also

Table 6.1 Market Observations from Jan. 2000—Jun. 2012

Statistic	Avg Volatility		Avg Correlation		Avg Dispersion	
	LC	SC	LC	SC	LC	SC
Avg:	36%	54%	32%	23%	36%	58%
Stdev:	16%	20%	15%	12%	16%	20%
Min:	18%	30%	3%	2%	16%	33%
Max:	119%	146%	81%	67%	118%	142%

illustrates the higher volatility period after the flash crash in May 2010 and again during the US debt crisis in fall 2011. Volatility appears to be time varying with a clustering effect. Thus, traditional models that give the same weighting to all historical observations may not be the most accurate representation of actual volatility. The volatility clustering phenomenon was the reason behind the advanced volatility modeling techniques such as ARCH, GARCH, and EWMA. These models are further discussed below. Table 6.1 provides a summary of our market observations over the period from January 2000 through June 2012.

FORECASTING STOCK VOLATILITY

In this section, we describe various volatility forecasting models as well as appropriate techniques to estimate their parameters. We also introduce a new model that incorporates a historical measure coupled with insight from the derivatives market. These models are:

- Historical Moving Average (HMA)
- Exponential Weighted Moving Average (EWMA)
- Autoregressive Models (ARCH and GARCH)
- HMA-VIX Adjustment

Some of these descriptions and our empirical findings have been disseminated in the Journal of Trading's "Intraday Volatility Models: Methods to Improve Real-Time Forecasts," Fall 2012. Below we follow the same outline and terminology as in the journal.

Volatility Models

We describe four different volatility models: the historical moving average (HMA), the exponential weighted moving average (EWMA) introduced by JP Morgan (1996), an autoregressive heteroscedasticity (ARCH) model introduced by Engle (1982), a generalized autoregressive conditional heteroscedasticity (GARCH) model introduced by Bollerslev (1986), and an HMA-VIX adjustment model that combines the stock's current realized volatility with an implied volatility measure[1].

We fit the parameters of these models to the SP500 and R2000 indexes over the one-year period July 1, 2011 through June 30, 2012 and compare the performance of the models. Readers are encouraged to experiment with these techniques to determine which model work best for their needs.

Price Returns

Price returns for each index were computed as the natural log of close-to-close price change:

$$y_t = \ln(p_t/p_{t-1}) \tag{6.13}$$

A general short-term model of price returns is:

$$y_t = C + \sigma_t \varepsilon_t \tag{6.14}$$

where C is a constant, ε_t is random noise with distribution $\varepsilon_t \sim N(0, 1)$, and σ_t is the time varying volatility component. In practice, the short-term constant term C is rarely known in advance and analysts often use a simplifying assumption of $C = 0$. Then the general short-term price returns model simplifies to:

$$y_t = \sigma_t \varepsilon_t \tag{6.15}$$

Data Sample   We used the SP500 and R2000 index returns over the one-year period July 1, 2011 through June 30, 2012 as our data sample. Figure 6.3a shows the daily price returns for the SP500. The standard deviation of daily returns over this period was 1.5% and the SP500 index had price swings as high as $+4.6\%$ on August 9, 2011 and as low as -6.9% on August 8, 2011. These large swings occurred during the US debt crisis (Aug.–Sep. 2011). Figure 6.3b shows the price returns for the R2000 index over the same period. The R2000 index had a standard deviation of daily returns of 2.1%, and price swings as high as $+6.7\%$ on August 9, 2011 and as low as -9.3% on August 8, 2011. As is evident in both figures, volatility was fairly low through middle of August 2011 when there was a sudden spike in volatility that caused it to remain high through year

---

[1]The HMA-VIX volatility model was presented at Curt Engler's CQA/SQA Trading Seminar (February 2009), "Volatility: Is it safe to get back in the water?" and taught as part of the volatility section in Cornell University's Graduate Financial Engineering Program, "Introduction to Algorithmic Trading," Fall 2009 (Kissell and Malamut). The HMA-VIX model was also published in Journal of Trading, "Intraday Volatility Models: Methods to Improve Real-Time Forecasts," Fall 2012.

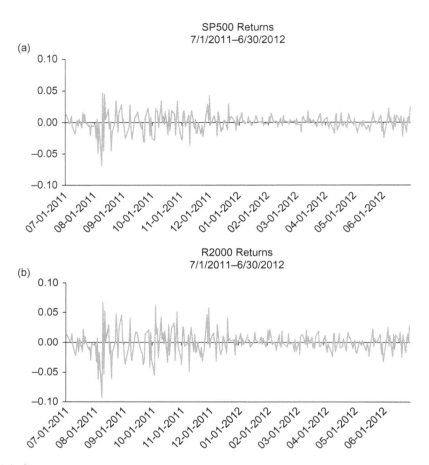

■ **Figure 6.3** Daily Index Returns.

end. This was then followed by a relatively low volatility period from January 2012 through the end of May 2012 when markets started to experience increasing volatility levels. The most important finding over the time period investigated is that the volatility of both indexes appears to be time varying with clustering. The volatility model needs to be able to quickly adjust to these types of sudden regime shifts.

Historical Moving Average (HMA)

The historical moving average volatility measure is computed by definition:

$$\overline{\sigma}_t^2 = \frac{1}{n-1} \sum_{k=1}^{n} y_{t-k}^2 \tag{6.16}$$

This is a simple unbiased average of squared returns (since we are taking the trend term to be $C = 0$). The advantage of this approach is that the calculation straightforward. The disadvantage is that the HMA assumes returns are independent and identically distributed with constant variance. However, this assumption does not seem to hold true over the period analyzed (see Figure 6.3a and Figure 6.3b). Since this model applies equal weightings to all historical data points it has been found to be slow to adapt to changing volatility regimes, such as the US debt crisis (Aug.−Sep. 2011) and during the financial crisis (Sep. 2008−Mar. 2009).

Exponential Weighted Moving Average (EWMA)

The exponential weighted moving average (EWMA) is computed as follows:

$$\hat{\sigma}_t^2 = (1 - \lambda)\gamma_{t-1}^2 + \lambda\hat{\sigma}_{t-1}^2 \tag{6.17}$$

EWMA applies weights to the historical observations following an exponential smoothing process with parameter λ where $0 \le \lambda \le 1$. The value of the smoothing parameter is determined via maximum likelihood estimation (MLE). JP Morgan (1994) first introduced this model as part of their Risk Metrics offering.

The advantage of the EWMA is that it places more emphasis on the recent data observations. This allows the model to quickly update in a changing volatility environment. Additionally, its forecasts only require the previous period price change and the previous volatility forecast. We do not need to recalculate the forecast using a long history of price returns.

Arch Volatility Model

The ARCH volatility model was introduced by Engle (1982) and consists of the "p" previous returns. We can formulate it as follows:

$$\hat{\sigma}_t^2 = \omega + \sum_{i=1}^{p} \alpha_i r_{t-i}^2 \tag{6.18}$$

where, $\omega > 0, \alpha_1, \ldots, \alpha_p \ge 0, \sum \alpha_i < 1$.

The parameters of the model are determined via ordinary least squares (OLS) regression analysis. The model differs from the HMA in that it does not apply the same weightings to all historical observations. To the extent that the more recent observations have a larger effect on current returns, the ARCH model will apply greater weight to the more recent observations. This allows the model to update quickly in a changing volatility environment.

A simple ARCH(1) model consists of the previous day's price return. This is formulated as:

$$\hat{\sigma}_t^2 = \omega + \alpha_1 y_{t-i}^2 \tag{6.19}$$

where, $\omega > 0, 0 \leq \alpha_1 < 1$.

GARCH Volatility Model

The GARCH volatility model was introduced by Bollerslev (1986) and is an extension of the ARCH model (Engle, 1982). A GARCH(p,q) model consists of "p" previous returns and "q" previous volatility forecasts as follows:

$$\hat{\sigma}_t^2 = \omega + \sum_{i=1}^{p} \alpha_i r_{t-i}^2 + \sum_{j=1}^{q} \beta_j \hat{\sigma}_{t-j}^2 \tag{6.20}$$

where, $\omega > 0, \alpha_1, \ldots, \alpha_p, b_1, \ldots, b_q \geq 0, \sum \alpha_i + \sum \beta_j < 1$.

The GARCH model applies more weight to the more recent observations thus allowing the model to quickly adapt to changing volatility regimes. The parameters of the model are determined via maximum likelihood estimation.

A simple GARCH(1,1) model consists of only the previous day's price return and previous day's volatility forecast and is formulated as:

$$\hat{\sigma}_t^2 = \omega + \alpha_1 y_{t-i}^2 + \beta_1 \hat{\sigma}_{t-j}^2 \tag{6.21}$$

where, $\omega > 0, \alpha_1, b_1 \geq 0, \alpha_1 + b_1 < 1$.

HMA-VIX ADJUSTMENT MODEL

The HMA-VIX volatility forecasting model is an approach that combines the stock's current volatility with an implied volatility estimate. We formulate this model as:

$$\hat{\sigma}_t = \overline{\sigma}_{t-1} \cdot \frac{VIX_{t-1}}{\overline{\sigma}_{SPX,t-1}} \cdot AdjFactor \tag{6.22}$$

where $\overline{\sigma}_{t-1}$ is the HMA stock trailing volatility, $\overline{\sigma}_{SPX,t-1}$ is the SP500 index trailing volatility, VIX_{t-1} is the VIX implied volatility index, and *AdjFactor* is the adjustment factor needed to correct for the risk premium embedded in the VIX contract.

Over the years the options market has proven to be a valuable, accurate, and timely indicator of market volatility and changing regimes. Options traders are able to adjust prices quickly based on changing volatility expectations. Analysis can easily infer these expectations through the options prices. This is known as the implied volatility. The question arises then if implied volatility is an accurate and timely estimate of volatility then why cannot analysts just use implied volatility from the options market rather than use results from these models? The answer is simple. Unfortunately, implied volatility estimates do not exist for all stocks. The options market at the stock level is only liquid for the largest stocks. Accurate implied volatility estimates do not exist across all stocks. Fortunately, the options market still provides valuable information that could be extended to the stock level and help provide accurate forward looking estimates, and in a more timely manner than the other historical techniques. This also provides ways for algorithms to quickly adjust to changing expectations in real-time and provide investors with improved trading performance.

The HMA-VIX technique consists of adjusting the stock's trailing volatility by the ratio of the VIX index to the SP500 trailing volatility plus a correction factor. The ratio of the VIX to the SP500 realized shows whether the options market believes that volatility will be increasing or decreasing. However, since the VIX usually trades at a premium of 1.31 to the SP500 trailing volatility we need to include an adjustment factor to correct for this premium. If the VIX Index/SP500 realized volatility >1.31 then we conclude that the options market believes volatility will be increasing and if the VIX Index/SP500 realized volatility <1.31 then we conclude that the options market believes volatility will be decreasing.

The increasing/decreasing expectation obtained from the options market is then applied to individual stocks. A comparison of the VIX implied to the SP500 trailing is shown in Figure 6.4.

The advantage of incorporating the implied expectations into our real-time volatility estimator is that if there is a sudden market event that will affect volatility it will almost immediately be reflected in the HMA-VIX measure. The historical models (HMA, EWMA, ARCH and GARCH) will not react to the sudden market event until after this event has affected stock prices. Thus the historical models will always be lagging behind the event to some degree. Furthermore, if the options market is anticipating an event that has not yet occurred, and has priced the uncertainty of the event into its prices, the HMA-VIX model will also reflect the anticipated event and increased uncertainty prior to that event taking

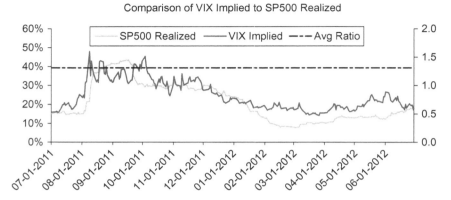

■ **Figure 6.4** Comparison of VIX Implied Volatility to the SP500 Realized Volatility.

place. Just the worry of a potential event taking place will be reflected in the HMA-VIX model. Models updated nightly will miss this event and will not necessarily provide timely accurate volatility estimates.

Determining Parameters via Maximum Likelihood Estimation

Parameters of the GARCH and EWMA volatility models are computed via maximum likelihood estimation (MLE). An overview of the estimation process follows.

Likelihood Function

Let log price returns be normally distributed with mean zero and time varying volatility, that is, $y_t \sim N(0, \hat{\sigma}_t^2)$. Then the probability density function (pdf) of these returns at any time is:

$$f_t(y_t; \hat{\sigma}_t) = \frac{1}{\sqrt{2\pi\hat{\sigma}_t^2}} \cdot e^{-\frac{y_t^2}{2\hat{\sigma}_t^2}} \qquad (6.23)$$

The likelihood of achieving the observed series of returns is:

$$L = \prod_{t=1}^{n} \frac{1}{\sqrt{2\pi\hat{\sigma}_t^2}} \cdot e^{-\frac{y_t^2}{2\hat{\sigma}_t^2}} \qquad (6.24)$$

The log likelihood function ln(L) of achieving this sequence of returns is:

$$\ln(L) = \sum_{t=1}^{n} \left(-\frac{1}{2}\ln(2\pi) - \frac{1}{2}\ln(\hat{\sigma}_t^2) - \frac{1}{2}\frac{\gamma_t^2}{\hat{\sigma}_t^2} \right) \qquad (6.25)$$

The parameters of the EWMA and GARCH models are found by maximizing the ln(L) where $\hat{\sigma}_t^2$ is defined from the corresponding volatility models. This maximization process can be simplified as:

$$\text{Max: } \ln(L) = \sum_{i=1}^{n} -\ln(\hat{\sigma}_i^2) - \frac{\gamma_i^2}{\hat{\sigma}_i^2} \qquad (6.26)$$

Many times optimization packages will only minimize an equation. In these situations, the parameters are found by minimizing the negative of the log likelihood function as follows:

$$\text{Min: } -\ln(L) = \sum_{i=1}^{n} \ln(\hat{\sigma}_t^2) + \frac{\gamma_i^2}{\hat{\sigma}_i^2} \qquad (6.27)$$

Estimation Results

Our parameter estimation results are shown in Table 6.2. For the EWMA model, we found $\lambda = 0.88$ for both the SP500 and R2000. For the GARCH model we found $\beta = 0.84$ for the SP500 and $\beta = 0.82$ for the R2000. These findings are slightly lower than what has been previously reported for other times where λ and β are closer to 0.95 and indicate a stronger persistence. This difference is likely due to the cause of volatility persistence over our timeframe—high volatility regime caused by the debt issues followed by a lower volatility regime after the issues were resolved. The Aug. 2011 through Sep. 2011 debt issue in the US was relatively short-lived and was resolved relatively quickly.

MEASURING MODEL PERFORMANCE

We compared the HMA-VIX technique to the historical moving average (HMA), exponential weighted moving average (EWMA), and generalized autoregressive conditional heteroscedasticity (GARCH) models. We evaluated the performance of the volatility models using three different criteria: root mean square error (RMSE), root mean z-score squared error (RMZSE), and an outlier analysis. Menchero, Wang and Orr (2012) and Patton (2011) provide an in-depth discussion of alternative volatility

Table 6.2 Estimated Parameters

	EWMA		GARCH	
Index	λ	Ω	α	β
SP500	0.8808	3.42E-06	0.1519	0.8420
R2000	0.8758	9.92E-06	0.1614	0.8222

evaluation statistics that can be used to further critique the accuracy of these models. Our usage of these aforementioned performance statistics is to provide a point of comparison across techniques. These procedures are:

Root Mean Square Error (RMSE)

$$RMSE = \sqrt{\frac{1}{n}\sum(\hat{\sigma}_t - \sigma_t)^2} \tag{6.28}$$

The RMSE is simply the difference squared between the estimated volatility $\hat{\sigma}_t$ and realized volatility. Realized volatility was measured as the square root of squared return (e.g., absolute value of return), that is, $\sigma_t = \sqrt{y_t^2}$. This follows along the lines of the more traditional statistical tests such as minimizing sum of squares used in regression analysis.

Root Mean Z-Score Squared Error (RMZSE)

$$RMZSE = \sqrt{\frac{1}{n}\sum\left(\frac{y_t^2}{\hat{\sigma}_t^2} - 1\right)^2} \tag{6.29}$$

The RMZSE is a measurement of the squared difference between our test statistic z and one.

This test is derived as follows. Let, $z = \frac{y - \mu}{\sigma}$. Then we have $E[z] = 0$ and $Var[z] = 1$.

Since we have $y_t \sim N(0, \hat{\sigma}_t^2)$, our test statistic z can be written as $z_t = \frac{y_t}{\hat{\sigma}_t}$. And the variance of z is simply, $Var[z_t] = \frac{y_t^2}{\hat{\sigma}_t^2} = 1$. The root mean z-score squared error is then a test of how close the test statistic is to its theoretical value.

Outlier Analysis

The outlier analysis was used to determine the number of times actual returns exceeded a predicted three standard deviation movement. That is:

$$\text{Outlier if } \left| \frac{\gamma_t}{\hat{\sigma}_t} \right| > 3$$

The outlier analysis consists of determining the total number of outliers observed based on the predicted volatility from each model. We chose three standard deviations as the criteria for outliers. If the absolute value of price return for the index was greater than 3 times the forecasted standard deviation for the index on that day the observation was counted as an outlier. The goal of the outlier analysis was to determine which model resulted in the fewest number of surprises.

Results

The HMA-VIX model was found to be a significant improvement over the historical moving average model. It also performed better than the GARCH and EWMA models under various test statistics, and as well as the GARCH and EWMA models in the other tests.

RMSE performance criteria: The HMA-VIX volatility model was the best performing model. The EWMA was the second best model for the SP500 index followed by the GARCH. The GARCH was the second best model for the R2000 index followed by the EWMA. So while the HMA-VIX was the best there was no clear second best. As expected, the simple HMA standard deviation model was the worst performing model for both indexes.

RMZSE performance criteria: The GARCH model was the best performing model for both the SP500 and R2000 indexes. The HMA-VIX was the second best performing model for the R2000 index followed by the EWMA. The EWMA was the second best performing model for the SP500 index followed closely by the HMA-VIX. Therefore, while the GARCH was the best there was no clear second best. The HMA-VIX model performed as well as the EWMA based on the RMZSE analysis.

These testing metrics for the RMSE and RMZSE are shown in Table 6.3.

Outlier performance criteria: The HMA-VIX and GARCH models had the fewest number of outliers (surprises) for predicted SP500 index returns. There were two surprises with these models. The EWMA approach resulted in three outliers and the HMA resulted in five. The GARCH and EWMA models resulted in two outliers each for the R2000 index. The HMA-VIX had three outliers and the HMA had six. Overall, the GARCH model had

Table 6.3 Performance Results

Model	SP500		R2000	
	RMSE	RMZSE	RMSE	RMZSE
HMA	0.0106	2.6310	0.0135	4.9373
HMA-VIX	0.0095	2.2965	0.0129	4.1787
EWMA	0.0102	2.1019	0.0131	4.1824
GARCH	0.0103	1.6825	0.0131	3.2585

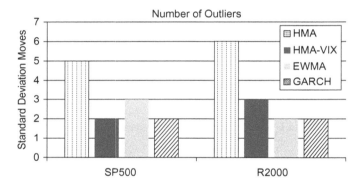

■ **Figure 6.5** Number of Outliers

fewest outliers with four in total, followed by the HMA-VIX and EWMA with five each. The HMA had the most outliers with 11 in total. The total number of outliers from each model is shown in Figure 6.5.

Figure 6.6 illustrates the ratio of price return divided by predicted HMA-VIX volatility for the SP500 index over our sample period. As shown in this figure, there were two outliers that occurred towards the end of August 2011 during the US debt ceiling crisis.

Under a conservative set of testing criteria, there appears to be compelling statistical evidence that the HMA-VIX adjustment model is as accurate as the EWMA and GARCH models. These are extremely encouraging results since only the HMA-VIX model is capable of reacting in real-time to market events or potential market events that will impact volatility but have not yet affected realized market prices. The historical models will not react to these events until that information is captured in the prices. This

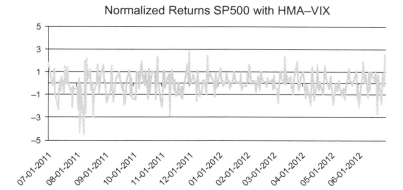

■ Figure 6.6 Normalized SP500 Returns using HMA-VIX

makes the HMA-VIX model a potentially powerful volatility forecasting model for intraday trading applications and electronic trading algorithms.

Some of the advantages of the HMA-VIX volatility model over other techniques are:

■ Reacts to new information sets prior to those events affecting prices. Historical models will only react to new information after it has already affected prices. There is always some degree of lag when using historical models or models based on realized prices.

■ Incorporates real-time information from the options market, e.g., forward looking implied volatility, across the full universe of stock. Implied stock volatility is only available for a very limited number of stocks.

■ Provides necessary real-time volatility estimates that can be incorporated into trading applications and electronic trading algorithms.

■ Allows algorithms to make real-time revision to their execution strategies, limit order model, and smart order routing logic in real-time.

■ Performed as well as, and in some cases better than, some of the more traditional volatility forecasting models.

As a follow-up exercise, we propose further research that combines alternative historical volatility measures with forward looking implied volatility terms. For example, combine the GARCH or EWMA models with an implied volatility term. The implied volatility term can be based on the ratio of the VIX index to S&P trailing volatility (as proposed above), or possibly based on the net change or log change in the VIX index from one period to the next. It is often said that execution performance will only be as good as the models that are used to manage the executions, and that those models are only as good as the accuracy of the forecasted explanatory factors. Since real-time volatility is often a key explanatory factor in these models, having improved volatility forecasts on hand will likely lead to more accurate models. And having more accurate models will allow investors to better react to changing market conditions. And most importantly, this ensures consistency between trading decisions and investing objectives of the fund.

Table 6.4 provides a comparison of the different volatility forecasting models.

Table 6.4 Volatility Forecasting Models

Volatility Model	Formula	Parameter(s)	Calculation
HMA	$\overline{\sigma}_t^2 = \frac{1}{N-1}\sum_{i=1}^{N} r_{t-i}^2$	n/a	By Definition
EWMA	$\widehat{\sigma}_t^2 = (1-\lambda)r_{t-1}^2 + \lambda\widehat{\sigma}_{t-1}^2$	λ	MLE
ARCH(1)	$\widehat{\sigma}_t^2 = \omega + \alpha r_{t-1}^2$	ω, α	OLS
GARCH(1,1)	$\widehat{\sigma}_t^2 = \omega + \alpha r_{t-1}^2 + \beta\widehat{\sigma}_{t-1}^2$	ω, α, β	MLE
VIX Adj.	$\widehat{\sigma}_t^2 = \overline{\sigma}_{t-1}^2 \cdot \left(\frac{VIX_{t-1}}{SP_{500}\overline{\sigma}_{t-1}}\right)^2$	n/a	By Definition
	$\widehat{\sigma}_t^2 = \beta_0 + \beta_1 \cdot VIX_{t-1}^2$	β_0, β_1	OLS

Problems Resulting from Relying on Historical Market Data for Covariance Calculations

Next we want to highlight two issues that may arise when relying on historical data for the calculation of covariance and correlation across stocks using historical price returns. These issues can have dire results on our estimates. They are:

- False Relationships.
- Degrees of Freedom.

False Relationships

It is possible for two stocks to move in the same direction and have a negative calculated covariance and correlation measure and it is possible for two stocks to move in opposite directions and have a positive calculated covariance and correlation measure. Reliance on market data to compute covariance or correlation between stocks can result in false measures.

Following the mathematical definition of covariance and correlation we find that the covariance of price change between two stocks is really a measure of the co-movement of the "error terms" of each stock not the co-movement of prices. For example, the statistical definition of covariance between two random variables x and y is:

$$\sigma_{xy} = E[(x - \bar{x})(y - \bar{y})]$$

It is quite possible for two stocks to have the same exact trend but whose errors (noise term) are on opposite sides of the trend lines. For example, if $\bar{x} = \bar{y} = z$, $x = d$, $y = -d$ and $d > z$, our covariance calculation is:

$$E[(x - \bar{x})(y - \bar{y})] = E[(d - z)(-d - z)] = E[-d^2 + z^2]$$

Since $d > z$, we have $E[-d^2 + z^2] < 0$ which is a negative measured covariance term indicating the stocks trend in opposite directions. But these two stocks move in exactly the same direction, namely, z.

It is also possible for two stocks to move in opposite directions but have a positive covariance measure. For example, if $\bar{x} = z$ and $\bar{y} = -z$, $x = y = d$, and $d > z$, the covariance calculation is:

$$E[(x - \bar{x})(y - \bar{y})] = E[(d - z)(d - - z)] = E[d^2 - z^2]$$

Since $d > z$ we have $E[d^2 - z^2] > 0$ which is a positive measured covariance term indicating the stocks trend in the same direction. But these two stocks move in the exact opposite direction.

The most important finding above is that when we compute covariance and correlation on a stock by stock basis using historical returns and price data it is possible that the calculated measure is opposite of what is happening in the market. These "false positive" and/or "false negative" relationships may be due to the error term about the trend rather than the trend or possibly due to too few data points in our sample.

Example 1: False negative signal calculations

Table 6.5a contains the data for two stocks A and B that are moving in the same direction. Figure 6.7a illustrates this movement over 24 periods. But when we calculate the covariance between these stocks we get a negative correlation, $\rho = -0.71$. How can stocks that move in the same direction have a negative covariance term? The answer is due to the excess terms being on opposite sides of the price trend (Figure 6.7b). Notice that these excess returns are now on opposite sides of the trend which results in a negative covariance measure. The excess returns are indeed negatively correlated but the direction of trend is positively correlated.

Table 6.5a False Negative Signals

	Market Prices		Period Returns		Excess Returns	
Period	A	B	A	B	A	B
0	$10.00	$20.00				
1	$11.42	$22.17	13.3%	10.3%	7.0%	5.3%
2	$11.12	$25.48	−2.6%	13.9%	−8.8%	8.9%
3	$12.60	$28.62	12.5%	11.6%	6.3%	6.6%
4	$12.96	$33.56	2.8%	15.9%	−3.4%	10.9%
5	$16.91	$30.59	26.6%	−9.3%	20.4%	−14.3%
6	$17.63	$33.58	4.2%	9.3%	−2.0%	4.3%
7	$17.78	$37.86	0.8%	12.0%	−5.4%	7.0%
8	$19.93	$38.93	11.4%	2.8%	5.2%	−2.2%
9	$23.13	$38.94	14.9%	0.0%	8.7%	−5.0%
10	$24.21	$39.64	4.6%	1.8%	−1.6%	−3.2%
11	$23.39	$46.32	−3.5%	15.6%	−9.7%	10.6%
12	$23.92	$49.59	2.3%	6.8%	−3.9%	1.8%
13	$25.50	$51.45	6.4%	3.7%	0.2%	−1.3%
14	$23.97	$56.96	−6.2%	10.2%	−12.4%	5.2%
15	$27.35	$56.60	13.2%	−0.6%	7.0%	−5.6%
16	$31.27	$57.37	13.4%	1.3%	7.2%	−3.7%
17	$30.03	$61.26	−4.0%	6.6%	−10.2%	1.6%
18	$36.04	$61.02	18.2%	−0.4%	12.0%	−5.4%
19	$32.01	$67.66	−11.9%	10.3%	−18.1%	5.3%
20	$33.16	$69.90	3.5%	3.3%	−2.7%	−1.7%
21	$37.32	$66.33	11.8%	−5.2%	5.6%	−10.2%
22	$34.71	$73.60	−7.3%	10.4%	−13.5%	5.4%
23	$39.08	$71.58	11.9%	−2.8%	5.7%	−7.8%
24	$44.33	$66.43	12.6%	−7.5%	6.4%	−12.5%
Avg:			6.2%	5.0%	0.0%	0.0%
Correl:			−0.71		−0.71	

Example 2: False positive signal calculations

Table 6.5b contains the data for two stocks C and D that are moving in opposite directions. Figure 6.7c illustrates this movement over 24 periods. But when we calculate the covariance between these stocks we get a negative correlation, $\rho = +0.90$. How can stocks that move in the same direction have a negative covariance term? The answer is due to the excess terms being on the same side of the price trend. Figure 6.7d illustrates the excess return in each time period. Notice that these excess returns are now on opposite sides of the trend which results in a negative covariance measure.

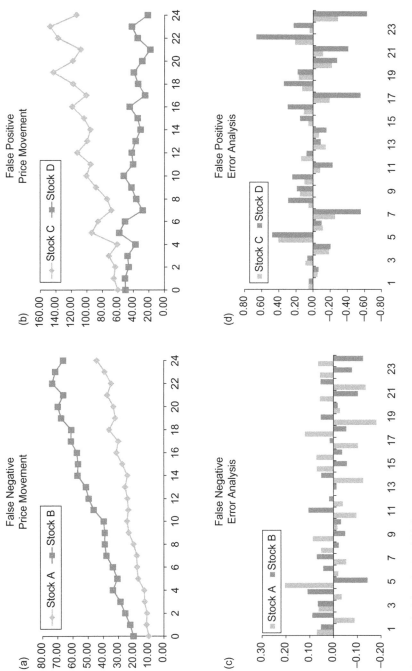

■ **Figure 6.7** False Relationship Signal Calculations.

Table 6.5b False Positive Signals

Period	Market Prices		Period Returns		Excess Returns	
	C	**D**	**C**	**D**	**C**	**D**
0	$60.00	$50.00				
1	$65.11	$50.82	8.2%	1.6%	5.5%	5.1%
2	$63.43	$45.93	−2.6%	−10.1%	−5.3%	−6.6%
3	$71.51	$47.43	12.0%	3.2%	9.3%	6.7%
4	$60.90	$37.31	−16.1%	−24.0%	−18.7%	−20.5%
5	$93.93	$58.09	43.3%	44.3%	40.7%	47.8%
6	$85.83	$50.77	−9.0%	−13.5%	−11.7%	−10.0%
7	$68.19	$28.10	−23.0%	−59.2%	−25.7%	−55.7%
8	$73.95	$36.34	8.1%	25.7%	5.5%	29.2%
9	$88.56	$42.51	18.0%	15.7%	15.4%	19.2%
10	$100.69	$52.41	12.8%	20.9%	10.2%	24.4%
11	$95.29	$40.31	−5.5%	−26.3%	−8.2%	−22.8%
12	$112.56	$42.10	16.7%	4.3%	14.0%	7.8%
13	$99.59	$37.12	−12.2%	−12.6%	−14.9%	−9.1%
14	$95.56	$30.63	−4.1%	−19.2%	−6.8%	−15.7%
15	$103.88	$34.49	8.3%	11.9%	5.7%	15.4%
16	$119.10	$44.81	13.7%	26.2%	11.0%	29.7%
17	$100.88	$24.90	−16.6%	−58.7%	−19.3%	−55.3%
18	$117.90	$33.90	15.6%	30.9%	12.9%	34.3%
19	$143.46	$39.28	19.6%	14.7%	17.0%	18.2%
20	$118.28	$28.70	−19.3%	−31.4%	−22.0%	−27.9%
21	$108.05	$18.39	−9.0%	−44.5%	−11.7%	−41.0%
22	$137.49	$34.52	24.1%	63.0%	21.4%	66.5%
23	$147.63	$41.95	7.1%	19.5%	4.4%	23.0%
24	$113.77	$21.63	−26.1%	−66.2%	−28.7%	−62.7%
Avg:			2.7%	−3.5%	0.0%	0.0%
Correl:			0.90		0.90	

The excess returns are indeed positively correlated but the direction of trend is negatively correlated.

To correct for the calculation of covariance and correlation it is advised to compare stock price movement based on a common trend (such as the market index) or a multi-factor model. Factor models are discussed further below.

Degrees of Freedom

A portfolio's covariance matrix consists of stock variances along the diagonal terms and covariance terms on the off diagonals. The covariance matrix is a symmetric matrix since the covariance between stock A and stock B is identical to the covariance between stock B and stock A.

If a portfolio consists of n stocks the covariance matrix will be $n \times n$ and with n^2 total elements. The number of unique variance terms in the matrix is equal to the number of stocks, n. The number of covariance terms is equal to $(n^2 - n)$ and the number of unique covariance terms is $(n^2 - n)/2 = n \cdot (n - 1)/2$.

The number of unique covariance parameters can also be determined from:

$$Unique\ Covariances = \binom{n}{2} = \frac{n(n-1)}{2} \tag{6.30}$$

The number of total unique elements k in the $n \times n$ covariance matrix is equal to the total number of variances plus total number of unique covariances. This is:

$$k = n + \frac{n(n-1)}{2} = \frac{n \cdot (n+1)}{2} \tag{6.31}$$

In order to estimate these total parameters we need a large enough set of data observations to ensure that the number of degrees of freedom is at least positive (as a starting point). For example, consider a system of m equations and k variables. In order to determine a solution for each variable we need to have $m \geq k$ or $m - k \geq 0$. If $m < k$ then the set of equations is underdetermined and no unique solution exists. Meaning, we cannot solve the system of equations exactly.

The number of data points, d, that we have in our historical sample period of time is equal to $d = n \cdot t$ since we have one data point for each stock n. If there are t days in our historical period we will have $n \cdot t$ data points. Therefore, we need to ensure that the total number of data points, d, is greater than or equal to the number of unique parameters, k, in order to be able to solve for all the parameters in our covariance matrix. This is:

$$d \geq k$$
$$n \cdot t \geq \frac{n \cdot (n+1)}{2}$$
$$t \geq \frac{(n+1)}{2}$$

Therefore, for a 500 stock portfolio there will be 125,250 unique parameters. Since there are 500 data points per day we need just over 1 year of data (250 trading days per year) just to calculate each parameter in the covariance matrix.

But now, the problem associated with estimating each parameter entry in the covariance matrix is further amplified because we are not solving for a deterministic set of equations. We are seeking to estimate the value of each parameter. A general rule of thumb is that there needs to be at least 20 observations for each parameter to have statistically meaningful results.

The number of data points required is then:

$$d \geq 20 \cdot k$$
$$n \cdot t \geq 20 \cdot \frac{n \cdot (n+1)}{2}$$
$$t \geq 10 \cdot (n+1)$$

Therefore, for a 500 stock portfolio (the size of the market index) we need 5010 days of observations, which is equivalent to over 20 years of data! Even if we require only 10 data points per parameter this still results in over 10 years of data! Figure 6.8b shows the number of days of data that is required to estimate the parameters of the covariance matrix for different numbers of stocks.

It has been suggested by some industry pundits that it is possible to estimate all unique parameters of the covariance matrix using the same number of observations as there are unique parameters. However, these pundits also state that in order for this methodology to be statistically correct we need to compute the covariance terms across the entire universe of stocks and not just for a subset of stocks. But even if this is true, the relationship across companies in the methodology needs to be stable. The reasoning is that if we do use the entire universe of stocks with enough data points we will uncover the true intrarelationship across all subgroups of stocks and have accurate variance and covariance measures.

In the US there are over 7000 stocks and thus over 24.5 million parameters. This would require over 14 years of data history! We are pretty confident in the last 14 years that many companies have changed main lines of products (e.g., Apple), changed their corporate strategy (e.g., IBM), and thus these relationships have changed. So even if we had enough data points we know that companies do change, violating the requirements for this approach.

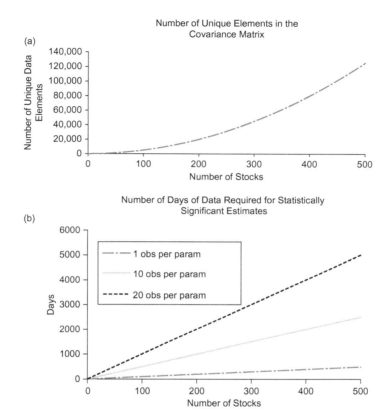

■ Figure 6.8 Errors Associated with Estimating Stock Price Movement Covariance

The last point to make is that for a global covariance matrix with a global universe of over 50,000 companies (at least 50,000!) there would be over 1.25 billion unique parameters and we would need a historical prices series of over 100 years! Think about how much has changed in just the last 10 years, let alone 100 years.

FACTOR MODELS

Factor models address the two deficiencies we encountered when using historical market data to compute covariance and correlation. First, these models do not require the large quantity of historical observations that are needed for the sample covariance approach in order to provide accurate risk estimates. Second, factor models use a set of common explanatory factors across all stocks and comparisons are made to these factors across all stocks. However, proper statistical analysis is still required to ensure accurate results.

Factor models provide better insight into the overall covariance and correlation structure between stocks and across the market. Positive correlation means that stocks will move in the same direction and negative correlation means that stocks will move in opposite direction.

A factor model has the form:

$$r_t = \alpha_0 + f_{1t}b_1 + f_{2t}b_2 + \cdots + f_{kt}b_k + e_t \tag{6.32}$$

where,

r_t = stock return in period t.
α_0 = constant term.
f_{kt} = factor k value in period t.
b_k = exposure of stock i to factor k. This is also referred to as beta, sensitivity, or factor loadings.
e_t = noise for stock i in period t. This is the return not explained by the model.

Parameters of the model are determined via ordinary least squares (OLS) regression analysis. Some analysts apply a weighting scheme so more recent observations have a higher weight in the regression analysis. These weighting schemes are often assigned using a smoothing function and "half-life" parameter. Various different weighting schemes for regression analysis can be found in Green (2000).

To perform a statistically correct regression analysis the regression model is required to have the following properties. (see Green (2000); Kennedy (1998); Mittelhammer (2000), etc.).

Regression properties:

1. $E(e_t) = 0$
2. $Var(e_t) = E(e'e) = \sigma_e^2$
3. $Var(f_k) = E[(f_k - \bar{f}_k)^2] = \sigma_{f_k}^2$
4. $E(ef_k) = 0$
5. $E(f_{kt}, f_{lt}) = 0$
6. $E(e_t e_{t-j}) = 0$
7. $E(e_{it} e_{jt}) = 0$

Property (1) states that the error term has a mean of zero. This will always be true for a regression model that includes a constant term $\hat{b}_{0k}$ or for a model using excess returns $E(r_{it}) = 0$. Property (2) states that the variance of the error term for each stock is σ_{ei}^2. Properties (1) and (2) are direct byproducts of a properly specified regression model. Property (3) states that the variance of each factor is σ_{fk}^2 and is true by definition. Property (4) states that the error term (residual) and each factor are independent. Analysts need to test to ensure this property is satisfied. Property (5) states that the explanatory factors are independent. Analysts need to properly select factors that are independent or make adjustments to ensure that they are independent. If the factors are not truly independent the sensitivities to these factors will be suspect. Property (6) states that the error terms are independent for all lagged time periods, i.e., no serial correlation or correlation of any lags across the error terms. Property (7) states that the error terms across all stocks are independent, i.e., the series of all error terms are uncorrelated. Since the error term in a factor model indicates company specific returns or noise that is not due to any particular market force these terms need to be independent across companies. If there are stocks with statistically significant correlated error terms then it is likely that there is some market force or some other explanatory variable that is driving returns that we have not accounted for in the model. In this case, while the sensitivities to the selected variables may be correct, some of our risk calculations may be suspect because we have not fully identified all sources of risk. For example, company specific risk, covariance and correlation, and portfolio risk may be suspect due to an incomplete model and may provide incorrect correlation calculations.

When constructing factor models analysts need to test and ensure that all properties are satisfied.

Matrix Notation

In matrix notation our single stock factor model is:

$$r_i = \alpha_i + Fb_i + e_i \tag{6.33}$$

where,

$$r_i = \begin{bmatrix} r_{i1} \\ r_{i2} \\ \vdots \\ r_{in} \end{bmatrix}, \ \alpha_i = \begin{bmatrix} \alpha_{i1} \\ \alpha_{i2} \\ \vdots \\ \alpha_{in} \end{bmatrix}, \ F = \begin{bmatrix} f_{11} & f_{21} & \cdots & f_{k1} \\ f_{12} & f_{22} & \cdots & f_{k2} \\ \vdots & \vdots & \ddots & \vdots \\ f_{1n} & f_{2n} & \cdots & f_{kn} \end{bmatrix}, \ b_k = \begin{bmatrix} b_{i1} \\ b_{i2} \\ \vdots \\ b_{ik} \end{bmatrix} \ e_i = \begin{bmatrix} e_{i1} \\ e_{i2} \\ \vdots \\ e_{in} \end{bmatrix}$$

r_i = vector of stock returns for stock i

r_{it} = return of stock i in period t

α_i = vector of the constant terms

F = column matrix of factor returns

f_{jt} = factor j in period t

b_i = vector of risk exposures
b_{ij} = risk sensitivity of stock i to factor j
e_i = vector of errors (unexplained return)
e_{it} = error term of stock i in period t
n = total number of time periods
m = total number of stocks
k = total number of factors

Constructing Factor Independence

Real-world data often results in factors that are not independent which violates regression property (5). This makes it extremely difficult to determine accurate risk exposures to these factors. In these situations, analysts can transform the set of dependent original factors into a new set of factors that are linearly independent (Kennedy, 1988).

This process is described as follows:

Let F, G, and H represent three explanatory factors that are correlated.

First, sort the factors by explanatory power. Let F be the primary driver of risk and return, let G be the secondary driver, and let H be the tertiary driver.

Second, remove the correlation between F and G. This is accomplished by regressing the secondary factor G on the primary factor F as follows:

$$G = \tilde{v}_0 + \tilde{v}_1 F + e_G$$

The error term in this regression e_G is the residual factor G that is not explained by the regression model and by definition (property (4)) is independent of F. Then let $\tilde{G}$ be simply e_G from the regression. That is:

$$\tilde{G} = G - \tilde{v}_0 - \tilde{v}_1 F$$

Third, remove the correlation between factor H and factor F and the new secondary factor $\tilde{G}$. This is accomplished by regressing H on F and $\tilde{G}$ as follows:

$$H = \hat{\gamma}_0 + \hat{\gamma}_1 F + \hat{\gamma}_2 \tilde{G} + e_H$$

The error term in this regression e_H is the residual factor H that is not explained by the regression model and by definition (property (4)) is independent of F and G. This process can be repeated for as many factors as are present.

The factor model with uncorrelated factors is finally re-written as:

$$r = \alpha_o + Fb_f + \tilde{G}b_{\tilde{g}} + \tilde{H}b_{\tilde{h}} + \varepsilon \qquad (6.34)$$

This representation now provides analysts with a methodology to calculate accurate risk exposures to a group of predefined factors which are now independent.

Estimating Covariance Using a Factor Model

A factor model across a universe of stocks can be written as:

$$R = \alpha + F\beta + \varepsilon \qquad (6.35)$$

where,

$$
R = \begin{bmatrix} r_{11} & r_{21} & \cdots & r_{m1} \\ r_{12} & r_{22} & \cdots & r_{m2} \\ \vdots & \vdots & \ddots & \vdots \\ r_{1n} & r_{2n} & \cdots & r_{mn} \end{bmatrix} \quad
F = \begin{bmatrix} f_{11} & f_{21} & \cdots & f_{k1} \\ f_{12} & f_{22} & \cdots & f_{k2} \\ \vdots & \vdots & \ddots & \vdots \\ f_{1n} & f_{2n} & \cdots & f_{kn} \end{bmatrix} \quad
\alpha' = \begin{bmatrix} \alpha_1 \\ \alpha_2 \\ \vdots \\ \alpha_n \end{bmatrix}
$$

$$
\beta = \begin{bmatrix} b_{11} & b_{21} & \cdots & b_{m1} \\ b_{12} & b_{22} & \cdots & b_{m2} \\ \vdots & \vdots & \ddots & \vdots \\ b_{1k} & b_{2k} & \cdots & b_{mk} \end{bmatrix} \quad
\varepsilon = \begin{bmatrix} \varepsilon_{11} & \varepsilon_{21} & \cdots & \varepsilon_{m1} \\ \varepsilon_{12} & \varepsilon_{22} & \cdots & \varepsilon_{m2} \\ \vdots & \vdots & \ddots & \vdots \\ \varepsilon_{1n} & \varepsilon_{2n} & \cdots & \varepsilon_{mn} \end{bmatrix}
$$

This formulation allows us to compute the covariance across all stocks without the issues that come up when using historical market data. This process is described following Elton and Gruber (1995) as follows:

The covariance matrix of returns C is calculated as:

$$C = E[(R - E[R])'(R - E[R])]$$

From our factor model relationship we have:

$$R = \alpha + F\beta + \varepsilon$$

The expected value of returns is:

$$E[R] = \alpha + \overline{F}\beta$$

Now we can determine the excess returns as:

$$R - E[R] = (F - \overline{F})\beta + \varepsilon$$

Now substituting in the above we have:

$$
\begin{aligned}
C &= E[((F-\overline{F})^2\beta + \varepsilon)'((F-\overline{F})\beta + \varepsilon)] \\
&= E[\beta'(F-\overline{F})^2\beta + 2\beta'(F-\overline{F})\varepsilon + \varepsilon'\varepsilon] \\
&= \beta'E[(F-\overline{F})^2]\beta + 2\beta'E[(F-\overline{F})\varepsilon] + E[\varepsilon'\varepsilon]
\end{aligned}
$$

By property (4):

$$E[2\beta'(F - \overline{F})\varepsilon] = 0$$

By property (2) and property (7) we have,

$$E[\varepsilon'\varepsilon] = \begin{bmatrix} \sigma_{e1}^2 & 0 & \cdots & 0 \\ 0 & \sigma_{e2}^2 & \cdots & 0 \\ \vdots & \vdots & \ddots & \vdots \\ 0 & 0 & \cdots & \sigma_{en}^2 \end{bmatrix} = \Lambda$$

which is the idiosyncratic variance matrix and is a diagonal matrix consisting of the variance of the regression term for each stock.

By property (3) and property (5) the factor covariance matrix is:

$$E[(F - \overline{F})^2] = \begin{bmatrix} \sigma_{f1}^2 & 0 & \cdots & 0 \\ 0 & \sigma_{f2}^2 & \cdots & 0 \\ \vdots & \vdots & \ddots & \vdots \\ 0 & 0 & \cdots & \sigma_{fk}^2 \end{bmatrix} = \mathrm{cov}(F)$$

The factor covariance matrix will be a diagonal matrix of factor variances. In certain situations there may be some correlation across factors. When this occurs, the off-diagonal entries will be the covariance between the factors. Additionally, the beta sensitivities may be suspect meaning that we may not know the true exposures to each factor and we will have some difficulty determining how much that particular factor contributes to returns. However, the covariance calculation will be correct providing we include the true factor covariance matrix.

Finally, we have our covariance matrix derived from the factor model to be:

Covariance Matrix

$$C = \beta' \mathrm{cov}(F)\beta + \Lambda \tag{6.36}$$

This matrix can be decomposed into the systematic and idiosyncratic components. Systematic risk component refers to the risk and returns that are explained by the factors. It is also commonly called market risk or factor risk. The idiosyncratic risk component refers to the risk and returns that are not explained by the factors. This component are also commonly called stock specific risk, company specific, and diversifiable risk. This is shown as:

$$C = \underbrace{\beta' \mathrm{cov}[F]\beta}_{\substack{\text{Systematic} \\ \text{Risk}}} + \underbrace{\Lambda}_{\substack{\text{Idiosyncratic} \\ \text{Risk}}} \tag{6.37}$$

TYPES OF FACTOR MODELS

Factor models can be divided into four categories of models: index models, macroeconomic models, cross-sectional or fundamental data models, and statistical factor models. These are described below.

Table 6.6 provides a comparison of the common risk factor models used in the industry. A description of each is provided below.

Table 6.6 Risk Factor Models

Volatility Model	Types	Formula	Parameter	Calculation
Time Series	Index, Macro, Technical	$R = \hat{B}F + \varepsilon$	$\hat{B}$	OLS
Cross-Sectional	Fundamental	$R = B\hat{F} + \varepsilon$	$\hat{F}$	OLS
Statistical/PCA	Implicit Factors	$R = \hat{B}\hat{F} + \varepsilon$	$\hat{B}, \hat{F}$	OLS, MLE
	Data Driven	$\overline{C}$		Eigenvalue, SVD Factor Analysis

Index Model

There are two forms of the index model commonly used in the industry: single index and multi-index models. The single index model is based on a single major market index such as the SP500. The same index is used as the input factor across all stocks. The multi-index model commonly incorporates the general market index, the stock's sector index, and additionally, the stock's industry index. The market index will be the same for all stocks but the sector index and industry index will be different based on the company's economic grouping. All stocks in the same sector will use the same sector index, and all stocks in the same industry will use the same industry index.

Single Index Model

The simplest of all the multi-factor models is the single index model. This model formulates a relationship between stock returns and market movement. In most situations, the SP500 index or some other broad market index is used as a proxy for the whole market.

In matrix notation, the single factor model has general form:

$$r_i = \alpha_i + \hat{b}_i R_m + e_i \tag{6.38}$$

r_i = column vector of stock returns for stock i
R_m = column vector of market returns
e_i = column vector of random noise for stock i
$\hat{b}_i$ = stock return sensitivity to market returns

In the single index model we need to estimate the risk exposure $\hat{b}_i$ to the general index R_m. In situations where the index used in the single index model is the broad market index and the constant term is the risk free rate, the single index model is known as the CAPM model (Sharpe, 1964) and the risk exposure $\hat{b}_i$ is the stock beta, β.

Multi-Index Models

The multi-index factor model is an extension of the single index model that captures additional relationships between price returns and corresponding sectors and industries. There have been numerous studies showing that the excess returns (error) from the single index model are correlated across stocks in the same sector, and with further incremental correlation across stocks in the same industry (see Elton and Gruber, 1995).

Let R_m = market returns, S_k = the stock's sector returns, and I_k = the stock's industry return. Then the linear relationship is:

$$r_i = \alpha_i + b_{im} R_m + b_{ik} S_k + b_{il} I_i + e_i$$

where b_{im} is the stock's sensitivity to the general market movement, b_{ik} is the stock's sensitivity to its sector movement, and b_{il} is the stock's sensitivity to its industry movement.

There is a large degree of correlation, however, across the general market, sectors, and industry. These factors are not independent and analysts need to make appropriate adjustment following the process outlined above.

The general multi-index model after multicollinearity now has form:

$$r_i = \alpha_i + \hat{b}_{im} R_m + \hat{b}_{isk}^* \tilde{S}_k + \hat{b}_{iIl}^* \tilde{I}_l + \varepsilon \tag{6.39}$$

Macroeconomic Factor Models

A macroeconomic multi-factor model defines a relationship between stock returns and a set of macroeconomic variables such as GDP, inflation, industrial production, bond yields, etc. The appeal of using macroeconomic data as the explanatory factors in the returns model is that these variables are readily measurable and have real economic meaning.

While macroeconomic models offer key insight into the general state of the economy they may not sufficiently capture the most accurate correlation structure of price movement across stocks. Additionally, macroeconomic models may not do a good job capturing the covariance of price movement across stocks in "new economies" or a "shifting regime" such as the sudden arrival of the financial crisis beginning in Sep. 2008.

Ross, Roll, and Chen (1986) identified the following four macroeconomic factors as having significant explanatory power with stock return. These strong relationships still hold today and are:

1. Unanticipated changes in inflation.
2. Unanticipated changes in industrial production.
3. Unanticipated changes in the yield between high-grade and low-grade corporate bonds.
4. Unanticipated changes in the yield between long-term government bonds and t-bills. This is the slope of the term structure.

Other macroeconomic factors have also been incorporated into these models include change in interest rates, growth rates, GDP, capital investment, unemployment, oil prices, housing starts, exchange rates, etc. The parameters are determined via regression analysis using monthly data over a five-year period, e.g., 60 observations.

It is often assumed that the macroeconomic factors used in the model are uncorrelated and analysts do not make any adjustment for correlation across returns. But improvements can be made to the model following the adjustment process described above.

A k-factor macroeconomic model has the form:

$$r_i = \alpha_{i0} + \hat{b}_{i1}f_1 + \hat{b}_{i2}f_2 + \cdots + \hat{b}_{ik}f_k + e_i \qquad (6.40)$$

Analysts need to estimate the risk exposures $b_{ik}s$ to these macroeconomic factors.

Cross-Sectional Multi-Factor Models

Cross-sectional models estimate stock returns from a set of variables that are specific to each company rather than through factors that are common across all stocks. Cross-sectional models use stock specific factors that are based on fundamental and technical data. The fundamental data consists of company characteristics and balance sheet information. The technical data (also called market driven) consists of trading activity metrics such as average daily trading volume, price momentum, size, etc.

Because of the reliance on fundamental data, many authors use the term "fundamental model" instead of cross-sectional model. The rationale behind the cross-sectional model is similar to the rationale behind the macroeconomic model. Since managers and decision-makers incorporate fundamental and technical analysis into their stock selection process it is only reasonable that these factors provide insight into return and risk for those stocks. Otherwise why would they be used?

Fama and French (1992) found that three factors consisting of (1) market returns, (2) company size (market capitalization), and (3) book to market ratio have considerable explanatory power. While the exact measure of these variables remains a topic of much discussion in academia, notice that the last two factors in the Fama-French model are company specific fundamental data.

While many may find it intuitive to incorporate cross-sectional data into multi-factor models, these models have some limitations. First, data requirements are cumbersome requiring analysts to develop models using company specific data (each company has its own set of factors). Second, it is often difficult to find a consistent set of robust factors across stocks that provide strong explanatory power. Ross and Roll had difficulty determining a set of factors that provided more explanatory power than the macroeconomic models without introducing excessive multicollinearity into the data (Figure 6.9).

The cross-sectional model is derived from company specific variables, referred to as company factor loadings. The parameters are typically

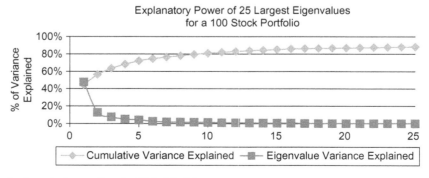

■ **Figure 6.9** Eigenvalue-Eigenvector Decomposition of a 100 Stock Portfolio

determined via regression analysis using monthly data over a longer period of time, e.g., a five-year period, with 60 monthly observations.

The cross-sectional model is written as:

$$r_{it} = x_{i1}^* \hat{f}_{1t} + x_{i2}^* \hat{f}_{2t} + \cdots + x_{ik}^* \hat{f}_{kt} + e_{it} \qquad (6.41)$$

where x_{ij}^* is the normalized factor loading of company i to factor j. For example,

$$x_{kl}^* = \frac{x_{kl} - E(x_k)}{\sigma(x_k)}$$

where $E(x_k)$ is the mean of x_k across all stocks and $\sigma(x_k)$ is the standard deviation of x_k across all stocks.

And unlike the previous models where the factors were known in advance and we estimate the risk sensitivities, here we know the factor loadings (from company data) and we need to estimate the factors.

Statistical Factor Models

Statistical factor models are also referred to as implicit factor models and principal component analysis (PCA). In these models neither the explanatory factors nor sensitivities to these factors are known in advance and they are not readily observed in the market. However, both the statistical factors and sensitivities can be derived from historical data.

There are three common techniques used in statistical factor models: eigenvalue-eigenvector decomposition, singular value decomposition, and factor analysis. Eigenvalue-eigenvector is based on a factoring scheme of the sample covariance matrix and singular value decomposition is based on a factoring scheme of the matrix of returns (see Pearson, 2002). Factor analysis (not to be confused with factor models) is based on a maximum likelihood estimate of the correlations across stocks. In this section we discuss the eigenvalue-eigenvector decomposition technique.

The statistical factor models differs from the previously mentioned models in that analysts estimate both the factors (F_ks) and the sensitivities to the factors (b_{ik}s) from a series of historical returns. This model does not make any prior assumptions regarding the appropriate set of explanatory factors or force any preconceived relationship into the model.

This approach is in contrast to the explicit modeling approaches where analysts must specify either a set of explanatory factors or a set of company specific factor loadings. In the explicit approaches analysts begin with either a set of specified factors and estimate sensitivities to those factors (i.e., index models and macroeconomic factor model) or begin with the factor loadings (fundamental data) and estimate the set of explanatory factors (cross-sectional model).

The advantage of a statistical factor model over the previously described explicit approaches is that it provides risk managers with a process to uncover accurate covariance and correlation relationships of returns without making any assumptions regarding what is driving the returns. Any preconceived bias is removed from the model. The disadvantage of this statistical approach is that it does not provide portfolio managers with a set of factors to easily determine what is driving returns since the statistical factors do not have any real-world meaning.

To the extent that analysts are only interested in uncovering covariance and correlation relationships for risk management purposes, PCA has proven to be a viable alternative to the traditional explicit modeling approaches. Additionally, with the recent growth of exchange traded funds (ETFs) many managers have begun correlating their statistical factors to these ETFs in much the same way Ross and Roll did with economic data to better understand these statistical factors.

The process to derive the statistical model is as follows:

Step 1. Compute the sample covariance matrix by definition from historical data. This matrix will likely suffer from spurious relationships due the data limitations (not enough degrees of freedom and potential false relationships). But these will be resolved via principal component analysis.

Let $\overline{C}$ represent the sample covariance matrix.

Step 2. Factor the sample covariance matrix. We based the factorization scheme on eigenvalue-eigenvector decomposition. This is:

$$\overline{C} = VDV' \tag{6.42}$$

where D is the diagonal matrix of eigenvalues sorted from largest to smallest, $\lambda_1 > \lambda_2 > \cdots > \lambda_n$ and V is the corresponding matrix of eigenvectors and these eigenvalues are determined by computing the percentage of total variance that is explained by each eigenvector. In finance, the terminology that is most often used when referring to determining the eigenvalue with the strongest predictive power is principal component analysis. These matrices are as follows:

$$D = \begin{bmatrix} \lambda_1 & 0 & \cdots & 0 \\ 0 & \lambda_2 & \cdots & 0 \\ \vdots & \vdots & \ddots & \vdots \\ 0 & 0 & \cdots & \lambda_n \end{bmatrix} \quad V = \begin{bmatrix} v_{11} & v_{21} & \cdots & v_{n1} \\ v_{12} & v_{22} & \cdots & v_{n2} \\ \vdots & \vdots & \ddots & \vdots \\ v_{1n} & v_{2n} & \cdots & v_{nn} \end{bmatrix}$$

Since D is a diagonal matrix, we have $D = D^{1/2}D^{1/2}$, $D = D'$, and $D^{1/2} = (D^{1/2})'$

Then, our covariance matrix C can be written as:

$$\overline{C} = VDV' = VD^{1/2}D^{1/2}V' = VD^{1/2}(VD^{1/2})'$$

Step 3. Compute β in terms of the eigenvalues and eigenvectors:

$$\beta = (VD^{1/2})'$$

Then the full sample covariance matrix expressed in terms of β is:

$$\beta'\beta = VD^{1/2}(VD^{1/2})' \qquad (6.43)$$

Step 4. Remove spurious relationship due to data limitation.

To remove the potential spurious relationship we only use the eigenvalues and eigenvectors with the strongest predictive power.

How many factors should be selected?

In our eigenvalue-eigenvector decomposition each eigenvalue λ_k of the sample covariance matrix explains exactly $\lambda_k / \sum \lambda$ percent of the total variance. Since the eigenvalues are sorted from highest to lowest, a plot of the percentage of variance explained will show how quickly the predictive power of the factors declines. If the covariance matrix is generated by say 10 factors then the first 10 eigenvalues should explain the large majority of the total variance.

There are many way to determine how many factors should be selected to model returns. For example, some analysts will select the minimum number of factors that explain a pre-specified amount of variance, some will select the number of factors up to where there is a break-point or fall-off in explanatory power. And others may select factors so that the variance $>1/n$. Assuming that each factor should explain at least $1/n$ of the total. Readers can refer to Dowd (1998) for further techniques.

If it is determined that there are k factors that sufficiently explain returns, the risk exposures are determined from the first k risk exposures for each stock since our eigenvalues are sorted from highest predictive power to lowest.

$$\beta = \begin{bmatrix} \beta_{11} & \beta_{21} & \cdots & \beta_{m1} \\ \beta_{12} & \beta_{22} & \cdots & \beta_{m2} \\ \vdots & \vdots & \ddots & \vdots \\ \beta_{1k} & \beta_{2k} & \cdots & \beta_{mk} \end{bmatrix}$$

The estimated covariance matrix is then:

$$C = \underset{nxk\ kXn}{\beta'\ \beta} + \underset{nXn}{\Lambda} \tag{6.44}$$

In this case the idiosyncratic matrix Λ is the diagonal matrix consisting of the difference between the sample covariance matrix and $\beta'\beta$. That is,

$$\Lambda = diag(\overline{C} - \beta'\beta) \tag{6.45}$$

It is important to note that in the above expression $\overline{C} - \beta'\beta$ the off-diagonal terms will often be non-zero. This difference is considered to be the spurious relationship caused by the data limitation and degrees of freedom issue stated above. Selection of an appropriate number of factors determined via eigenvalue decomposition will help eliminate these false relationships.

Advanced Algorithmic Forecasting Techniques

INTRODUCTION

This chapter introduces readers to advanced algorithmic forecasting techniques. We begin by reformulating our transaction cost equations in terms of the various trading strategy definitions, such as percentage of volume, trade rate, and trade schedules, and calibrate the parameters for these model variations. Estimated market impact costs for each approach are compared for the different data samples.

Readers are next introduced to the various sources of algorithmic trading risk including price volatility, liquidity risk, and parameter estimation uncertainty. We derive algorithmic forecasting techniques to estimate daily volume and monthly ADVs. The daily volume forecasting model is based on an autoregressive moving average (ARMA) time series which incorporates a median metric with a day of week effect adjustment factor, and the monthly ADV model incorporates previous volume levels, momentum, and market volatility.

The chapter concludes with an overview of the various transaction equations that are utilized to construct the efficient trading frontier and to develop optimal "best execution" strategies. All of which are essential building blocks for traders and portfolio managers interested in improving portfolio returns through best in class transaction costs management practices[1].

---

[1]We would like to thank Connie Li, M.S. from Cornell Financial Engineering and Quantitative Analyst at Numeric Investors, for providing invaluable insight into the proper formulation of these mathematical techniques and for testing and verifying these equations.

TRADING COST EQUATIONS

Our market impact and timing risk equations expressed in terms of percentage of trading volume POV are:

$$I^*_{bp} = \hat{a}_1 \cdot \left(\frac{X}{ADV}\right)^{\hat{a}_2} \cdot \sigma^{\hat{a}_3} \qquad (7.1)$$

$$MI_{bp} = \hat{b}_1 \cdot I^* \cdot POV^{\hat{a}_4} + (1 - \hat{b}_1) \cdot I^* \qquad (7.2)$$

$$TR_{bp} = \sigma \cdot \sqrt{\frac{1}{3} \cdot \frac{1}{250} \cdot \frac{X}{ADV} \cdot \frac{1 - POV}{POV}} \cdot 10^4 bp \qquad (7.3)$$

where,

X = total shares to trade
ADV = average daily volume
σ = annualized volatility (expressed as a decimal, e.g., 0.20)
$POV = \frac{X}{X + V_t}$ = percentage of trading volume rate
V_t = expected market volume during trading period (excluding the order's shares X)
$\hat{a}_1, \hat{a}_2, \hat{a}_3, \hat{a}_4, \hat{b}_1$ = model parameters estimated via non-linear estimation techniques

Model Inputs

On the surface, the cost estimation process seems fairly straightforward, especially after having already estimated the model parameters. Investors simply need to enter their shares X and preferred POV execution strategy, and the model will determine cost estimates for these inputs.

Although, is the process really this simple and straightforward? Will the model provide accurate cost forecasts?

To answer these questions, let's take a closer look at our equations. Our transaction cost model actually consists of three different sets of input information:

1. User specified inputs: X, POV
2. Model parameters: $\hat{a}_1, \hat{a}_2, \hat{a}_3, \hat{a}_4, \hat{b}_1$
3. Explanatory factors: σ, ADV, V_t

The first set of input information is entered by the user and is based on the investment decision and the investor's urgency preference. In Chapter 5, Estimating I-Star Model Parameters, we provided non-linear regression techniques to estimate the parameters of the model and test the model's sensitivity. In Chapter 6, Price Volatility, we provided techniques to forecast price volatility and price covariance. What about the volume statistics—*ADV* and V_t? How are these variables determined?

During the parameter estimation phase we showed how to measure these variables over the historical sample period to incorporate them into the non-linear regression model. What about our forecasting needs? Are the volume patterns expected to be exactly the same in the future as they were in the past? The answer is likely no. If historical variable values are used for the forward-looking forecast with differing market conditions then the actual costs could be much different than the forecasted estimates even with the correct model form and actual parameters.

For example, suppose there was a sudden spike in volatility and a decrease in liquidity, similar to what was experienced during the financial crisis and debt crisis. If the cost model did not incorporate these new input variables, a higher cost execution strategy would be selected as a result of dramatically underestimating costs. Furthermore, many times portfolio managers are interested in future expected costs under various market conditions. Managers typically buy shares under the most favorable conditions, such as low volatility and high liquidity, and sell shares under very dire circumstances, such as high volatility, low liquidity, and decreasing prices. It is actually these market circumstances at times that help managers decide whether to buy or sell shares. So even if the model is formulated correctly and the parameters are exact, incorporating the incorrect volume and volatility values will lead to inaccurate cost estimates. This chapter will provide the necessary techniques to forecast daily V_t and monthly *ADV* conditions.

TRADING STRATEGY

Algorithmic trading makes use of three types of trading strategies: percentage of volume *POV*, trading rate α, and trade schedule x_k

Let,

X = total shares to trade
V_t = expected volume during the trading horizon (excluding shares from the order)

The trading strategy variables are:

Percentage of Volume

$$POV = \frac{X}{X + V_t} \quad 0\% \leq POV \leq 100\% \tag{7.4}$$

The percentage of volume POV variable measures the amount of market volume the order participated with over the trading period. For example, if a trader executes 20,000 shares of stock over a period where 100,000 shares traded in the market (including the order) the POV rate is $20,000/100,000 = 20\%$. POV is a very intuitive measure. For example, $POV = 25\%$ means the order participated with 25% of market volume, and $POV = 100\%$ means that the trader accounted for all the market volume during this period. POV is the preferred trading strategy metric when monitoring current and historical activity.

The disadvantage of the POV strategy is that it contains a decision variable in the denominator, which creates an additional layer of mathematical complexity during trade strategy optimization and increases the solution time.

Trading Rate

$$\alpha = \frac{X}{V_t} \quad \alpha \geq 0 \tag{7.5}$$

The trading rate variable α is the ratio of the shares traded X to the market volume V_t during the trading period, excluding its own traded shares. For example, if a trader executed 20,000 shares in the market over a period of time when 100,000 shares traded in the market; 20,000 shares were from the investor's order and 80,000 shares from other participants, so the trading rate is $\alpha = 20,000/80,000 = 25\%$. If a trader executed 20,000 shares in the market over a period of time when 30,000 shares traded in the market; 20,000 shares were from the investor's order and 10,000 shares from other participants, so the trading rate is $\alpha = 20,000/10,000 = 200\%$.

Trading rate, unfortunately, is not as intuitive as POV rate. A trade rate of $\alpha = 100\%$ does not mean that the trader participated with 100% of market volume but rather the investor participated with 50% of market volume. The advantage of the trade rate is that it does not have a decision variable in the denominator so trading solution calculations are less complex and optimization processing time is much quicker. Trading rate is the preferred metric when forecasting costs and developing single stock optimal trading strategies.

Trade Schedule

The trade schedule x_k strategy defines exactly how many shares to transact in a given trading period. For example, the trade schedule for an order executed over n-period is

$$x_1, x_2, x_3, \ldots, x_n \tag{7.6}$$

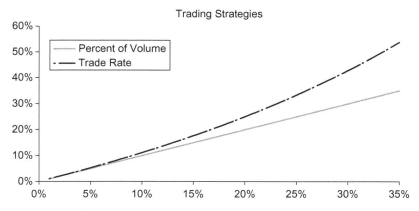

■ Figure 7.1 Trading Strategies.

and represents the number of shares to trade in periods $1, 2, 3, \ldots, n$. The total number of shares executed over this period is $X = \sum x_i$. The advantage of the trade schedule is that it allows front-loading and/or back-loading of trades to take advantage of anticipated price movement, volume conditions, as well as effective risk management during a basket trade (these are further discussed in Chapter 9, Portfolio Algorithms).

Comparison of *POV* rate to Trade Rate

There is a direct relationship between the trading rate α and *POV* rate:

$$POV = \frac{\alpha}{1+\alpha} \text{ and } \alpha = \frac{POV}{1-POV}$$

A comparison of *POV* rate to α is shown in Figure 7.1. For *POV* less than 15% there is minimal difference in these two calculations. However, as we start increasing these rates, the measures start to deviate.

TRADING TIME

We define trading time in terms of volume time units. The value represents the percentage of a normal day's volume that would have traded at a given point in time. For example, if 1,000,000 shares trade on an average trading day, the volume time when 250,000 shares trade is: $t^* = 250,000/1,000,000 = 0.25$. The volume time when 1,250,000 shares trade is: $t^* = 1,250,000/1,000,000 = 1.25$.

Volume time t^* is expressed as:

$$t^* = \frac{V_t}{ADV} \qquad (7.7)$$

Trading time can also be written in terms of trade rate α and POV rate. This calculation is as follows. Suppose the order is comprised of X shares. Then we can write trading time as:

$$t^* = \frac{V_t}{ADV} \cdot \left(\frac{X}{X}\right) = \frac{X}{ADV} \cdot \frac{V_t}{X}$$

In terms of trade rate α we have:

$$t^* = \frac{X}{ADV} \cdot \alpha^{-1} \qquad (7.8)$$

In terms of POV rate we have:

$$t^* = \frac{X}{ADV} \cdot \frac{1 - POV}{POV} \qquad (7.9)$$

TRADING RISK COMPONENTS

The timing risk (TR) measure is a proxy for the total uncertainty surrounding the cost estimate. In other words, it is the standard error of our forecast. This uncertainty is comprised of three components: price uncertainty, volume variance, and parameter estimation error. These are further described as follows:

Price Volatility: price volatility refers to the uncertainty surrounding price movement over the trading period. It will cause trading cost (ex-post) to be either higher or lower depending upon the movement and side of the order. For example, if the price moves up $0.50/share, this movement results in a higher cost for buy orders but a lower cost (savings) for sell orders. For a basket of stock, price volatility also includes the covariance or correlation across all names in the basket. Price volatility is the most commonly quoted standard error for market impact analysis. It is also very often the only standard error component.

Volume Variance: volume variance refers to the uncertainty in volumes and volume profiles over the trading horizon which could be less than, equal to, or more than a day. For example, if an investor trades an order over the full day, the cost will be different if total volume is 1,000,000 shares, 5,000,000 shares, or only 200,000 shares.

Parameter Estimation Error: parameter estimation error is the standard error component from our non-linear regression models. As shown in Chapter 5, Estimating I-Star Model Parameters, there is some degree of uncertainty surrounding the parameters which will affect market impact estimates. For simplicity, we define the timing risk measure to only include the price volatility term when quoting the standard error of the market impact estimate but analysts conducting advanced sensitivity analysis may want to incorporate these additional components into the timing risk estimate. We have found the easiest way to determine the overall uncertainty is via Monte-Carlo simulation where volumes, intraday profile, price movement, and parameter values are sampled from historical observations and their estimated distribution. Investors performing this type of analysis may find that corresponding market impact uncertainty is much larger than simply the standard deviation of price movement.

TRADING COST MODELS—REFORMULATED
Market Impact Expression

Our market impact equations can be restated in terms of our trading strategies as follows:

I-Star

The I-Star calculation written in basis point and total dollar units is:

$$I_{bp}^* = a_1 \cdot \left(\frac{Q}{ADV} \right)^{a_2} \cdot \sigma^{a_3} \qquad (7.10)$$

$$I_{\$/Share}^* = a_1 \cdot \left(\frac{Q}{ADV} \right)^{a_2} \cdot \sigma^{a_3} \cdot 10^{-4} \cdot P_0 \qquad (7.11)$$

$$I_{\$}^* = a_1 \cdot \left(\frac{Q}{ADV} \right)^{a_2} \cdot \sigma^{a_3} \cdot 10^{-4} \cdot X \cdot P_0 \qquad (7.12)$$

Market Impact for a Single Stock Order

The units of the market impact cost will be the same units as the instantaneous cost I-Star. Market impact cost for the three different trading strategy definitions is:

$$MI(POV) = \hat{b}_1 \cdot I^* \cdot POV^{\hat{a}_4} + (1 - \hat{b}_1) \cdot I^* \qquad (7.13)$$

$$MI(\alpha) = \hat{b}_1 \cdot I^* \cdot \alpha^{\hat{a}_4} + (1 - \hat{b}_1) \cdot I^* \qquad (7.14)$$

$$MI(x_k) = \sum_{k=1}^{t} \left(b_1 \cdot I^* \cdot \frac{x_k^2}{X \cdot v_k} \right) + (1 - b_1) \cdot I^* \tag{7.15}$$

with $\sum x_k = X$

The derivation of the trade schedule market impact formula above (Equation 7.15) is:

Start with the instantaneous cost estimate I^*. This value is allocated to each trade period based on the percentage of the order transacted in that period. If x_k shares of the total order X were executed in period k then the percentage I^* allocated to period k is:

$$I^* \cdot \frac{x_k}{X}$$

Therefore, the percentage of temporary impact allocated to period k is $b_1 \cdot I^* \cdot \frac{x_k}{X}$ and the percentage of permanent impact allocated to period k is $(1 - b_1) \cdot I^* \cdot \frac{x_k}{X}$.

The temporary impact cost is allocated to the investor based on the percentage of volume of the trade in that period. This is:

$$b_1 \cdot I^* \cdot \frac{x_k}{X} \cdot \frac{x_k}{x_k + v_k}$$

For simplicity, however, we rewrite temporary impact cost in terms of the trade rate as follows:

$$b_1 \cdot I^* \cdot \frac{x_k}{X} \cdot \frac{x_k}{v_k}$$

Finally, the total market impact cost of a trade schedule over all periods is determined by summing the cost over all periods. That is:

$$MI(x_k) = \sum_{k=1}^{n} b_1 \cdot I^* \cdot \frac{x_k}{X} \cdot \frac{x_k}{v_k} + \sum_{k=1}^{n} \frac{x_k}{X} \cdot (1 - b_1) \cdot I^*$$

This formulation is then simplified as:

$$MI(x_k) = \sum_{k=1}^{n} \left(b_1 \cdot I^* \cdot \frac{x_k^2}{X \cdot v_k} \right) + (1 - b_1) \cdot I^*$$

The units of the market impact cost formula will be the same as the units used in the I^* formula shown in Equations 7.10 to 7.12. Kissell and Glantz (2003) and Kissell, Glantz, and Malamut (2004) provide alternative derivations of the trade schedule market impact formulation.

Important note:

Notice that the market impact formulation for a one-period trade schedule reduces to:

$$MI_{bp} = \hat{b}_1 \cdot I^* \cdot \left(\frac{X}{V_t}\right) + (1 - \hat{b}_1) \cdot I^* \qquad (7.16)$$

This is the same formulation as Equation 7.15 with $a_4 = 1$. The importance of this equation is that it will be used to calibrate the market impact parameters for the trade schedule solution (shown below). Recall that this was also the simplified version of the model described in the two-step regression process shown in Chapter 5, Estimating I-Star Model Parameters.

Market impact cost across stock is an additive function. Therefore, the impact for a basket of stock is the sum of impacts for the entire basket. The addition problem is simplified when market impact is expressed in dollar units so that we do not need to worry about trade value weightings across stocks. These are:

Market Impact for a Basket of Stock

$$MI_\$(POV) = \sum_{i=1}^{m} (\hat{b}_1 \cdot I_i^* \cdot POV_i^{\hat{a}_4} + (1 - \hat{b}_1) \cdot I_i^*) \qquad (7.17)$$

$$MI_\$(\alpha) = \sum_{i=1}^{m} (\hat{b}_1 \cdot I_i^* \cdot \alpha_i^{\hat{a}_4} + (1 - \hat{b}_1) \cdot I_i^*) \qquad (7.18)$$

$$MI_\$(x_k) = \sum_{i=1}^{m} \sum_{k=1}^{n} \left(b_1 \cdot I_i^* \cdot \frac{x_{ik}^2}{X_i \cdot v_{ik}} \right) + (1 - b_1) \cdot I_i^* \qquad (7.19)$$

TIMING RISK EQUATION

The timing risk for an order executed over a period of time t^* following a constant trading strategy is as follows:

$$TR(t^*)_{bp} = \sigma \cdot \sqrt{\frac{1}{250} \cdot \frac{1}{3} \cdot t^*} \cdot 10^4 bp \qquad (7.20)$$

This equation simply scales price volatility for the corresponding trading period t^* and adjusts for the trade strategy (e.g., decreasing portfolio size). For example, σ is first scaled to a one-day period by dividing by $\sqrt{250}$, then this quantity is scaled to the appropriate trading time period by multiplying by $\sqrt{t^*}$. Recall that t^* is expressed in volume time units where $t^* = 1$ represents a one-day time period (volume-time). And since the order size is decreasing in each period, timing risk needs to be further adjusted downward by the $\sqrt{1/3}$ factor (see derivation below). This value is converted to basis points by multiplying by $10^4 bp$.

Therefore, timing risk is expressed in terms of POV and α following Equations 7.8 and 7.9 respectively. This is:

$$TR_{bp}(POV) = \sigma \cdot \sqrt{\frac{1}{250} \cdot \frac{1}{3} \cdot \frac{X}{ADV} \cdot \frac{1 - POV}{POV}} \cdot 10^4 bp \qquad (7.21)$$

$$TR_{bp}(\alpha) = \sigma \cdot \sqrt{\frac{1}{250} \cdot \frac{1}{3} \cdot \frac{X}{ADV} \cdot \alpha^{-1}} \cdot 10^4 bp \qquad (7.22)$$

These values expressed in terms of dollars follow directly from above:

$$TR_{\$}(POV) = \sigma \cdot \sqrt{\frac{1}{250} \cdot \frac{1}{3} \cdot \frac{X}{ADV} \cdot \frac{1 - POV}{POV}} \cdot X \cdot P_0 \qquad (7.23)$$

$$TR_{\$}(\alpha) = \sigma \cdot \sqrt{\frac{1}{250} \cdot \frac{1}{3} \cdot \frac{X}{ADV} \cdot \alpha^{-1}} \cdot X \cdot P_0 \qquad (7.24)$$

The reason the timing risk equations simplify so nicely is that the POV and α strategies assume a constant trading rate. Timing risk for a trade schedule, however, is not as nice. It is slightly more complicated since we need to estimate the risk for each period. This is as follows:

Let,

> $r_k =$ number of unexecuted shares at the beginning of period k
> $r_k = \sum_{j=k}^{n} x_j$
> $d =$ number of trading periods per day
> $v_k =$ expected volume in period k excluding the order shares
> $\left(\sigma^2 \cdot \frac{1}{250} \cdot \frac{1}{d}\right) =$ price variance scaled for the length of a trading period
> $P_0 =$ stock price at the beginning of the trading period

Timing risk for a trade schedule is the sum of the dollar risk in each trading period. That is:

$$TR_{\$}(x_k) = \sqrt{\sum_{k=1}^{n} r_k^2 \cdot \sigma^2 \cdot \frac{1}{250} \cdot \frac{1}{d} \cdot P_0^2} \qquad (7.25)$$

In this notation, σ^2 is expressed in (\$/share)^2 units and scaled for the length of the trading period. We divide by 250 to arrive at the volatility for a day and then further divide by the number of periods per day d. For example, if we break the day into 10 equal periods of volume we have $d = 10$. Finally, multiplying by P_0^2 converts volatility from (return)^2 units to ($/share)^2. Timing risk (variance) is now the sum of each period's variance over n trading horizons. Taking the square root gives timing risk value in total dollars.

Now suppose that we follow a constant trade rate. That is, the portfolio will be decreasing in a constant manner.

Derivation of the 1/3 factor

As shown above, risk for a specified trade rate is:

$$\Re(\alpha) = \sqrt{X^2 \cdot \frac{1}{250} \cdot \frac{X}{ADV} \cdot \frac{1}{\alpha} \cdot \sigma^2 \cdot \frac{1}{3} \cdot P_0^2} \qquad (7.26)$$

The derivation of the 1/3 adjustment factor is as follows:

Let R represent the vector of shares and C represent the covariance matrix scaled for a single period expressed in $\left(\$/Share\right)^2$. Then the one-period portfolio risk is:

$$\Re(1) = \sqrt{R'CR} \qquad (7.27)$$

For simplicity, we proceed forward using variance (risk squared). This is:

$$\Re^2(1) = R'CR \qquad (7.28)$$

The total variance over n periods is an additive function:

$$\Re^2(n) = \underbrace{R'CR}_{1} + \underbrace{R'CR}_{2} + \cdots + \underbrace{R'CR}_{n} = n \cdot R'CR \qquad (7.29)$$

For a constant portfolio R, variance scales with the square root of the number of trading periods:

$$\Re(n) = \sqrt{n \cdot R'CR} = \sqrt{n} \cdot \sqrt{R'CR} = \sqrt{n} \cdot \Re(1) \qquad (7.30)$$

This is often shown using the time notation as follows:

$$\Re(t) = \sqrt{t} \cdot \Re(1) \qquad (7.31)$$

For a portfolio where the share quantities change from period to period, the risk calculation will not simplify as it does above. Risk will need to be computed over all periods. This is:

$$\Re^2(n) = \sqrt{\underbrace{R_1'CR_1}_{1} + \underbrace{R_2'CR_2}_{2} + \ldots + \underbrace{R_n'CR_n}_{n}} \qquad (7.32)$$

Where R_k is the vector of portfolio shares in period k, this reduces to:

$$\Re(n) = \sqrt{\sum_{k=1}^{n} R'_k CR_k} \tag{7.33}$$

Trading risk for a trade schedule for a single stock execution is calculated as follows:

$$\Re^2(r_k) = \sum_{j=1}^{n} r_j^2 \cdot \sigma^2 \cdot P_0^2 \tag{7.34}$$

where σ^2 is the corresponding one-period variance expressed in $(\$/\text{Share})^2$, P_0 is the current price, and $\Re^2(r_k)$ is the total dollar variance for the strategy. Notice that we are simply summing the variance in each period.

For a continuous trade rate strategy where we execute the same number of shares in each period, the number of unexecuted shares at the beginning of each trade period is calculated as follows:

$$r_j = X - \frac{X}{n} \cdot (j-1) = X \left(1 - \frac{(j-1)}{n} \right) \tag{7.35}$$

where X is the total number of shares in the order.

Then the number of unexecuted shares at the beginning of each period squared is:

$$r_j^2 = X^2 \left(1 - \frac{(j-1)}{n} \right)^2 \tag{7.36}$$

Now let,

$\sigma^2 =$ the annualized variance
$t^* =$ total time to trade in terms of a year (same units as volatility), e.g., $t = 1$ is one year, $t = 1/250 = 1$ day, etc.
$n =$ number of periods in the trading interval

Then we have,

$t^* \cdot \sigma^2 =$ variance scaled for the time period
$t^* \cdot \frac{\sigma^2}{n} =$ variance scaled for a trading interval

For example, if the trading time is one day and the day is segmented into ten periods then we have:

$$\sigma^2(trading\ period) = \frac{1}{250} \cdot \frac{\sigma^2}{10} \tag{7.37}$$

The variance of the trade schedule is:

$$\mathfrak{R}^2(r_k) = \sum_{j=1}^{n} R_j^2 \cdot t^* \cdot \frac{\sigma^2}{n} \cdot P_0^2 \tag{7.38}$$

By Equations 7.36 and 7.38 we have:

$$\mathfrak{R}^2(r_k) = \sum_{j=1}^{n} X^2 \left(1 - \frac{(j-1)}{n}\right)^2 \cdot t^* \cdot \frac{\sigma^2}{n} \cdot P_0^2 \tag{7.39}$$

By factoring we have:

$$\mathfrak{R}^2(r_k) = X^2 \cdot P_0^2 \cdot t^* \cdot \frac{\sigma^2}{n} \cdot \sum_{j=1}^{n} \left(1 - \frac{(j-1)}{n}\right)^2 \tag{7.40}$$

And by expansion we have:

$$\mathfrak{R}^2(r_k) = X^2 \cdot P_0^2 \cdot t^* \cdot \frac{\sigma^2}{n} \cdot \sum_{j=1}^{n} \left(1 - \frac{2(j-1)}{n} + \frac{(j-1)^2}{n^2}\right) \tag{7.41}$$

Using the following identities:

$$\sum_{j=1}^{n} 1 = n$$

$$\sum_{j=1}^{n} x = \frac{n(n+1)}{2}$$

$$\sum_{j=1}^{n} x^2 = \frac{n(n+1)(2n+1)}{6}$$

Equation 7.41 is now:

$$\mathfrak{R}^2 = X^2 \cdot P_0^2 \cdot t^* \cdot \sigma^2 \cdot \frac{1}{n} \left(n - (n-1) + \frac{(n-1)(2n-1)}{6n^2}\right) \tag{7.42}$$

This further reduces to:

$$\mathfrak{R}^2 = X^2 \cdot P_0^2 \cdot t^* \cdot \sigma^2 \cdot \left(\frac{1}{3} + \frac{1}{n} + \frac{1}{2n} + \frac{1}{6n^2}\right) \tag{7.43}$$

Now if we let the number of trading periods over the defined trading time increase, the size of the trading interval becomes infinitely small and our trade schedule strategy approaches a continuous trade rate strategy.

To show this we take the limit as n approaches infinity in Equation 7.43:

$$\lim_{n \to \infty} X^2 \cdot P_0^2 \cdot t^* \cdot \sigma^2 \cdot \left(\frac{1}{3} + \frac{1}{n} + \frac{1}{2n} + \frac{1}{6n^2}\right) = X^2 \cdot t^* \cdot \sigma^2 \cdot \frac{1}{3} \tag{7.44}$$

Therefore, the timing risk for a continuous strategy trading over a period of time t^* is:

$$\Re = \sqrt{X^2 \cdot P_0^2 \cdot t^* \sigma^2 \cdot \frac{1}{3}} \qquad (7.45)$$

Substituting back for $t^* = \frac{1}{250} \cdot \frac{X}{ADV} \cdot \frac{1}{\alpha}$ we get:

$$\Re(\alpha) = \sqrt{X^2 \cdot P_0^2 \cdot \frac{1}{250} \cdot \frac{X}{ADV} \cdot \frac{1}{\alpha} \cdot \sigma^2 \cdot \frac{1}{3}} \qquad (7.46)$$

Simplifying we have, timing risk for a single stock order:

$$\Re(\alpha) = \sigma \cdot X \cdot P_0 \sqrt{\frac{1}{250} \cdot \frac{1}{3} \cdot \frac{X}{ADV} \cdot \frac{1}{\alpha}} \qquad (7.47)$$

QED

Timing Risk for a Basket of Stock

The timing risk for a basket of stock expressed in total dollars is:

$$TR_\$(x_k) = \sqrt{\sigma^2 \cdot \sum_{k=1}^{n} r_k' \, \tilde{C} r_k} \qquad (7.48)$$

where,

$r_k =$ column vector of unexecuted shares at the beginning of the period k

$$r_k = \begin{pmatrix} r_{1k} \\ r_{2k} \\ \vdots \\ r_{mk} \end{pmatrix}$$

$r_{ik} =$ unexecuted shares of stock i at the beginning of period k
$\tilde{C} =$ covariance matrix expressed in terms of $(\$/share)^2$ and scaled for a trading period.

To express the timing risk for the basket of stock in terms of basis points we simply divide the timing risk dollar amount by the initial value of the trade list $V_\$ = \sum X \cdot P_0 \cdot 10^4$.

COMPARISON OF MARKET IMPACT ESTIMATES

Market impact parameters are computed for the different trading strategy representations of the model (Equation 7.13, Equation 7.14 and

Table 7.1 Market Impact Parameters by Trade Strategy Definition

	a_1	a_2	a_3	a_4	b_1	non-R^2
All Data						
POV	708	0.55	0.71	0.50	0.98	0.41
Trade Rate	534	0.57	0.71	0.35	0.96	0.41
Trade Schedule	656	0.48	0.45	1	0.90	0.38
Large Cap Sample						
POV	687	0.70	0.72	0.35	0.98	0.42
Trade Rate	567	0.72	0.73	0.25	0.96	0.42
Trade Schedule	707	0.59	0.46	1	0.90	0.37
Small Cap Sample						
POV	702	0.47	0.69	0.60	0.97	0.42
Trade Rate	499	0.49	0.69	0.40	0.97	0.42
Trade Schedule	665	0.42	0.47	1	0.90	0.39

Equation 7.16) and for each of the data samples: all data, large cap, and small cap categories. These results are shown in Table 7.1.

As expected, the non-linear R^2 statistics are almost equivalent for the *POV* and trade rate strategies since there is a near one-to-one relationship between *POV* and α, especially for realistic percentages of volume levels (e.g., *POV* < 40%). Additionally, the trade schedule non-linear R^2 is just slightly lower than *POV* and trade rate strategies, which implies the trade schedule formulation provides reasonable results.

Comparison of parameter values, however, from the different models is not the preferred process to evaluate models. As we showed in Chapter 5, models could have seemingly different parameter sets yet provide the same cost estimates. Then the easiest way to compare models is through cost estimates for various sizes and strategies.

Our analysis consisted of comparing costs for sizes from 1% ADV to 35% ADV for a full day VWAP strategy and an equivalent *POV* = 20% strategy. We used the parameters for the full universe category and a volatility = 30% for the comparison test. The results are shown in Figure 7.2 and show that these results are consistent under the various model forms. Readers are encouraged to verify these calculations and to compare the models using the parameters from large cap and small cap data sets for different strategies.

Figure 7.2a compares market impact estimates for a VWAP strategy using *POV*, trade rate, and the trade schedule. Notice that the *POV* rate

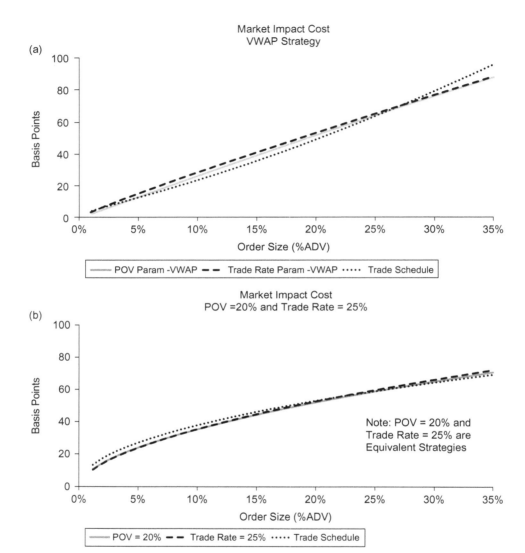

■ **Figure 7.2** Comparison of Market Impact Estimates. (a) Market Impact Cost-VWAP Strategy; (b) Market Impact Cost-Aggressive Strategy.

and trade rate estimates are virtually indistinguishable and the trade schedule estimates only have a slight difference.

Figure 7.2b compares market impact estimates for $POV = 20\%$, and trade rate $\alpha = 25\%$. The trade schedule cost estimates also corresponded to $\alpha = 25\%$ but with $a_4 = 1$. Again there is minimal difference between the three trade strategy definitions.

VOLUME FORECASTING TECHNIQUES
Daily Volumes

Our daily volume forecasting approach is based on an autoregressive moving average (ARMA) technique. Our research finds daily volumes to be dependent upon: (1) either a moving average (ADV) or a moving median (MDV) daily volume, (2) a historical look-back period of 10 days, (3) a day of week effect, or (4) a lagged daily volume term. Additional adjustments can also be made to the volume forecasts on special event days such as earnings, index reconstitution, triple and quadruple witching days, fed day, etc. (see Chapter 2, Market Microstructure).

Our daily volume forecasting analysis is as follows:

Definitions

Historical Look-Back Period. The number of days (data points) to use in the forecasts. For example, should the measure be based on 66 days, 30 days, 20 days, 10 days, or 5 days of data?

Average Daily Volume (ADV). Average daily volume computed over a historical period. We will use a rolling average in our forecast.

Median Daily Volume (MDV). Median daily volume computed over a historical period. We use a rolling median in our forecast.

Day of Week. A measure of the weekly cyclical patterns of trading volumes. Stocks tend to trade different percentages per days. This cyclical effect has varied over time and differs across market cap categories.

Lagged Daily Volume Term. We found some evidence of persistence in market volume. Many times both high and low volume can persist for days. However, the persistence is more often associated with high volume days due to the effect of trading large orders over multiple days to minimize price impact. Thus, when an institution is transacting a multi-day order, there is likely to be excess volume.

Author's Note: It is important to differentiate between the ADV measure used to normalize order size in the market impact estimate and the ADV or MDV measure used to predict daily volume. The ADV used in the former model needs to be consistent with the definition used by traders to quantify size. For example, if traders are using a 30-day ADV measure as a reference point for size, the market impact model should use the same metric. It is essential that the ADV measure that is used to quote order size by the trader be the exact measure that is used to calibrate the market impact parameters in the estimation stage. The daily volume forecast, however, is used to determine costs for the underlying trading strategy—whether it be a trade schedule, a POV based strategy, or a trading rate based strategy. An order for 100,000 shares or 10% ADV will have different expected costs if the volume on the day is 1,000,000 shares or 2,000,000 shares. In this case, a more accurate daily volume estimate will increase precision in the cost estimate and lead to improved trading performance.

Daily Forecasting Analysis—Methodology

Time Period: Jan. 1, 2011 through Dec. 31, 2011.

Sample Universe: SP500 (large cap) and R2000 (small cap) indexes on Dec. 31, 2011. We only included stocks where we had complete trading history over the period Nov. 10, 2010 through Dec. 31, 2011. The days from Nov. 11, 2010 to Dec. 31, 2010 were used to calculate the starting point for the historical average daily volume (ADV) and historical median daily volume (MDV) on Jan. 1, 2011.

Variable Notation

$V(t)$ = actual volume on day t.
$\hat{V}(t)$ = forecasted volume for day t.
$MDV(n)$ = median daily volume computed using previous n trading days.
$ADV(n)$ = Average daily volume computed using previous n trading days.
$Day\ Of\ Week(t)$ = The percentage of weekly volume that typically trades on the given weekday.
$\hat{\beta}$ = Autoregressive sensitivity parameter—estimated via OLS regression analysis.
$e(t)$ = forecast error on day t.

ARMA Daily Forecasting Model

$$\hat{V}(t) = \overline{V}_t(n) \cdot Day\ Of\ Week(t) + \hat{\beta} \cdot e(t-1) \qquad (7.49)$$

where $\overline{V}_t(n)$ is either the n-day moving ADV or n-day moving MDV, and $e(t-1)$ is the previous day's volume forecast error (actual minus estimate). That is

$$e(t-1) = V(t-1) - (\overline{V}_{t-1}(n) \cdot Day\ Of\ Week(t-1)) \qquad (7.50)$$

The error term above is calculated as the difference between actual volume on the day and estimated volume only using the day of week adjustment factor. The theoretical ARMA model will cause persistence of the error term as it includes the previous day's error in the forecast, e.g., $e(t - 2)$. However, our analysis has found that we could achieve more accurate estimates defining the error term only as shown above. Additionally, computation of daily volume estimates is also made easier since we do not need to maintain a series of forecast errors.

Analysis Goal

The goal of our daily volume forecasting analysis is to determine:

- Which is better ADV or MDV?
- What is the appropriate number of historical days?
- Day of week adjustment factor.
- Autoregressive volume term.

The preferred form of the ARMA model is determined via a three step process; the forecasting model should be re-examined at least on a monthly basis and recalibrated when necessary.

Step 1. Determine which is more appropriate: ADV or MDV and the historical look-back number of days.

- Compute the ADV and MDV simple forecast measure for various look-back periods, e.g., let the historical look-back period range from t = 1 to 30.
- Compute the percentage error between the actual volume on the day and simple forecast measure. That is:
  - $\varepsilon(t) = \ln(V(t)/\overline{V}(n))$
  - The percentage error is used to allow us to compare error terms across stocks with different liquidity.
- Calculate the standard deviation of the error term for each stock over the sample period.
- Calculate the average standard deviation across all stocks in the sample.
- Repeat the analysis for look-back periods from 1 to 30 days.
- Plot the average standard deviation across stocks for each day (from 1 to 30).

A plot of our forecast error analysis for each measure is shown in Figure 7.3a for large cap stocks and in Figure 7.3b for small cap stocks. Notice that for both large and small cap stocks the MDV measure has a lower error than the ADV. This is primarily due to the positive skew of daily volume which causes the corresponding ADV measure to be higher. Next, notice that the error term for both market cap categories follows a

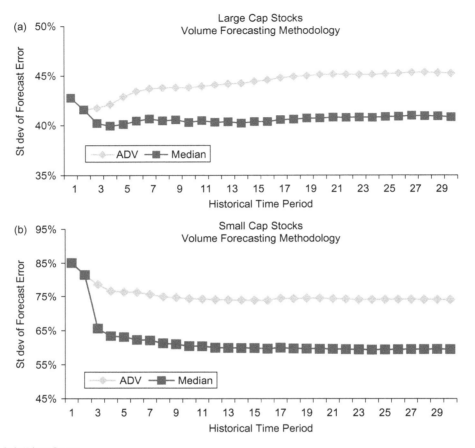

■ **Figure 7.3** Daily Volume Forecast.

convex shape with a minimum error point. For large cap stocks the minimum error is around 5–10 days and for small cap stocks the minimum error is close to 10 days.

Conclusion 1

● We conclude that the median daily volume using a historical period of 10 days, i.e., MDV(10), has the lowest forecast error across stocks and market cap during our analysis period.

Author's Note: As shown above, the ADV measure will more often be higher than the actual volume due to the positive skew of the volume distribution. Volume distributions tend to have more above average than below average outliers. This will result in actual costs being higher than the predicted cost. For example, if we trade 200,000 shares out of a total ADV of 1,000,000 shares, we may be tempted to state that a full day strategy corresponds to a trading rate of 20%. However, if the actual volume on the day is only 800,000 shares, the actual trading rate will be 25%, resulting in higher than predicted costs. The market impact forecasting error will be biased (to the high side) when using the ADV measure to predict daily volume. In our sample, we found the ADV to be higher than the actual volume on the day 65% of the time.

Step 2. Estimate the *Day Of Week*(t) parameter.

We analyzed whether or not there is a cyclical trading pattern during the week. To avoid bias that may be caused by special event days such as FOMC, triple witching, index reconstitution, earnings, month end, etc., we adjusted for these days in our analysis. It is important to note that if month end is not excluded from the data there may be a strong bias suggesting that Friday is the heaviest trading day of the week, since three out of seven month ends occur on Fridays (due to weekends). Many investors trade more often on the last day of the month.

Our day of week process is as follows:

- For each stock compute the percentage of actual volume traded on the day compared to the average volume in the week.
- Exclude the special event days that are historically associated with higher traded volume.
- Compute the average percentage traded on each day across all stocks in the sample.
- It is important to use a large enough sample in the analysis. We used one full year trading period to compute the day of week effect.

The result of our day of week analysis is shown in Figure 7.4. Monday is consistently the lowest volume day in the week for large and small cap stocks. After adjusting for month end volume, we found that small cap volume increases on Fridays but large cap volume decreases. The effect may be due to investors not being willing to hold an open position in small cap stocks over the weekend for fear there is too much market exposure for small cap stocks and therefore they may elect to pay higher market impact before weekend to ensure completion.

Conclusion 2

- Stock trading patterns exhibit a cyclical weekly pattern.
- The cyclical pattern is different for large cap and small cap stocks.

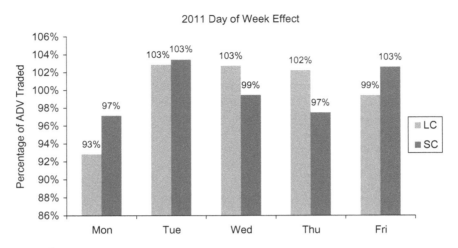

■ **Figure 7.4** Day of Week Effect.

Step 3. Estimate the autoregressive parameter $\hat{\beta}$.

The autoregressive parameter is used to correct for persistence of volume over consecutive days. We found above average volume days were more likely to be followed by above average volume days and below average volume days were more likely to be followed by below average volume days. But the relationship was much more significant for above average volume days than for the below average volume days. This process is as follows:

- Estimate volume for the day based on the 10-day median plus the day of week adjustment.
- Compute the forecast error term as the difference between the actual volume on the day and the estimated volume. This difference is:
  - $\varepsilon(t) = V(t) - Median(t) \cdot Day\ Of\ Week(t)$
- Run a regression of the error term on its one-day lagged term, that is,
  - $\varepsilon(t) = \alpha + \beta \cdot \varepsilon(t-1)$
- Compute the slope term β for large and small cap stocks.

Large cap stocks had a much larger degree of autocorrelation than small cap stocks. The average correlation of errors was $\beta_{large} = 0.452$ for large cap stocks and $\beta_{small} = 0.284$ for small cap stocks. After correcting for autocorrelation there was still a very slight amount of autocorrelation present, but this will have a negligible effect on our forecasts due to the constant term in our regression model.

Forecast Improvements

We next compared the results from our preferred ARMA model (shown above) to a simple 30-day ADV measure (e.g., ADV30) to determine the extent of the forecasting improvement. The preferred ARMA model reduced forecast error 17% for large cap stocks and 21% for small cap stocks.

Daily Volume Forecasting Model

Our daily volume forecasting models for large and small cap stocks can finally be formulated as:

Large Cap $\quad \hat{V}(t) = MDV(10) \cdot Day\ Of\ Week(t) + 0.450 \cdot e(t-1)$ (7.51)

Small Cap $\quad \hat{V}(t) = MDV(10) \cdot Day\ Of\ Week(t) + 0.283 \cdot e(t-1)$ (7.52)

Conclusion 3

- There is statistical evidence that there is persistence of volume trading.
- Forecasts can be improved through incorporation of an autoregressive term.
- Table 7.2 shows the one period lag analysis results and the error correlation results. Notice how dramatically the correlation across successive period error terms decreases. The correlation of the error terms decrease −0.311 by using the ARMA model Table 7.3 shows

Table 7.2 One-Period Lag—Error Correlation

Category	Beta	Correlation Before Adj.	Correlation After Adj.	Net Chg. Improvement
LC	0.450	0.452	− 0.015	− 0.437
SC	0.283	0.284	− 0.008	− 0.276
All	0.319	0.320	− 0.010	− 0.311

Table 7.3 Comparison of Error Terms Between ARMA Model and ADV Model

Mkt Cap	ADV30	ARMA	Net Change Improvement	Percent Improvement
LC	· 40.9%	33.9%	− 7.0%	− 17.1%
SC	56.0%	44.3%	− 11.7%	− 20.9%
All	53.2%	42.8%	− 10.4%	− 19.5%

Forecasting Model: Y(t) = Median(10)*DOW + AR*Y(t − 1)

the ARMA volume forecasting model improvement over the naïve thirty day average daily volume measure (e.g., ADV30). The error between actual volume and forecasting volume decreases -19.5% with the ARMA model over the naïve ADV30 measure.

Author's Note:

- In theory, the beta term in an ARMA model is often shown to be forecasted without the constant term alpha, but since we are using a moving median it is not guaranteed that the mean error will be zero and thus a constant term is needed.
- The ARMA forecast with the "beta" autoregressive terms can be computed both with and without special event days. Since it is important that this technique be continuous, unlike the day of week adjustment, we need to include all days. As an adjustment, we can: (1) Treat the special event day and day after the special event as any other day and include an adjustment for the previous day's forecasted error, (2) define the forecast error to be zero on a special event day (this way it will not be included in the next day's forecast), or (3) use a dummy variable for special event days.
- Our analysis calculated an autoregressive term across all stocks in each market cap category. Users may also prefer to use a stock specific autoregressive term instead. We did not find statistical evidence that a stock specific beta is more accurate for large cap stocks but there was some evidence supporting the need for a stock specific beta for the small cap stocks. Readers are encouraged to experiment with stock specific forecasts to determine what works best for their specific needs.

FORECASTING MONTHLY VOLUMES

In this section we describe a process to forecast average monthly volume levels. This process could also be extended to estimate annual volume levels. Having a forward looking ADV estimate can be very helpful for the portfolio manager who is looking to rebalance his/her portfolio at some future point in time when volumes may look much different than they do now.

Methodology:

Period: Ten years of data: Jan. 2002 through Dec. 2011.
Universe: Average daily volumes for large cap (SP500) and small cap (R2000) stocks by month.

Definitions:

$V(t)$ = average daily volume across all stocks per day in corresponding market cap category.

$\sigma(t)$ = average stock volatility in the month.

SPX = SP500 index value on last day in month.

$\Delta V(t)$ = log change in daily volume (MOM).

$$\Delta V(t) = \ln\{V(t)\} - \ln\{V(t-1)\}$$

$\Delta V(t-1)$ = previous month's log change in daily volume (MOM) to incorporate an autoregressive term. This is also a proxy for momentum.

$$\Delta V(t-1) = \ln\{V(t-1)\} - \ln\{V(t-2)\}$$

$\Delta V(t-12)$ = log change in daily volume (MOM) one year ago to incorporate a monthly pattern.

$$\Delta V(t-12) = \ln\{V(t-12)\} - \ln\{V(t-13)\}$$

$\Delta\sigma_{large}(t)$ = log change in large cap volatility (MOM).

$$\Delta\sigma_{large}(t) = \ln\{\sigma_{large}(t)\} - \ln\{\sigma_{large}(t-1)\}$$

$\Delta\sigma_{small}(t)$ = log change in small cap volatility (MOM).

$$\Delta\sigma_{small}(t) = \ln\{\sigma_{small}(t)\} - \ln\{\sigma_{small}(t-1)\}$$

$\Delta Spx(t)$ = log change in SP500 index value (MOM). We used the change in SP500 index values for both large cap and small cap forecasts.

$$\Delta Spx(t) = \ln\{spx(t)\} - \ln\{spx(t-1)\}$$

Monthly Volume Forecasting Model:

$$\Delta V(t) = b_0 + b_1 \cdot \Delta V(t-1) + b_2 \cdot \Delta V(t-12) + b_3 \cdot \Delta\sigma + b_4 \cdot \Delta Spx$$

Figure 7.5 shows the average daily volume per stock in each month for large cap and small cap stocks over the period Jan. 2002 through Dec. 2011 (10 years of monthly data). For example, in Jun. 2011 the average daily volume for a large cap stock was 5,766,850 per day and the average daily volume for a small cap stock was 555,530 per day. It is important to note that historical volume levels will change based on stock splits and corporate actions.

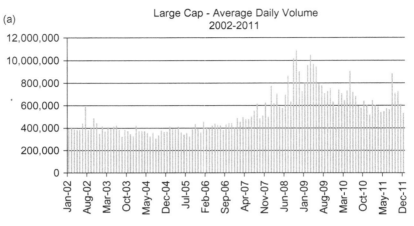

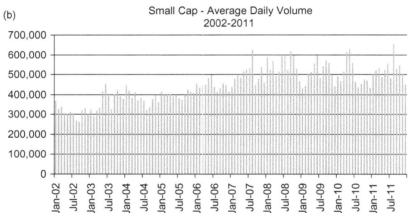

■ **Figure 7.5** (a) Average Daily Volumes by Month for Large Cap Stocks; (b) Average Daily Volumes by Month for Small Cap Stocks.

Analysis:

The monthly volume forecasting analysis is to determine an appropriate relationship to predict the expected change in monthly volume levels. Since the number of trading days will differ in each month due to weekends, holidays, etc., it is important that we adjust for the number of trading days in order to make a fair comparison across time. Our analysis included a one and twelve month autoregressive terms, the change in monthly volatility levels for each market cap category, and the MOM change in SP500 index for both large and small cap stocks.

Table 7.4 Monthly Volume Forecast Regression Results

Large Cap Stocks (SP500)

Ten Years: 2002–2011

	const	$\Delta V(-1)$	$\Delta V(-12)$	$\Delta\sigma$	ΔSpx
beta	0.0056	−0.2990	0.1354	0.5461	−0.1877
se	0.0064	0.0435	0.0461	0.0328	0.1466
t-stat	0.8806	−6.8785	2.9352	16.6699	−1.2798
R^2	0.79				

Recent Five Years: 2007–2011

	const	$\Delta V(-1)$	$\Delta V(-12)$	$\Delta\sigma$	ΔSpx
beta	−0.001	−0.313	0.088	0.529	−0.432
se	0.009	0.060	0.061	0.042	0.196
t-stat	−0.072	−5.245	1.461	12.601	−2.201
R^2	0.83				

Small Cap Stocks (SP500)

Ten Years: 2002–2011

	const	$\Delta V(-1)$	$\Delta V(-12)$	$\Delta\sigma$	ΔSpx
beta	0.0029	−0.2212	0.2385	0.3569	0.3437
se	0.0077	0.0732	0.0784	0.0475	0.1717
t-stat	0.3773	−3.0216	3.0423	7.5084	2.0016
R^2	0.39				

Recent Five Years: 2007–2011

	const	$\Delta V(-1)$	$\Delta V(-12)$	$\Delta\sigma$	ΔSpx
beta	−0.001	−0.225	0.143	0.362	0.140
se	0.012	0.100	0.109	0.058	0.222
t-stat	−0.103	−2.254	1.310	6.259	0.629
R^2	0.49				

We estimated our regression coefficients for large and small cap stocks using a ten year horizon. We further defined three periods to evaluate the stability of these relationships. These periods are:

Ten Years: 1/2002–12/2011
Five Years: 1/2007–12/2011
Five Years: 1/2002–12/2006

Regression results:

The result of our regression study is shown in Table 7.4. In total, there are six scenarios that were analyzed—three for LC and three for SC. The results show the estimated betas, corresponding standard errors and t-stat, and the R^2 statistic. Overall our regression model had a very strong fit. The model did explain a larger percentage of the variation for large cap stocks than for small cap stocks (as is expected due to the trading stability of the smaller companies). The R^2 using 10 years of data was $R^2 = 0.70$ for large cap stocks and $R^2 = 0.39$ for small cap stocks. Overall, our formulation is a very statistically sound model.

Our monthly volume forecasting models for large and small cap stocks are:

Large Cap: $\Delta V(t) = 0.0056 - 0.2990 \cdot \Delta V(t-1) + 0.1354 \cdot \Delta V(t-12)$
$$+ 0.5461 \cdot \Delta\sigma_{large}(t) - 0.1877 \cdot \Delta Spx$$

$$(7.53)$$

Small Cap: $\Delta V(t) = 0.0029 - 0.2212 \cdot \Delta V(t-1) + 0.2385 \cdot \Delta V(t-12)$
$$+ 0.3569 \cdot \Delta\sigma_{small}(t) + 0.3437 \cdot \Delta Spx$$

$$(7.54)$$

Main Observations:

- Monthly volumes exhibit trend reversion. The sign of the $\Delta V(t-12)$ variable was negative across both large and small cap stocks in each of the scenarios. If volume levels were up in one month, they were more likely to be down in the following month. If volume levels were down in one month, they were more likely to be up in the following month. This relationship is more significant for large cap than small cap stocks. Monthly volumes exhibit a positive seasonal trend. The sign of the $\Delta V(t-12)$ variable was positive and significant in most cases indicating a monthly pattern exists, although monthly volume levels vary. For example, December and August are consistently the lowest volume months during the year. October and January are the two highest volume months of the year (measured over our 10 year period). The relationship is stronger for small cap than for large cap stocks.
- Volumes are positively correlated with volatility. One way of thinking about this is that volume causes the volatility. Another explanation is that portfolio managers have a better opportunity to differentiate themselves and earn a higher return in times of increasing volatility. Hence they trade and rebalance their portfolios more often. The relationship here is slightly stronger for large cap stocks.
- The connection between volume and price level (SPX index) is the only factor that produces different relationships for large and small cap stocks. Small cap volume has always been positively related to price level. As the market increases, volume in small cap stocks increases, likely due to high investor sentiment during times of increasing market levels. Investors will put more in small cap stocks in a rising market hoping to earn higher returns, but will trade small stocks less often in a decreasing market. The strength of this relationship, has declined between the 2002−2006 and 2007−2011 periods. During 2002−2006 large cap stock volumes were also positively related to market price levels. But that relationship has reversed during 2007−2011.

- Currently, large cap stock volume is inversely related to prices. The relationship could be due to the current investor sentiment (since the financial crises). Investors are very weary of the market and fear further sharp declines. A cash investment of a fixed dollar amount will purchase fewer shares in a rising market but more shares in a falling market. Redemption of a fixed dollar amount will require few shares be traded in a rising market and more shares be traded in a declining market. We expect that this trend will stay constant and may additionally become negative for small cap volumes until investor sentiment and overall market confidence increases.
- Our analysis did not uncover any relationships between volume levels and correlation. However, correlation still remains a favorite indicator for portfolio managers. We suggest readers experiment with alternative correlation measures such as log change and actual level. This may improve the accuracy of our volume forecast model.

FORECASTING COVARIANCE

In this section we discuss a technique to construct a short-term risk model based on our price volatility forecasting model and multi-factor model to estimate covariance. In Kissell and Glantz (2003) we provide a detailed process to construct a short-term trading risk model based on a principal component risk model and a GARCH volatility estimate. In this section we provide a more general process that can incorporate any risk model combined with any volatility estimate (e.g., the HMA-VIX approach).

This process is as follows:

Let,

$\overline{C}$ = covariance matrix constructed from our multi-factor model
D = diagonal matrix of historical volatilities (from risk model)
$\hat{D}$ = diagonal matrix of forecasted volatilities (e.g., HMA-VIX, Garch, etc.)
P = diagonal matrix of current prices

Step 1: Convert the covariance matrix C to a correlation matrix *Rho* by dividing by the corresponding volatility terms:

$$Rho = D^{-1}CD^{-1}$$

Step 2: Incorporate the forecasted volatility from the preferred forecasting model, e.g., HMA-VIX, GARCH, EWMA, etc., into the new covariance matrix $\hat{C}$:

$$\hat{C} = \hat{D}(Rho)\hat{D} = \hat{D}D^{-1}CD^{-1}\hat{D}$$

This covariance matrix will now be scaled to the same time period as the price volatility term. For example, if the volatility forecast is a one-day forecast then the covariance matrix $\hat{C}$ is also a one-day estimate. If we are interested in a time period that is different than the time scale of the price volatility estimate we simply divide by that appropriate value. For example, if we break the day into n trading periods the covariance matrix for the time horizon is:

$$\hat{C} = \frac{1}{n} \cdot \hat{D}(Rho)\hat{D} = \frac{1}{n} \cdot \hat{D}D^{-1}CD^{-1}\hat{D}$$

Step 3: Convert the covariance matrix expressed in (returns)^2 into (\$/share)^2. Here we simply multiply by our diagonal price matrix from P above:

$$\tilde{C} = P\hat{C}P = \frac{1}{n} \cdot P\hat{D}D^{-1}CD^{-1}\hat{D}P$$

This covariance matrix is now scaled for the appropriate length of time for our trading period and is expressed in (\$/*share*)^2 for our trade schedule timing risk calculations. This matrix will also be extremely important for portfolio optimization.

The general form of our trading risk model is:

$$\tilde{C} = \frac{1}{n} \cdot P\hat{D}D^{-1}CD^{-1}\hat{D}P \tag{7.55a}$$

Many times investors will need the covariance matrix to be adjusted for a one-sided portfolio. In this case, we adjust the entries in the covariance matrix based on the side of the order. For example,

$$c_{ij}^* = side(i) \cdot side(j) \cdot c_{ij} \tag{7.55b}$$

Where c_{ij} is the computed covariance scaled for the length of the trading period and in (\$/*share*)^2. The full side adjusted covariance is computed via matrix multiplication following techniques above as follows:

$$\tilde{C} = \frac{1}{n} \cdot (Side)P\hat{D}D^{-1}CD^{-1}\hat{D}P(Side) \tag{7.55c}$$

Where (*Side*) is a diagonal matrix consisting of either a 1 if a buy order or -1 if a sell order. We make use of the side adjusted trading risk covariance in Chapter 8 through Chapter 10.

EFFICIENT TRADING FRONTIER

The efficient trading frontier (ETF) is the set of all optimal trading strategies. These are the strategies that contain the least risk for a specified cost and have the lowest cost for a specified risk. A rational investor is someone who will only trade via an optimal trading strategy. If an investor is trading via a strategy that is not on the efficient trading frontier it is unlikely that they will achieve best execution regardless of their actual execution costs.

If a strategy is not optimal (e.g., it is above the ETF) then there exists a strategy with either (1) a lower cost for the same level of risk, (2) less risk for the same cost, or (3) a strategy with lower cost and less risk.

The efficient trading frontier is constructed via an optimization process. The general equation is:

$$Min \; L = Cost + \lambda \cdot Risk \qquad (7.56)$$

where cost represents both market impact and alpha cost. In situations where investors do not have an alpha forecast or believe the natural price drift over the trading horizon to be zero they will only include the market impact cost component in the optimization.

Analysts then solve this equation for all values of $\lambda > 0$ and plot the sets of cost and risk. An example of the efficient trading frontier is shown in Figure 7.6. This figure illustrates the trade-off between market impact and timing risk. As the strategy becomes more aggressive timing risk decreases but market impact increases. As the strategy becomes more passive timing risk increases but market impact decreases. Market impact and timing risk are conflicting terms. Decreasing one term results in an increase in the other term. Unfortunately there is no way to simultaneously minimize both terms.

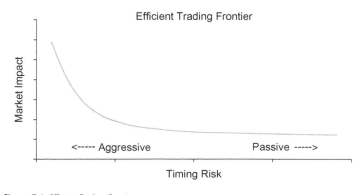

■ **Figure 7.6** Efficient Trading Frontier.

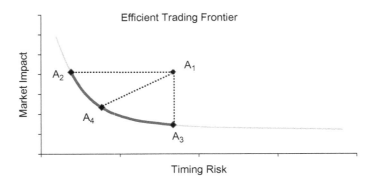

■ **Figure 7.7** Illustration of Different Trading Strategies.

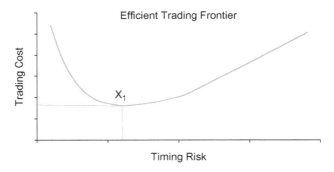

■ **Figure 7.8** Illustration of Trading Cost in Presence of Alpha.

Figure 7.7 illustrates various optimal trading strategies. Strategy A_1 in the figure is not an optimal strategy because it does not contain the least cost for the level of risk or the lower risk for the corresponding cost. For example, strategy A_2 has the same market impact as A_1 but reduced risk. Strategy A_3 has the same timing risk as A_1 but lower impact. Strategy A_4 has both lower market impact and less timing risk that A_1. All of these strategies would be preferred over A_1.

Figure 7.8 illustrates the efficient trading frontier in the presence of alpha momentum. Notice in this case that market impact is decreasing until strategy X_1 is reached. After this time, however, market impact begins to increase again due to the alpha cost of the trade. If the trader executes too passively the increased alpha cost will become greater than the reduced market impact cost. Hence, in these situations traders waiting too long to trade will incur increased risk and increased cost. The most passive trader should execute the trade represented by strategy X_1.

The optimization process for a single stock order and trade portfolio is shown below. The single stock process is further discussed in Chapter 8, Algorithmic Decision Making Framework and the portfolio optimization process is further discussed in Chapter 9, Portfolio Algorithms.

Single Stock Trade Cost Objective Function

$$Min(b_1 I^* \alpha^{a4} + (1 - b_1)I^*) + \lambda \cdot \left(\sigma \cdot \sqrt{\frac{1}{250} \cdot \frac{1}{3} \cdot \frac{S}{ADV} \cdot \frac{1}{\alpha}} \cdot 10^4 bp \right) \quad (7.57)$$

Portfolio Trade Cost Objective Function

$$Min \sum_{i=1}^{m} \sum_{j=1}^{n} x_{ij} \cdot \left(b_1 \cdot I_i^* \cdot \left(\frac{x_{ij}}{X_i} \right) \cdot \frac{x_{ij}}{v_{ij}} + \left(\frac{x_{ij}}{X_i} \right) \cdot (1 - b_1) \cdot I_i^* \right) + \lambda \cdot \sqrt{\sigma^2 \cdot \sum_{j=1}^{n} r_k' \, Cr_k}$$

$$(7.58)$$

Both of these optimizations will also contain user specified constraints.

Author's Note:

It is important to mention that the parameter λ is used to specify the investor's level of risk aversion. This represents how much market impact cost the investor is willing to incur to reduce timing risk by an additional unit. In this formulation lambda can take on any value greater than zero. That is, $\lambda \geq 0$.

In a cost-risk optimization the value of lambda is directly related to the resulting optimal trading strategy plotted on the efficient trading frontier. The tangent on the ETF at this point will be equal to the negative value of specified lambda. If $\lambda = 1$ then the tangent of the ETF at this point will have a slope equal to $m = -1$.

In general, setting lambda high will result in an aggressive strategy with higher market impact but lower timing risk. Setting lambda low will result in a passive strategy with lower market impact but higher timing risk. Unfortunately, there is no universal convention for the meaning of lambda in the optimization process.

Some brokers will optimize the trade-off between cost and variance rather than cost and standard deviation (as we show above). In optimization, the meaning of lambda will be much different using variance rather than standard deviation (the square root of lambda). Additionally, the value of lambda used in cost-variance optimization will not be the negative value of the tangent of the strategy on the efficient trading frontier.

Additionally, some brokers specify a mapping between the strategies on the efficient trading frontier and a qualitative term. For example, using Low, Med, and High, or Passive, Normal, and Urgent, where each of these qualitative labels are mapped to different values of lambda. Other brokers may map values of lambda to be between $1 \leq \lambda \leq 10$ where $1 =$ passive and $10 =$ aggressive, or between $1 \leq \lambda \leq 3$. Some only allow values between say $0 \leq \lambda \leq 1$ in a slightly reformulated optimization such as:

$$MinL = \lambda \cdot Cost + (1 - \lambda) \cdot Risk$$

It is important to point out that there is not enough consistency in the industry to compare results based on the selected value of lambda. There are large differences across the meaning of algorithmic parameters. Investors need to understand the optimization process used by their brokers and vendors and its meaning on the cost-risk trade-off in order to make an informed trading decision.

Algorithmic Decision Making Framework

INTRODUCTION

We introduce readers to the algorithmic decision making framework. The process includes macro- and micro-level decisions specified prior to trading to ensure consistency between the "the trading goal" and the "investment objective." The macro-level decision refers to the best execution strategy (BES) most consistent with the investment objective. The micro-level decision refers to how the algorithm will behave in real-time and how it will adapt to changing market conditions. Subsequently, it is the goal of the limit order model and smart order router to ensure that order placement and actual executions adhere to the investor's specifications. Only investors who possess full knowledge and proper specification of these criteria will be positioned to achieve best execution.

Before we discuss our algorithmic decision making framework, it is important to restate a few important concepts. Algorithmic trading is the computerized execution of financial instruments following pre-specified rules and guideline. Algorithmic trading provides many benefits. They do exactly what they are instructed to do—and do it well. However, one of the more unfortunate aspects of algorithmic trading is that they do exactly what they are instructed to do. If they are not provided with instructions that are in the best interest of the fund over all possible sets of market events, the results will likely be unfavorable execution and subpar performance.

Algorithmic decision frameworks have been previously studied from the perspective of the macro and micro viewpoint. For example, macro decisions have been studied by Barra (1997); Bertsimas and Lo (1998); Almgren and Chriss (1999, 2000); Cox (2000); and Kissell, Glantz, and Malamut (2004). Micro decisions have been studied more recently such as in Journal of Trading's "Algorithmic Decision Making Framework,"

The Science of Algorithmic Trading and Portfolio Management. DOI: http://dx.doi.org/10.1016/B978-0-12-401689-7.00008-8

(Kissell and Malamut, 2006), and Institutional Investor's Guide to Algorithmic Trading, "Understanding the P&L Distribution of Trading Algorithms," (Kissell and Malamut, 2005). Additionally, Almgren and Lorenz analyzed real-time adaptive strategies in Institutional Investor's Algorithmic Trading III: Precision, Control, Execution, "Adaptive Arrival Price," (Spring 2007), and also in Journal of Trading's "Bayesian Adaptive Trading with a Daily Cycle," (Fall 2006).

We now expand the previous research findings and provide an appropriate algorithmic framework that incorporates both macro and micro decisions. Our focus is with regards to single stock trading algorithms and single stock algorithmic decisions.

In Chapter 9, Portfolio Algorithms, we provide an algorithmic decision making framework for portfolio and program trading.

EQUATIONS

The equations used to specify macro and micro trading goals are stated below. Since we are comparing execution prices to a benchmark price in $/share units our transaction cost analysis will be expressed in $/share units. For single stock execution (in the US) this is most consistent with how prices and costs are quoted by traders and investors. Additionally, our process will incorporate the trading rate strategy α but readers are encouraged to examine and experiment with this framework for the percentage of volume and trade schedule strategies. These transaction cost models were presented in Chapter 7, Advanced Algorithmic Forecasting Techniques.

Variables

$I' = $ *Instantaneous Impact in $/Share*

$MI' = $ *Market Impact in $/Share*

$TR' = $ *Timing Risk in $/Share*

$PA' = $ *Price Appreciation Cost in $/Share*

$X = $ *Order Shares*

$Y = $ *Shares Traded (Completed)*

$(X - Y) = Unexecuted\ Shares\ (Residual)$

$\theta = \dfrac{Y}{X} = Percentage\ of\ Shares\ Traded$

$(1 - \theta) = \dfrac{X - Y}{X} = Percentage\ of\ Shares\ Remaining$

$ADV = Average\ Daily\ Volume$

$V_t = Volume\ over\ Trading\ Horizon\ (excluding\ the\ order)$

$\sigma = Annualized\ Volatility$

$\alpha = Trade\ Rate\ at\ Time = t$

$POV_t = Percentage\ of\ Volume\ at\ Time = t$

$P_t = Market\ Price\ at\ Time = t$

$\bar{P}_t = Realized\ Average\ Execution\ Price\ at\ Time = t$

$\mu = Natural\ Price\ Appreciation\ of\ the\ Stock\ (not\ caused\ by\ trading\ imbalance)$

Important Equations

I-Star $\quad I'_{\$/Share} = a_1 \cdot \left(\dfrac{X}{ADV} \right)^{a2} \cdot \sigma^{a3} \cdot P_0 \cdot 10^{-4}$ $\hspace{2cm}$ (8.1)

Market Impact $\quad MI'_{\$/Share}(\alpha) = b_1 \cdot I' \cdot \alpha + (1 - b_1) \cdot I'$ $\hspace{1cm}$ (8.2)

Timing Risk $\quad TR'_{\$/Share}(\alpha) = \sigma \cdot \sqrt{\dfrac{1}{250} \cdot \dfrac{1}{3} \cdot \dfrac{X}{ADV} \cdot \alpha^{-1}} \cdot P_0$ $\hspace{1cm}$ (8.3)

Price Appreciation Cost $\quad PA'_{\$/Share}(\alpha) = \dfrac{X}{ADV} \cdot \dfrac{1}{\alpha_t} \cdot \mu_t$ $\hspace{1cm}$ (8.4)

Future Price $\quad E(P_n) = P_0 + (1 - b_1) \cdot I'$ $\hspace{2cm}$ (8.5)

Benchmark Cost $\quad Cost = (\bar{P} - P_b) \cdot Side$ $\hspace{2cm}$ (8.6)

$$\text{Trade Rate} \quad \alpha_t = \frac{Y}{V_t} \tag{8.7}$$

$$POV_t \quad \alpha_t = \frac{Y}{Y + V_t} \tag{8.8}$$

Important Note: for our analysis we are using the market impact formulation with the trade rate strategy with $a_4 = 1$. This is equivalent to the single period trade schedule formulation discussed in Chapter 7, Advanced Algorithmic Forecasting Techniques.

ALGORITHMIC DECISION MAKING FRAMEWORK

The algorithmic decision making framework is about traders instructing the algorithm to behave in a manner consistent with the investment objectives of the fund. If traders enter orders into an algorithm without any pre-specified rules, or with rules that are not consistent with their investment objective, the only thing we can be certain of is that the algorithm will not achieve best execution. Of course the algorithm may realize favorable prices at times, but this would be due to luck rather than actual intentions. Best execution is only achieved through proper planning.

Best execution is evaluated based on the information set at the time of the trading decision (e.g., ex-ante). Anything else is akin to playing Monday morning quarterback.

The algorithmic decision making framework consists of:

1. Select Benchmark Price
2. Specify Trading Goal (Best Execution Strategy)
3. Specify Adaptation Tactic

1) Select Benchmark Price

Investors need to first select their benchmark price. This could be the current price which is also known as the arrival price, a historical price such as the previous day's closing price, or a future price such as the closing price on the trade day. These are described as follows:

Arrival Price Benchmark

The arrival price benchmark is often selected by fundamental managers. These are managers who determine what to buy and what to sell based on company balance sheets and long-term growth expectations. These managers may also use a combination of quantitative and fundamental (e.g., "quant-timental" managers) information to construct portfolios based on what stocks are likely to outperform their peer group over time. These managers often have a long term view on the stocks.

The arrival price benchmark is also an appropriate benchmark price for situations where a market event triggers the portfolio manager or trader to release an order to the market.

The arrival price benchmark is:

$$E_0[\text{Arrival Cost}] = (\overline{P} - P_0) \cdot Side$$

The E_0 notation here is used to denote that this is the expected cost at the beginning of trading. The expected cost for a buy order with strategy expressed in terms of trading rate is:

Let,

$$P_0 = \text{Arrival Price}$$

$$\overline{P} = P_0 + (b_1 \cdot I' \cdot \alpha) + (1 - b_1) \cdot I'$$

Then the expected cost is:

$$E_0[\overline{P} - P_0] = (b_1 \cdot I' \cdot \alpha) + (1 - b_1) \cdot I' = MI'$$

And we have,

$$E_0[\text{Arrival Cost}] = (b_1 \cdot I' \cdot \alpha) + (1 - b_1) \cdot I' \tag{8.9}$$

For the arrival price benchmark the expected cost is equal to the market impact of the trade.

Historical Price Benchmark

Quantitative managers who run optimization models may select a historical price as their benchmark if this represents the price used in the optimization process. Until recently, many quant managers would run optimizers after the close incorporating the closing price on the day. Optimizers determine the mix of stocks and shares to hold in the portfolio and the corresponding trade list for the next morning. These orders are then submitted to the market at the open the next day. The overnight price movement represents a price jump or discontinuity in the market that the trader is not able to participate with and represents either a sunk cost or a saving to the fund at the time trading begins. If the manager is looking to buy shares in a stock that closed at \$30.00 but opened at \$30.05, the \$0.05/share move represents a sunk cost to the manager. But if the stock opened at $29.95, the $0.05/share move represents a saving to the manager. Depending upon this overnight price movement the manager may change the trading strategy to become more or less aggressive. Even in situations where the portfolio manager uses current market prices in the stock selection process there will likely be some delay (although it may be very short in duration) in determining what stocks and shares need to be purchased and/or sold and releasing those orders to the market. Thus by the time these orders are entered into the market the current market price will be different than the historical price benchmark that was used to determine what stocks to hold in the portfolio.

The historical benchmark price is:

$$E_0[Historical\ Cost] = (\overline{P} - P_{hist}) \cdot Side$$

The E_0 notation is used to denote that this is the expected cost at the beginning of trading. This cost for a buy order and a strategy expressed in terms of trade rate using our formulas above is described as follows:

Let,

$$P_{hist} = Historical\ Decision\ Price$$

$$\overline{P} = P_0 + (b_1 \cdot I' \cdot \alpha) + (1 - b_1) \cdot I'$$

Then the difference is:

$$E_0[\overline{P} - P_{hist}] = (b_1 \cdot I' \cdot \alpha) + (1 - b_1) \cdot I' + (P_0 - P_{hist})$$

If we define the delay cost to be $(P_0 - P_{hist})$ our historical cost is:

$$E_0[\overline{P} - P_{hist}] = MI' + Delay$$

And we have:

$$E_0[Historical\ Cost] = (b_1 \cdot I' \cdot \alpha) + (1 - b_1) \cdot I' + (P_0 - P_{hist}) \qquad (8.10)$$

Notice that this is the same cost function as the arrival cost above plus a delay component that is a constant. When a manager selects the previous night's closing price as the benchmark price and starts trading at the open, the "Delay" cost represents the overnight price movement and translates to either a sunk cost or savings to the fund.

Unfortunately, as managers' claim time and time again, this movement represents a sunk cost much more often than it represents a saving because managers as a group do a very good job at figuring out what stocks are mispriced. Thus, if there is any gap in trading, such as the overnight close, the rest of the market will usually learn of this mispricing and adjust their market quotes to reflect the proper pricing at the open the next day.

Future Price Benchmark

Index managers often select the closing price to be their benchmark because this is the price that will be used to value the fund. Any transaction that is different than the closing price on the day of the trade will cause the actual value of the fund to differ from the index benchmark price thus causing tracking error. In order to avoid incremental tracking error and potential subpar performance, index managers often seek to achieve the closing price on the day.

An interesting aspect of using a future price as the benchmark is that your performance will look better than it actually is because the future price will also include the permanent impact of the order. So while permanent impact will have an adverse effect on the arrival price or historical price benchmark, it will not have any effect on a future price benchmark (from a cost perspective). Investors are expected to perform better against a future price benchmark by the amount of the permanent impact than they will against the arrival price.

The future cost derivation is explained as follows:

$$E_0[Future\ Cost] = (\overline{P} - E(P_n))$$

The E_0 notation is used to denote that this is the expected cost at the beginning of trading. Thus,

$$E(P_n) = P_0 + (1 - b_1) \cdot I'$$

$$\overline{P} = P_0 + (b_1 \cdot I' \cdot \alpha) + (1 - b_1) \cdot I'$$

The expected future cost is:

$$E_0[\overline{P} - E(P_n)] = (b_1 \cdot I' \cdot \alpha) = Temporary\ Market\ Impact$$

And we have,

$$E_0[Future\ Cost] = (b_1 \cdot I' \cdot \alpha) \tag{8.11}$$

Notice that the future cost function only consists of temporary market impact. As stated, this is because permanent impact is reflected in the future price. So the future cost function that only includes temporary impact will be lower than the arrival cost function that includes both temporary and permanent impact.

Comparison of Benchmark Prices

A comparison of the efficient trading frontier for the different benchmark prices is shown in Figure 8.1. The arrival price frontier is the middle curve in the graph. The arrival price cost consists of both temporary and permanent impact measured from the price when trading begins. In this example, the previous close frontier has the highest cost for the corresponding level of timing risk due to adverse overnight price movement (sunk cost). In reality, the previous close frontier could be higher, lower, or the same as the arrival price frontier. The future price benchmark, such as the day's closing price, is the lowest frontier. It consists only of the temporary impact because the future price will be comprised of the permanent impact cost.

If an investor is ever given the choice of which benchmark to use to judge performance it behooves the investors to select a future price benchmark since this cost will likely be less than a historical or arrival price benchmark. The future price will always include the permanent impact of the order and the temporary impact that has not yet fully dissipated (see Chapter 4, Market Impact Models).

2) Specify Trading Goal

The next step in the process is to select the trading goal so that it is consistent with the underlying investment objective. To assist in the process we describe five potential best execution strategies for investors. While these may not comprise all possibilities they do address needs for many investment professionals. Techniques for specifying the macro-level trading goal have been previously studied by Kissell, Glantz, and Malamut (2004). We expand on those findings.

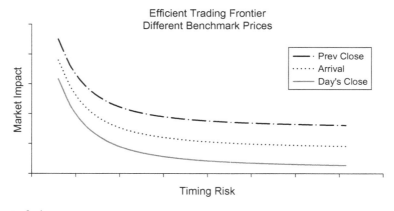

■ **Figure 8.1** Efficient Frontier Graphs.

These trading goals are: (1) minimize cost, (2) minimize cost with risk constraint, (3) minimize risk with cost constraint, (4) balance the trade-off between cost and risk, and (5) price improvement.

1. Minimize Cost

The first criterion "minimize cost" sets out to find the least-cost trading strategy. Investors may seek to find the strategy that minimizes market impact cost. If investors have an alpha or momentum expectation over the trading horizon, they will seek to minimize the combination of market impact cost and price appreciation cost. The solution of this goal is found via optimization.

$$Min\ MI' + Alpha' \qquad (8.12)$$

In situations where investors do not have any alpha view or price momentum expectation the solution to this optimization will be to trade as passively as possible. That is, participate with volume over the entire designated trading horizon because temporary impact is a decreasing function and will be lowest trading over the longest possible horizon. A VWAP strategy in this case is the strategy that will minimize cost. In situations where investors have an adverse alpha expectation the cost function will achieve a global minimum. If this minimum value corresponds to a time that is less than the designated trading time the order will finish early. If the minimum value corresponds to a time that is greater than the designated trading time then the solution will again be a VWAP strategy.

Example: An investor is looking to minimize market impact and price appreciation cost. The optimal trading rate to minimize this cost is found through minimizing the following equation:

$$Min: b_1 \cdot I' \cdot \alpha + (1 - b_1) \cdot I' + \frac{X}{ADV} \cdot \frac{1}{\alpha} \cdot \mu \qquad (8.13)$$

Solving for the optimal trading rate α^* we get:

$$\alpha^* = \sqrt{\frac{X \cdot \mu}{b_1 \cdot I \cdot ADV}} \qquad (8.14)$$

Figure 8.2a illustrates a situation where investors seek to minimize market impact and alpha. Notice in this case that the efficient trading frontier decreases and then increases and has a minimum cost at \$0.16 with a corresponding timing risk of \$0.65. The trading rate that will minimize total cost is 9%. This is denoted by strategy A1 in the diagram.

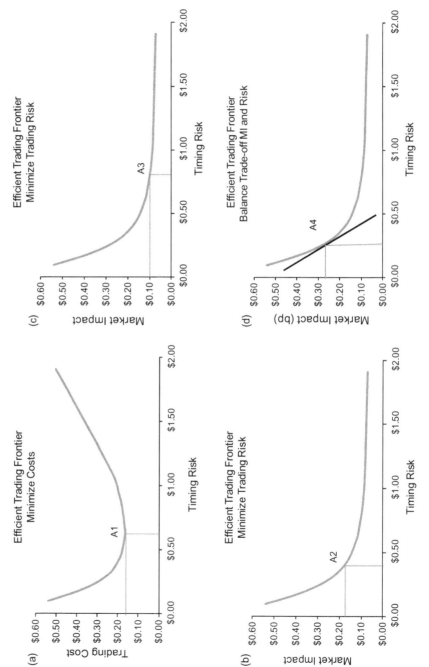

■ **Figure 8.2** (a–d) Best Execution Strategies.

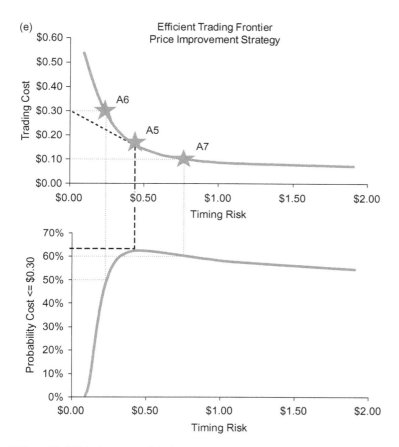

■ **Figure 8.2** (e) Price Improvement Derivation.

2. Minimize Cost with Risk Constraint

Our second criterion is to minimize cost for a specified quantity of risk. The risk constraint is often specified by the portfolio manager or by firm mandate that will not allow risk to exceed the specified level $\mathfrak{R}^*$. The optimization is:

$$
\begin{aligned}
&Min\ Cost \\
&s.t.\quad TR' = \mathfrak{R}^*
\end{aligned}
\tag{8.15}
$$

Example: An investor looking to minimize market impact cost (not including price appreciation) subject to a specified level of timing risk will determine the optimal trading rate through minimizing the following equation:

$$
Min:\ b_1 \cdot I' \cdot \alpha + (1 - b_1) \cdot I'
$$

$$
s.t.\quad \sigma \cdot \sqrt{\frac{1}{250} \cdot \frac{1}{3} \cdot \frac{X}{ADV} \cdot \alpha^{-1}} \cdot P_0 = \mathfrak{R}^*
\tag{8.16}
$$

The optimal trading rate is:

$$\alpha^* = \frac{X \cdot \sigma^2 \cdot P_0^2}{3 \cdot 250 \cdot ADV \cdot \mathfrak{R}^{*2}} \tag{8.17}$$

Figure 8.2b illustrates this trading goal through strategy A2. Here, A2 is the strategy that minimizes cost for a risk exposure of \$0.40/share. It has an expected cost of \$0.20/share and corresponds to a trading rate of 25% and a POV rate of 20%.

3. Minimize Risk with Cost Constraint

A portfolio manager's preferred investment stock is LMK with an expected return of 10% and the next most attractive stock is RLK with an expected return of 9.5%. The manager determines that X shares can be purchased at a cost of 50 bp (0.50%). Purchasing any more shares of LMK will cause the cost to be greater than the incremental return of 50 bp and the manager would be better off investing some portion of the dollars in the second most attractive stock. Therefore, the manager decides to transact the X shares using a strategy that will minimize risk for a cost of 50 bp. This optimization is as follows:

$$\begin{aligned} &Min\ TR' \\ s.t.\quad &Cost' = C^* \end{aligned} \tag{8.18}$$

Example: An investor looking to minimize timing risk for a specified level of market impact cost (not including price appreciation) will determine the optimal trading rate through minimizing the following equation:

$$\begin{aligned} Min: &\ \sigma \cdot \sqrt{\frac{1}{250} \cdot \frac{1}{3} \cdot \frac{X}{ADV} \cdot \alpha^{-1}} \cdot P_0 \\ s.t.\quad &b_1 \cdot I' \cdot \alpha + (1 - b_1) \cdot I' = C^* \end{aligned} \tag{8.19}$$

Solving for the optimal trading rate we get:

$$\alpha^* = \frac{C^* - (1 - b_1) \cdot I}{b_1 \cdot I} \tag{8.20}$$

Figure 8.2c illustrates the BES for an investor with a maximum cost of \$0.10/share. Here, strategy A3 has the lowest timing risk of \$0.73/share, and corresponds to a trading rate of 7% and a POV rate of 6.5%.

4. Balance Trade-off between Cost and Risk

The fourth criterion "Balance Trade-off between Cost and Risk" is used by investors with a certain level of risk aversion defined by the parameter λ. Risk adverse investors will set λ to be high to avoid market exposure and risk neutral investors set λ to be small. Setting lambda to be zero is equivalent to our first criterion—minimize cost—since the risk term would be ignored.

This trading goal is also known as the standard "cost-risk" optimization or algorithmic optimization objective function. It is formulated as follows:

$$Min\ \text{Cost}' + \lambda \cdot \text{Risk}' \tag{8.21}$$

Example: An investor looking to minimize the combination of market impact cost (not including price appreciation) and timing risk for a specified risk aversion value λ will determine the optimal trading rate through minimizing the following equation:

$$Min: (b_1 \cdot I' \cdot \alpha + (1 - b_1) \cdot I') + \lambda \cdot \left(\sigma \cdot \sqrt{\frac{1}{250} \cdot \frac{1}{3} \cdot \frac{X}{ADV}} \cdot \alpha^{-1} \cdot P_0 \right) \tag{8.22}$$

Solving for the optimal trading rate we get:

$$\alpha^* = \left(\frac{b_1 \cdot I}{\lambda \cdot \sigma \cdot P_0 \cdot \sqrt{\frac{1}{3} \cdot \frac{1}{250} \cdot \frac{X}{ADV}}} \right)^{\frac{2}{3}} \tag{8.23}$$

Notice that this optimal solution is in terms of the investor's risk aversion parameter.

Figure 8.2d illustrates the BES for an investor with a risk aversion $\lambda = 1$. In this case the solution is at the point where the tangent to the ETF is -1. This is noted by strategy A4 on the ETF and has cost = \$0.28/share and timing risk = \$0.25. The trading rate that achieves this optimal strategy is 57% and corresponds to POV = 36%.

5. Price Improvement

The fifth criterion "Price Improvement" is used by investors wishing to maximize the probability that they will execute more favorably than a specified cost. Usually, this is the goal of participants seeking to maximize short-term returns or exploit a pricing discrepancy. Additionally, it is often the goal used by agency traders seeking to maximize the likelihood of outperforming a cost such as a principal bid, or the strategy utilized by a principal trading desk looking to minimize chances of gamblers ruin and maximize profiting opportunity. The proof of the price improvement strategy was derived by Roberto Malamut (see *Optimal Trading Strategies,* p. 225, and Kissell, Glantz, and Malamut (2004, p 45).

The price improvement optimization is:

$$Max: Prob\ (Cost \leq C^*) \tag{8.24a}$$

where C^* is the specified target price, cost, or principal bid that the investor is seeking to outperform. Mathematically, this optimization can also be written as:

$$Max: \frac{C^* - E[Cost]}{TR'} \tag{8.24b}$$

Example: An investor seeking to maximize the probability of achieving price improvement over a market impact cost of C^* (not including price appreciation cost) will determine the optimal trading rate via the following optimization:

$$Max: \frac{C^* - (b_1 \cdot I' \cdot \alpha + (1 - b_1) \cdot I')}{\sigma \cdot \sqrt{\frac{1}{250} \cdot \frac{1}{3} \cdot \frac{X}{ADV} \cdot \alpha^{-1}} \cdot P_0} \tag{8.25}$$

Solving for the optimal trading rate we get:

$$\alpha^* = \frac{C^* - K}{3 \cdot b_1 \cdot I} \tag{8.26}$$

Figure 8.2e illustrates the process used to determine the price improvement strategy for a cost of $0.30/share. The top graph shows the efficient trading frontier with cost on the y-axis and timing risk on the x-axis. The bottom graph shows the probability that each of the strategies on the efficient trading frontier will incur a cost of less than $0.30/share. The probability was determined assuming a normal distribution with mean equal to the expected cost (y-axis) and standard deviation equal to the timing risk (x-axis). The strategy that maximizes the likelihood that the cost will be less than $0.30/share is found by drawing a line from the cost of $0.30 on the y-axis tangent to the efficient trading frontier. This is denoted by strategy A5 and has expected cost of $0.18/share and timing risk of $0.40/share. The strategy has a 62.1% chance of outperforming $0.30/share. Strategy A6 has an expected cost of $0.30/share and timing risk of $0.23/share and so a probability of 50% of outperforming $0.30/share. Obviously, any strategy to the left of A6 will have an expected cost higher than $0.30/share so the probability that the actual cost will be less than $0.30/share will be less than 50%. Strategy A7 has an expected cost of $0.10/share and timing risk of $0.84/share. The corresponding probability that this strategy will have a cost less than $0.30/share is 59.7%. Notice the shape of this probability curve. The probability of outperforming $0.30/share increases fast until strategy A5 (the highest probability of outperforming). The probability of outperforming $0.30/share decreases at a slower rate as we execute more passively. This tells us that it is more beneficial to trade more passively than more aggressively when seeking to outperform a specified price.

Further Insight

Many times traders will specify the strategy in simpler terms. In some cases the strategies are well thought out and developed, but in other cases, they are simply instructions for the algorithms to follow and may or may not be in the best interest of the fund. These strategies include:

Volume Based. A volume-based strategy will instruct the algorithm to follow a specified trading rate such as 15 or 20% of the volume. At times, these trading rate values are consistent with the investment objective of the fund and thus achieve best execution, but in other cases they will not be consistent with the investment objective and will not achieve best execution regardless of the actual performance. The volume-based strategy could be in terms of trade rate α or percentage of volume POV.

Price Based. A price-based algorithm will instruct the algorithm to trade at a certain rate based on market prices. As prices become more or less favorable these algorithms will trade either faster or slower. These include algorithms such as ladder and step functions.

Hyper-Aggressive. A hyper-aggressive algorithm is one that executes as many shares as possible within a specified price level. As long as actual prices are more favorable than the specified price these algorithms will transact as aggressively as they can.

Passive/Dark Pool. The passive or dark pool algorithms are those that will transact primarily in dark pools and crossing networks. Usually these algorithms do not have a specified maximum trading rate such as is associated with the volume-based algorithms. They can participate with as much volume as possible provided they only transact in dark pools. Investors believe that if they are trading only in dark pools then they are minimizing their market impact cost and information leakage. While this is a widely held belief it is not correct. Market impact cost is caused by buying/selling imbalance. If you are on the side of the imbalance you will incur a higher cost. If you enter shares into a dark pool and your entire order is traded then there is a counterparty with at least as many shares as your order and possibly more shares, otherwise you would not have had your entire order executed. If you enter shares into a dark pool and only a portion of your order is executed then there is counterparty with an order that was smaller than your order, otherwise your entire order would have been executed. If you enter an order into a dark pool but do not have any shares traded then there were not any counterparties in the dark pool at that time.

Get-Me-Done. The get-me-done type of algorithms will trade at a specified rate until the order is complete. However, if there is ample liquidity in the order book the algorithm will accelerate trading and sweep the book providing that doing so will complete the order. Usually traders will not want to accelerate trading and sweep the book complete, because by doing so they will likely signal their trading intentions to the market which may result in higher future prices (buy order), lower future prices (sell order), and higher permanent impact for all orders. But in the case where the order would be completed by sweeping the book the less favorable future prices will not affect the performance of the investor. One way to disguise trading intentions is to sweep all liquidity in the book except for the final 100 shares so that you do not affect the market price or NBBO. Many algorithms are set up to react to changing prices or quotes and utilizing this type of sweeping technique would keep those algorithms at bay—at least for the time being anyway.

When we analyze our trading goals it is important to point out that every BES has an expected cost (mean) and timing risk (uncertainty) component. Once the strategy on the efficient trading frontier is determined the expected cost and timing risk value are shown by drawing a horizontal line from the strategy to the y-axis to determine the expected cost and a vertical line from the strategy to the x-axis to determine the corresponding timing risk. Additionally, drawing a line that is tangent to the strategy on the efficient trading frontier to the y-axis results in the value where the strategy will have the highest probability of out-performing. In all case these values are unique except for the situation where we seek to minimize cost (market impact and alpha). In this case, the tangent from strategy A1 (Figure 8.2a) will be a horizontal line, and the expected cost of the strategy and the cost where the strategy will have the highest probability of outperforming will be the same. The strategy will have a 50% chance of outperforming that particular cost.

3) Specify Adaptation Tactic

The third step in the algorithmic decision making process is to specify how the algorithm is to adapt to changing market conditions. This is also commonly referred to as dynamic optimization or adaptive pricing. Mathematicians may cringe at phrase "dynamic optimization" in this instance since this process is really "real-time re-optimization" and not true mathematical "dynamic optimization." But regardless of the nomenclature used, the goal is for investors to define how the algorithm will react to changing market conditions.

There are often times when investors may not want the algorithm to adapt to changing market conditions. For example, investors seeking to achieve the day's closing price would not want to make any adjustments to the algorithm because it may cause the algorithm to finish early and increase tracking error compared to the closing price. Investors seeking to achieve the VWAP price would want to adhere to the intraday volume profile regardless of price movement or volatility. Additionally, investors trading hedged baskets or hedged portfolios may not want to deviate from their initial prescribed strategy regardless of market conditions since doing so may ruin the hedge and increase risk exposure. Portfolio adaptation tactics are further discussed in Chapter 9, Portfolio Algorithms.

Below we discuss three methodologies for revising intraday algorithmic trajectories based on expected total trading cost. These are: targeted cost, aggressive-in-the-money (AIM), and passive-in-the-money (PIM) strategies[1]. These studies also provide an in-depth analysis surrounding the underlying profit and loss distributions corresponding to these tactics as well as real-time solutions.

Adaptation tactics and how they influence algorithmic decision and trading performance were previously studied by Kissell and Malamut (2005, 2006) and Almgren and Lorenz (2006, 2007). We follow the approach from the Journal of Trading introduced by Kissell and Malamut (2005, 2006).

Projected Cost

At the beginning of trading the projected cost is equal to the initial cost estimate (described in step 2 above). But after trading begins the projected cost will be comprised of four components: realized cost, momentum cost, remaining market impact cost, and alpha trend cost.

---

[1] Tom Kane, former Managing Director at JP Morgan and Merrill Lynch, introduced the naming of the Aggressive-in-the-Money (AIM) and Passive-in-the-Money (PIM) adaptation tactics. Furthermore, unlike many of the names chosen for algorithms, the AIM and PIM provide investors with a description of their behavior.

This is explained as follows:

Suppose that X represents the total order shares, Y represents the shares that have traded, and $(X - Y)$ represents the shares that have not yet traded (unexecuted shares). Then,

$\theta = \frac{Y}{X}$ represents the percentage of shares traded.
$(1 - \theta) = \frac{(X - Y)}{X}$ represents the percentage of shares not yet traded.

Additionally, let,

$E_0[\cdot] =$ initial time expectation. The expected cost and prices at the beginning of trading.
$E_t[\cdot] =$ time expectation. The expected cost and prices at the current point in time.
$E_0[Cost] = C^*$ initial estimated cost.
$P_t =$ market price at time t.
$\overline{P}_t =$ average execution price of the Y shares at time t.

Then the cost components are:

Realized Cost: the actual cost of the Y traded shares. This is:

$$Realized(Cost(t)) = \theta \cdot (\overline{P}_t - P_0) \cdot Side \qquad (8.27)$$

Momentum Cost: the price movement in the stock from the time trading began to the current time. This price movement results in either a sunk cost or savings to the investor. Momentum cost is applied to the number of unexecuted shares since these are the shares that will realize the cost or realize the savings. This is:

$$E_t[Momentum(Cost)] = (1 - \theta) \cdot (P_t - P_0) \cdot Side \qquad (8.28)$$

Remaining Market Impact: the remaining market impact cost of the unexecuted shares. This is the expected price impact that will result from trading the unexecuted shares in the current market environment (liquidity and volatility) and with the current trading rate. Mathematically this is:

$$E_t[MI'] = (1 - \theta) \cdot (b_1 \cdot I' \cdot \alpha_t + (1 - b_1) \cdot I') \qquad (8.29)$$

Alpha Trend: the cost that will result due to the alpha trend over the trading horizon. Mathematically this is:

$$E_t[Alpha'] = (1 - \theta) \cdot \frac{X}{ADV} \cdot \frac{1}{\alpha_t} \cdot \mu_t \qquad (8.30)$$

where μ_t is the alpha trend over the trading horizon expressed in \$/share.

Notice that the remaining market impact cost and alpha trend are the only components that can be affected by the trading strategy. Investors will seek to manage projected costs by trading either faster or slower based on market conditions, price momentum, and desired adaptation tactic.

The timing risk of these unexecuted shares represents the uncertainty surrounding the market impact estimate for the remaining shares. This is:

$$E_t[TR'] = \sigma \cdot \sqrt{(1 - \theta) \cdot \frac{X}{ADV} \cdot \frac{1}{\alpha_t}} \cdot P_0 \tag{8.31}$$

The projected cost of the order is then,

$$E_t\left[Projected\,Cost\,(\$/share)\right] = Realized(Cost(t)) = + E_t[Momentum]$$
$$+ E_t[MI'] + E_t[\text{Alpha}'] \tag{8.32}$$

Written formulaically this is:

$$E_t\left[Projected\right] = \theta \cdot (\overline{P}_t - P_0) \cdot Side \cdot (1 - \theta)(((\overline{P}_t - P_0) \cdot Side)$$
$$+ (b_1 \cdot I' \cdot \alpha_t + (1 - b_1) \cdot I')) + (1 - \theta) \cdot \frac{X}{ADV} \cdot \frac{1}{\alpha_t} \cdot \mu_t \tag{8.33}$$

Notice that this projected cost expression consists of a component that is "sunk" and "unavoidable" and a component that is "controllable." The sunk cost component is comprised of the realized cost for the transacted shares. The unavoidable component is comprised of the momentum cost and permanent market impact cost. The controllable component is comprised of the price impact of the shares that are to be traded and alpha cost of these shares. This is also the component that can be managed through proper selection of the trading strategy.

Recall that our initial cost is denoted as $E_0[Cost]$ and is expressed in $/share but could also be expressed in total dollars or in basis points.

For simplicity, we proceed in the examples below without the alpha term in basis point units. We leave it as an exercise for our readers to work through the math including the alpha cost component.

Let K_t represent the costs that are unavoidable (realized, momentum, and permanent):

$$K_t = \theta \cdot (\overline{P}_t - P_0) \cdot Side + (1 - \theta) \cdot ((\overline{P}_t - P_0) \cdot Side) + (1 - \theta) \cdot (1 - b_1) \cdot I' \tag{8.34}$$

Then the projected cost can be simplified as follows:

$$E_t(Cost) = K_t + (1 - \theta)(b_1 \cdot I' \cdot \alpha_t)$$

where α_t is the trading rate that will be used from the current time through the completion of the order.

Target Cost Tactic

The targeted cost adaptation tactic will minimize the squared difference between the projected cost and original cost from the best execution strategy. This tactic will always revise the strategy to put us back on track to get as close as we can to the original expected trading cost. Here, the strategy becomes more aggressive in times of favorable price movement and more passive in times of adverse price movement.

Mathematically, the target cost optimization is found by minimizing the following:

$$Min \quad L = (E_0(Cost) - E_t(Cost))^2$$
$$s.t. \quad LB^* \leq \alpha_t \leq UB^*$$

The general optimization needs to include expectations for volume and volatility over the remainder of the trading period. Volatility can be adjusted by the HMA-VIX adjustment described in Chapter 6, Price Volatility. The upper and lower bounds are included to ensure that the optimal strategy will be within levels specified by the investor. Some investors request to trade no slower than some level (e.g., $\geq 5\%$) and no faster than another level (e.g., $\leq 40\%$). Furthermore, most investors do require completion of the order by some end time. Thus the optimal strategy needs to be at least fast enough to ensure that trading will be completed by the specified end time or market close.

For the trading rate strategy the targeted cost adaptation tactic objective function is:

$$Min \quad L = (C^* - (K_t + (1-\theta)(b_1 \cdot I' \cdot \alpha_t)))^2 \tag{8.35}$$

where C^* is the original expected cost from the BES, and K_t is the unavoidable cost at time t based on realized cost and market movement. Solving for α_t we get,

$$\alpha_t = \frac{C^* - K_t}{(1-\theta)(b_1 \cdot I')} \tag{8.36}$$

This rate is then adjusted to ensure it satisfies our boundary conditions. That is written mathematically as:

$$\alpha_t^* = min(max(LB, \alpha_t), UB)$$

Figure 8.3a compares the cost distribution from a targeted cost adaptation tactic to the cost distribution of a constant trading rate. The expected cost of the targeted rate C_2 will be lower than the constant rate C_1 since this tactic

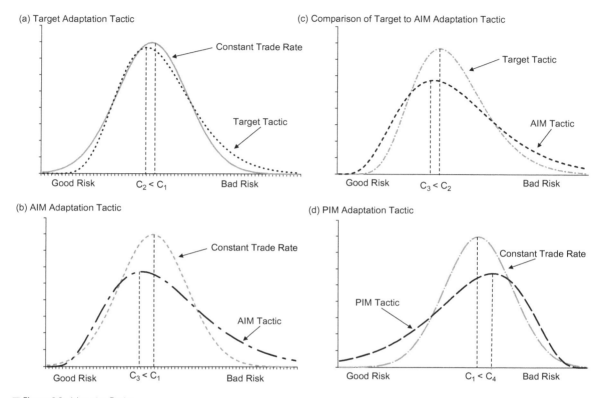

(a) Target Adaptation Tactic

Constant Trade Rate

Target Tactic

Good Risk $C_2 < C_1$ Bad Risk

(b) AIM Adaptation Tactic

Constant Trade Rate

AIM Tactic

Good Risk $C_3 < C_1$ Bad Risk

(c) Comparison of Target to AIM Adaptation Tactic

Target Tactic

AIM Tactic

Good Risk $C_3 < C_2$ Bad Risk

(d) PIM Adaptation Tactic

Constant Trade Rate

PIM Tactic

Good Risk $C_1 < C_4$ Bad Risk

■ **Figure 8.3** Adaptation Tactics.

takes advantage of favorable market conditions. However, since the trading rate executes slower in times of adverse price movement it will expose the trader to higher market risk. This is shown in the cost distribution with a fatter tail at the right signifying higher probability of these higher costs. Additionally, in times of continued favorable price momentum over the life of the defined trading horizon this tactic will not have the opportunity to transact with the most favorable market prices since it is likely that the order will be completed before the most favorable prices occur in the market.

Aggressive-in-the-Money

The aggressive-in-the-money (AIM) adaptation tactic maximizes the probability that the actual cost will be less than the original cost from the best execution strategy. This optimization is equivalent to maximizing the Sharpe Ratio of the trade where performance (return) is measured as the difference between original cost and projected cost. This type of optimization has also been defined as maximizing the information ratio of the trade (Almgren and Chriss, 2000). The AIM adaptation tactic becomes more aggressive in times of favorable price momentum and less aggressive in times of adverse price momentum.

If investors selected the price improvement best execution strategy in step 2 it is essential that we use the exact same cost in the AIM adaptation strategy, otherwise the resulting strategy will have a lower probability of executing more favorably than initially intended and will not be consistent with the trading goal.

Mathematically, the AIM tactic is found by maximizing the following equation:

$$Min \quad L = \frac{E_0(Cost) - E_t(Cost)}{E_t(TR)}$$

$$s.t. \quad LB^* \le \alpha_t \le UB^*$$

Here, $E_0(Cost)$ is either the original expected cost from the BES in step 2 or the cost used to generate the price improvement strategy in step 2. And $E_t(Cost)$ and $E_t(TR)$ are the expected projected cost and timing risk for the order at time t. Also, the expected cost term needs to include the alpha cost component in times when traders have a short-term alpha expectation.

For the trading rate strategy (without an alpha term) the AIM adaptation tactic objective function is:

$$Min \quad L = \frac{C^* - (K_t + (1 - \theta)(b_1 \cdot I' \cdot \alpha_t))}{\sigma \cdot \sqrt{(1 - \theta) \cdot \frac{X}{ADV} \cdot \frac{1}{\alpha_t} \cdot P_0}} \tag{8.37}$$

Where C^* is either the original expected cost from the BES or the cost used to generate the price improvement best execution strategy and K_t is the unavoidable cost at time t based on realized cost and market movement.

Solving for α_t we get,

$$\alpha_t = \frac{C^* - K_t}{3 \cdot (1 - \theta)(b_1 \cdot I')} \tag{8.38}$$

This trading rate then needs to ensure that it satisfies our boundary conditions (i.e., user specified maximum and/or minimum rates, or the minimum rate required to ensure completion of the order). That is written mathematically as:

$$\alpha_t^* = min(max(LB, \alpha_t), UB)$$

Compared to a constant trading rate, the AIM adaptation tactic will incur a lower cost on average but will have increased risk exposure. The cost distribution of the AIM tactic is shown in Figure 8.3b. Notice that its expected cost C_3 is lower than that of the constant rate C_1, but it does have higher bad risk exposure.

Compared to the targeted cost strategy the AIM tactic will trade at a slightly slower rate. Notice that the optimal AIM strategy is 1/3 of the optimal targeted cost solution. This results in a slightly lower cost than the targeted tactic and an increased potential for better prices if the favorable trend continues. But it is also associated with increased risk exposure and a potential for higher costs in times of adverse price movement. Figure 8.3c shows the cost distribution of the targeted cost tactic to the AIM tactic. Notice that the AIM cost C_3 is lower than the targeted cost C_2 but also has increased risk exposure as shown by the fatter tail on the right hand side (bad risk).

Passive-in-the-Money

The passive-in-the-money (PIM) adaptation tactic is a price-based scaling tactic intended to limit the potential losses and high costs in times of adverse price movement. It allows investors to better participate in gains to share in gains in times of favorable price movement.

The PIM tactic was originally designed with the Arrow-Pratt constant relative risk aversion (CRRA) formulation (see Pratt, 1964 and Arrow, 1971) in mind. But after conversations with a large number of investors we revised this adaptation tactic to be the mirror image of the AIM tactic. For example, investors may decelerate when trading prices are favorable because they believe the prices are going to continue to improve, and by trading slower they will further reduce market impact and continue to realize better prices. PIM will trade faster in times of adverse movement because investors believe that the adverse trend will continue to worsen. Thus we would rather pay higher market impact cost to avoid the most adverse prices and minimize the "bad" fat tail events. In other words, the PIM adaptation tactic minimizes potential bad outliers and increases our chances of achieving the good outliers.

The PIM tactic is found by maximizing the negative of the AIM adaptation tactic. Mathematically this formulated as:

$$Max \quad L = -\frac{E_0(Cost) - E_t(Cost)}{E_t(TR)}$$

or alternatively as the following minimization problem,

$$Min \quad L = \frac{E_t(Cost) - E_0(Cost)}{E_t(TR)}$$

In this case $E_0(Cost)$ is the original cost from the BES in step 2 and $E_t(TR)$ is the price uncertainty for the remainder of the order. If the investor has an alpha expectation this additional cost will need to be incorporated into this expression.

For the trading rate strategy (without an alpha term) the PIM adaptation tactic objective function is:

$$Min \quad L = \frac{(K_t + (1-\theta)(b_1 \cdot I' \cdot \alpha_t)) - C^*}{\sigma \cdot \sqrt{(1-\theta) \cdot \frac{X}{ADV} \cdot \frac{1}{\alpha_t} \cdot P_0}} \tag{8.39}$$

where C^* is the original expected cost, K_t is the unavoidable cost at time t, and the denominator is the price uncertainty for the remainder of the order.

Solving for α_t we get,

$$\alpha_t = \frac{K_t - C^*}{3 \cdot (1-\theta)(b_1 \cdot I')} \tag{8.40}$$

Incorporation of our boundary conditions yields,

$$\alpha_t^* = min(max(LB, \alpha_t), UB)$$

The optimal PIM solution is similar to the optimal AIM solution. The only difference is in the numerator where the PIM numerator is the negative of the AIM numerator. This ensures a mirror image between the two adaptation tactics. As prices become more favorable the PIM tactic will slow down and as prices become less favorable the PIM tactic will speed up. This results in a cost distribution with a higher mean cost than the AIM and targeted cost tactics, but with a much lower probability of incurring higher costs due to persistence of adverse price movement. The PIM tactic protects against the fat tail and "bad" risk events and provides increased opportunity to achieve better prices in times of favorable trends. But this comes with a slightly higher cost. Unfortunately in finance there is no free lunch. This is shown in Figure 8.3d. Notice that the expected cost for PIM, C_4, is higher than of the constant rate C_1. But it is associated with less bad risk and a much higher possibility of achieving the better prices from the favorable price trend.

Comparison across Adaptation Tactics

As a strategy deviates from the initial best execution strategy due to the specification of an adaptation tactic, it results in a new expected cost distribution. The targeted cost and AIM adaptation tactics result in a skewed distribution with a lower expected trading cost but with more "bad" risk. These strategies will take advantage of better market prices by trading at a quicker rate, which results in a lower expected trading cost, but they also trade at a slower rate when prices are less favorable, which increases the chances of incurring costly outliers.

Comparison of the targeted tactic (Equation 8.36) to the AIM tactic (Equation 8.38) shows that the AIM tactic trading rate is 1/3 of the targeted cost rate. This results in a lower expected cost for the AIM strategy since it allows the fund to participate with favorable price trends for a longer period of time, but since it trades slower it will also expose the order to more market risk which will result in a higher potential for costly outliers when adverse price trends persist.

One strategy that is being used by investors to overcome the "bad" risk issue associated with the targeted cost or AIM adaptation tactic is to set the lower bound or minimum trading rate equal to the original trading rate from step 2—specify trading goal. This then allows investors to take advantage of the better prices when they arise and it will not cause the fund to incur incremental market risk over the constant rate since it will trade no slower than the original trading rate.

The PIM adaptation tactic (Equation 8.40) also results in a skewed distribution, but, unlike the target and AIM tactics, PIM protects investors from "bad" risk by accelerating trading when the adverse momentum begins. The goal of PIM is to complete the order before prices can become too expensive to trade. An advantage of PIM is that it increases "good" risk exposure since it trades slower when there are favorable market prices. But this results in a slightly higher cost on average—but with bad risk protection.

Figure 8.4 provides an example of the difference between the three tactics for a buy order. At the current price the target rate is trading at 30% and both AIM and PIM are at 10%. As prices decline, the target and AIM tactics increase and the PIM tactic decreases. Notice that the target tactic increases three times as quickly (as we determined mathematically). The PIM tactic will decrease down to its minimum rate (which is specified by investor) and in this example is 5%. As prices increase, both the target and AIM tactics decrease and will continue to decrease down to the investor's specified minimum rate. Since the AIM tactic (10%) is

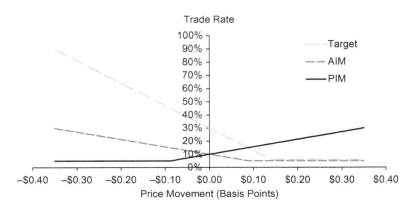

■ **Figure 8.4** Comparison of Adaptation Tactics.

lower than the target rate (30%) it will reach its minimum rate quicker than the target rate. But notice how the PIM tactic trading rate increases as prices increase. It is the exact mirror image to the AIM tactic.

Modified Adaptation Tactics

Many investors are proponents of specifying adaptation tactics based only on the current market prices and arrival price, or based on current market prices plus remaining market impact cost and arrival price. In this type of scenario, the algorithm ignores what has happened in the past, and only considers the current point in time and expected future prices.

While this type of tactic is preferred by some market participants, we are not proponents of this type of tactic. If an investor is not concerned with what happened in the past while they are trading they should not be concerned with what happened in the past after trading is completed. Otherwise, they exhibit inconsistent behavior. If an investor wants to trade a certain way halfway through the order that is different than at the beginning of the order, the chances are that they could have defined a different initial strategy and adaptation tactic at the beginning of trading and would have realized even better results. Consistency and pre-planning is paramount when it comes to algorithmic trading.

How Often Should We Re-Optimize Our Tactics?

There are many different theories for when and why an investor should re-optimize their trading rates. Some of these are quite appropriate and some are really quite silly. Suffice it to say, investors should not treat trading algorithms like video games and make changes only for the sake of making changes. Revisions to strategies should be made when appropriate from the perspective of the investment objective as well as to disguise trading intentions. Of course revisions could always be made because of the arrival of new information that was either not known at the beginning of trading or is different than what was believed at the beginning of trading.

There are four techniques that are actively being used by algorithms to revise trading strategies. These are continuous, trade-based, time period, and z-score. First, some algorithms continuously revise their trading rate. With today's computer power some algorithms are continuously revising rates based on what is observed in the market. Second, some algorithms revise their trading rate after each child order is executed or cancelled. Third, some trading algorithms will revise their trading rate based on a defined time period such as every 30 seconds (or faster or slower). However, some of the more sophisticated algorithms employing higher levels of sophistication and anti-game logic will revise their trading algorithms based on a z-score criterion. This is as follows:

The z-score is measured as the difference between the expected cost at the current point in time and the original expected cost divided by the remaining timing risk of the order. That is:

$$E_t(Z) = \frac{E_0(Cost) - E_t(Cost)}{E_t(TR)} \qquad (8.41)$$

This formulation of $E_t(Z)$ will be negative if we are underperforming our expected cost and positive if we are outperforming our expected cost. Negative is bad and positive is good.

The algorithm may determine to only revise the trading rate if the z-score is higher or lower than a specified value such as ± 1 or $\pm \frac{1}{2}$. For example, only revise the trading rate if the projected z-score at time t is greater than or less than a trader specified criteria. The z-score specifies the number of standard deviations the expected finishing cost will be from the original cost. Quite often in statistics we use a z-score within ± 1 to signify expected performance, but in trading, many investors prefer a more conservative measure of the z-score and use $\pm \frac{1}{2}$ as the reference value to re-optimize. Additionally, some investors elect to re-optimize only if the projected z-score is less than some value such as $Z < -1$ or $Z < -\frac{1}{2}$ indicating less than desirable performance. Additionally, and what is often most important for many investors, the z-score re-optimization logic makes it more difficult to uncover what the algorithm is doing

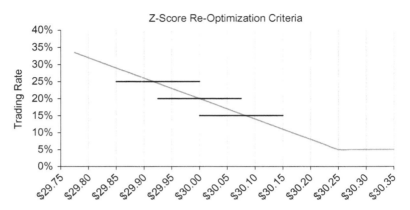

■ **Figure 8.5** Z-Score Re-Optimization Criteria.

because it is not updating continuously, after each trade, or based on a specified period of time.

An example of the z-score re-optimization criteria for a buy order is shown in Figure 8.5. At the current price of $30.00 the algorithm is transacting with a trading rate of 20%. The algorithm will continue to trade at this rate while price levels are between $29.92 and $30.08. If prices fall below $29.92 then the algorithm will increase to a trading rate of 25% and continue at this rate as long as prices are between $29.85 and $30.00. Notice that this logic does not return the trading algorithm to the original rate of 20% when prices move back above $29.92. If prices increase above $30.08 the trading rate will decrease to 15% and remain at this level while prices are between $30.00 and $30.15. Again, notice that the algorithm does not return to the original rate of 20% after the prices fall back below $30.08. It is important to point out in this example that at a market price of $30.00 it is possible for the algorithm to be transacting at three different trading rates: 15%, 20%, or 25%. This makes it increasingly difficult for any trader to decipher the intentions of the trading algorithm or goal of the trader.

The exact rate in use by the algorithm will be determined in part by current and forecasted market conditions, realized and projected trading costs, and the investor's z-score criteria. All of which makes it increasingly difficult to uncover the investor's execution strategy.

Portfolio Algorithms

INTRODUCTION

Portfolio algorithms and multi-period trade schedule optimization has gained momentum in the financial community due to the increase in program and algorithmic trading. By understanding how portfolio trading decisions influence returns, traders will be better prepared to make decisions consistent with the overall investment objectives of the fund. Unfortunately, traditional optimization techniques are not adequate for portfolio needs due to the non-linearity of the price impact function, the large number of decision variables, and the time it takes to calculate the answer.

Each time a trader is given a trade list to execute (e.g., basket, program, or portfolio) they face an inherent dilemma. Trading too quickly will result in too much price impact due to liquidity demands and information leakage but trading too slowly will result in too much risk which could lead to even higher costs in times of adverse price movement ("trader's dilemma").

In order to address the conflicting expressions of market impact cost and timing risk, traders derive a trade schedule ("slicing strategy") that balances the trade-off between price impact and risk based on a specified level of risk aversion. The appropriate computational technique to solve this problem for a portfolio is the multi-period trade schedule optimization. But unlike the portfolio manager who usually has ample time to run sophisticated optimization algorithms and perform thorough sensitivity analysis, a timely solution for the trader is mandatory especially considering that many times they are given the order just before the market open or during the trading day. Most currently available optimization routines take too much time to solve the trader's dilemma to be useful for investors. These packages can take several minutes, hours, or more, especially if the problem involves thousands of stocks over a long trading horizon. Traders require real-time solutions in seconds or less.

Trade schedule optimization to minimize total trading costs has been previously studied. For example, Bertsimas and Lo (1998) provided an approach to minimize price impact in the presence of expected future

The Science of Algorithmic Trading and Portfolio Management. DOI: http://dx.doi.org/10.1016/B978-0-12-401689-7.00009-X

information. The goal is to minimize total cost arising from price impact and price drift. Almgren and Chriss (1999, 2000) expanded on the idea of trade schedule optimization by incorporating a risk aversion parameter and set out to balance two conflicting terms (price impact and risk) based on the investor's risk appetite. Their proposed market impact formulation contains the right shape and market impact properties (e.g., convex shape with dollar value) but their objective function results in a path dependent stochastic process with a difficult and slow solution.

Malamut (2002, 2006) devised an approximated QP formulation and provided insight into parametric trade schedules to solve a non-linear impact formulation. Kissell, Glantz and Malamut (2004) incorporate a drift term into the objective functions and offer alternative goals to mean-variance optimization such as maximizing the probability of outperforming a specified cost (e.g., maximize the Sharpe ratio of the trade). Obizhaeva and Wang (2005) study an inter-temporal (not static) trade sequencing problem. They seek to solve a path dependent problem similar to Almgren and Chriss (2000) by understanding the half-life of a trade (e.g., the time for temporary impact to dissipate). Their techniques, however, are only presented for a single stock order.

In the financial literature, the mean-variance portfolio optimization of Markowitz (1952) clearly stands out as one of the more important quantitative approaches. The technique is widely used by portfolio managers and is an effective tool to manage risk and improve returns. However, mean-variance optimization is mostly used in the context of a one-period investment model. Li and Ng (2000) derive a solution to the multi-period mean-variance optimization problem where the allocation decision is reviewed in every period. Therefore, the proposed solution is dynamic since the decision to invest is reviewed after each period's results are known.

In this chapter, we present a multi-period trade schedule optimization approach for portfolio optimizers (Malamut, 2002). We offer four approaches that can be used to solve the trader's dilemma in an amount of time that can be useful for traders. These approaches expand on techniques presented in *Optimal Trading Strategies* (2003) and introduce real-time adaptation techniques to determine when it is appropriate to take advantage of market conditions given the overall risk composition of the trade basket.

TRADER'S DILEMMA

A typical trading situation is as follows: traders are provided with a basket of stock (e.g., program, trade list, portfolio, etc.) to transact in the market.

The basket may be one-sided (e.g., all buys or all sells) or two-sided (e.g., both buys and sells). Traders are then tasked with determining the most appropriate way to transact the order over a specified period of time. This is accomplished by balancing the trade-off between cost and risk based on a user specified level of risk aversion. Mathematically, this is stated as follows:

$$Min \; Cost(x_k) + \lambda \cdot Risk(x_k)$$

Where λ represents trader specified level of risk aversion, and x_k is used to denote the discrete trade schedule representing exactly how the shares are to be transacted in each period for each stock.

Variables

$X = $ *shares to trade*

$Y_t = $ *shares executed at time t*

$$Side(i) = \begin{cases} +1 & \textit{if buy order} \\ -1 & \textit{if sell order} \end{cases}$$

$ADV = $ *average daily volume*

$\sigma = $ *annualized volatility*

$C = $ *covariance matrix, scaled for the length of the trading period in* $(\$/Share)^2$

$x_{ij} = $ *shares of stock i to trade in period t*

$r_{ij} = $ *residual shares of stock i at beginning of period t*

$$r_{ij} = \sum_{k=j}^{n} x_{ik}$$

$v_{it} = $ *volume for stock i in period t*

$P_{i0} = $ *arrival price*

$P_{it} = $ *market price in period t*

$\overline{P}_{it} = $ *average execution price at time t*

$m = $ *number of stocks in the portfolio*

$n = $ *number of trading periods during the horizon*

$d = $ *number of trading periods per day*

TRANSACTION COST EQUATIONS

This section describes the trading cost equations that will be used to solve the portfolio trader's dilemma. When performing portfolio optimization it is most beneficial to express the trading strategy in terms of a trade schedule to allow us to most effectively manage total portfolio risk and express costs in total dollars to allow us to easily sum costs across stocks. The formulations used here are based on the equations provided in Chapter 7, Advanced Algorithmic Forecasting Techniques.

For a portfolio of m stocks that are to be executed over n trading periods we have:

The trade schedule x as an $m \times n$ matrix as follows:

$$
x = \begin{pmatrix} x_{11} & x_{12} & \cdots & x_{1n} \\ x_{21} & x_{22} & \cdots & x_{2n} \\ \vdots & \vdots & \ddots & \vdots \\ x_{m1} & x_{m2} & \cdots & x_{mn} \end{pmatrix}
$$

The residual schedule r as an $m \times n$ matrix as follows:

$$
r = \begin{pmatrix} r_{11} & r_{12} & \cdots & r_{1n} \\ r_{21} & r_{22} & \cdots & r_{2n} \\ \vdots & \vdots & \ddots & \vdots \\ r_{m1} & r_{m2} & \cdots & r_{mn} \end{pmatrix}
$$

where $r_{ij} = \sum_{k=j}^{n} x_{ik}$.

For simplicity of notation, we define x_k and r_k to be the column vectors of the trade and residual matrices, respectively, as follows:

$$
x_k = \begin{pmatrix} x_{1k} \\ x_{2k} \\ \vdots \\ x_{mk} \end{pmatrix} \qquad r_k = \begin{pmatrix} r_{1k} \\ r_{2k} \\ \vdots \\ r_{mk} \end{pmatrix}
$$

The covariance matrix C is an $m \times n$ matrix as follows:

$$
C = \begin{pmatrix} c_{11} & c_{12} & \cdots & c_{1m} \\ c_{21} & c_{22} & \cdots & c_{2m} \\ \vdots & \vdots & \ddots & \vdots \\ c_{m1} & c_{m2} & \cdots & c_{mm} \end{pmatrix}
$$

where C is scaled for the length of the trading horizon and expressed in terms of $(\$/\text{share})^2$, c_{ij} is the covariance between stock i and stock j, and c_{ii} is the variance of stock i.

Market Impact

$$I_{\$i}^* = a_1 \cdot \left(\frac{X_i}{ADV_i}\right)^{a_2} \cdot \sigma_i^{a_3} \cdot 10^{-4} \cdot P_{i0} \cdot X_i \tag{9.1}$$

$$MI_\$(x_k) = \sum_{i=1}^{m} \left(\sum_{t=1}^{n} \frac{b_1 \cdot I_i^* \cdot x_{it}^2}{X_i \cdot v_{it}}\right) + (1 - b_1)I_i^* \tag{9.2}$$

This formulation of market impact follows from the trade schedule formulation with parameter $a_4 = 1$. These parameters are given in Table 9.1.

Price Appreciation

$$PA_\$(x_k) = \sum_{i=1}^{n}\sum_{t=1}^{m} x_{ij} \cdot \Delta p_i^* \cdot t \tag{9.3}$$

where Δp_i is the per period price appreciation term expressed in $/share adjusted for the side of the order. For example, if the price is expected to increase $0.05/share per period and we are selling shares the price appreciation term is:

$$\Delta p_i^* = side(i) \cdot \Delta p_i = -1 \cdot \$0.05/share = -\$0.05/share$$

Table 9.1 Market Impact Parameters—Trade Schedule Strategy

Data Sample	a_1	a_2	a_3	a_4	b_1
All Data	656	0.48	0.45	1	0.90
Large Cap	707	0.59	0.46	1	0.90
Small Cap	665	0.42	0.47	1	0.90

Timing Risk

$$TR(r_k) = \sqrt{\sum_{k=1}^{n} r_k' C r_k} \qquad (9.4)$$

where C is the trading risk covariance matrix expressed in ($/share)^2, and is scaled for the length of the trading interval, and r_k is the residual vector of unexecuted shares at the beginning of period k.

One-Sided Optimization Problem

It is important to note that our I-Star market impact equation (Equation 9.1) requires that the trade size X be positive, i.e., $X_i > 0$ for all stocks. This creates a difficulty when optimizing a two-sided portfolio since we need to have a way to incorporate the negative market relationship between buy and sell orders. Similar to how we adjusted the price appreciation term by the side of the order, we adjust the covariance term by the side of each order and thereby convert the portfolio optimization problem to a one-sided optimization problem. That is:

$$c_{ij}^* = side(i) \cdot side(j) \cdot c_{ij}$$

Then our transaction cost equations will properly account for the sided covariance across stocks. Notice in this calculation that the variance of a stock will be positive, the covariance between two stocks on the same side (i.e., both buys or both sells) will be equal to the original covariance term and the covariance between two stocks with opposite side orders (i.e., one buy and one sell) will be the negative of the original covariance term.

OPTIMIZATION FORMULATION

Using the expressions above the complete portfolio trader's dilemma translates to:

$$\underset{x}{Min} \left\{ \sum_{i=1}^{m} \left(\sum_{t=1}^{n} \frac{b_1 \cdot I_i^* \cdot x_{it}^2}{X_i \cdot v_{it}} \right) + (1 - b_1) I_i^* + \sum_{i=1}^{n} \sum_{t=1}^{m} x_{ij} \cdot \Delta p_{ij} \cdot t \right\} + \lambda \cdot \sqrt{\sum_{k=1}^{n} r_k' C r_k} \qquad (9.5)$$

Subject to constraints:

i.	$\sum_{t=1}^{n} x_{it} = X_i$	Completion
ii.	$x_{it} \geq 0$	No Short Sales
iii.	$r_{it} - r_{it+1} \geq 0$	Shrinking Portfolio
iv.	$r_{it} = \sum_{k=t}^{n} x_{ik}$	Residual Schedule

v.	$x_{it} = r_{it} - r_{it+1}$	Trade Schedule
vi.	$\alpha^*_{i,min} \le \frac{x_{it}}{v_{it}} \le \alpha^*_{i,min}$	Trade Rate Bounds
vii.	$x^*_{i,min} \le x_{it} \le x^*_{i,max}$	Trade Size Bounds
viii.	$r^*_{i,min} \le r_{it} \le r^*_{i,max}$	Residual Size Bounds
ix.	$LB \le \sum_{i=1}^{m}\sum_{j=1}^{k}(side(i)\cdot x_{ij}\cdot p_{ik}) \le UB$	Self-Financing
x.	$LB \le (side(i)\cdot r_{ij}p_{ij}) \le UB$	Risk-Management

Constraint Description

Investors may include all or some of the constraints above. These constraints are fund specific and can be omitted if deemed unnecessary by the trader. These constraints are described as follows:

i. **Completion**: ensures that the optimization solution will executes all shares in all orders within the defined trading horizon.

ii. **No Short Sales**: ensures that the side of the order will not change. For example, the optimization solution will only buy shares for a buy order and sell shares for a sell order. Without this constraint, it is possible that the optimization may overbuy or oversell during the day and then have to offset the newly acquired position.

iii. **Shrinking Portfolio**: ensures that the size of the order keeps decreasing. For example, if the order is to buy 100,000 shares the positions will always be decreasing towards zero and will never increase. Without this constraint the optimization solution may determine it would be best to first sell 25,000 shares so that the order increases to 125,000 shares. While this type of strategy may be the best way to manage overall portfolio risk it may not be an acceptable solution for the investor. For example, it exposes the investor to short-term risk if the stock is halted after the position size increases to 125,000 shares.

iv. **Residual Schedule**: defines the residual share quantity in each period in terms of unexecuted shares at that point in time. Used if the decision variable is the trade share amount.

v. **Trade Schedule**: defines the shares to trade in terms of the residual shares. Used if the decision variable is the residual trade vector.

vi. **Trade Rate Bounds**: the defined maximum and minimum trading rates. For example, investors may wish to trade at least 1% of the total market volume in each period but no more than say 25% of total market volume in each period. These constraints are most often defined in terms of percentage of volume rate so may need to be converted to the trade rate definition.

vii. **Trade Size Bounds**: defines the maximum and/or minimum number of shares to execute in each period through completion of the order. For example, investors may wish to trade at least 100 shares in each period and no more than say 25,000 shares.

viii. **Residual Size Bounds**: defines the maximum and/or minimum position sizes (e.g., unexecuted shares) at different points in time. The investor may wish to give the optimizer some leeway on the solution but within a user specified tolerance band. For example, the investor may require one-half of the order to be executed within the first two hours of the trading day.

The last two constraints are often stated as *cash-balancing* constraints. However, the term cash balancing is a very vague term in the industry and has two different meanings. These cash-balancing constraints are "self-financing," and "risk-management."

ix. **Self-Financing**: The self-financing constraint is used by investors who are looking to have their sell orders finance their buy orders. This constraint manages the cash transactions throughout the day. For example, if this constraint is positive it indicates that they have bought more than they have sold, and therefore will need to pay incremental dollars. If this constraint is negative it indicates that they will have sold more than they have purchased, and will have incremental dollars that they will receive. Investors will often place tolerance bands on the cash position so that they will not have to provide too much additional cash for the purchases or receive too much cash back from sells. The self-financing constraint manages cash-flow from the perspective of shares already traded. It is often intended to keep the fund from having to raise cash at the end of the day in cases where the buy dollar amount was higher than the sell dollar amount.

x. **Risk-Management**: The risk-management constraint is used by investors to manage risk throughout the trading day. Here risk is managed by the net value of the unexecuted shares. These investors believe that as long as the value of the remaining shares to be purchased is equal to the value of shares to be sold the portfolio is hedged from market movement. This constraint, however, does not incorporate the sensitivity to the market. For example, if the investor is buying a list of high beta technology stocks and selling a list of low beta consumer staples they may not be hedged from market movement. If the market goes up the prices of the technology stocks are likely to go up more than the consumer staples stocks thus requiring the investor to provide additional cash at the end of the day and incur a higher trading cost. The risk-management constraint manages cash-flow from the perspective of unexecuted shares. Cash balancing for risk-management was originally implemented when investors did not have full confidence in the underlying intraday covariance model.

Objective Function Difficulty

The formulation of the objective function above presents many difficulties. First, the problem is not linear or quadratic, thus creating increased complexity for the optimization routine. Second, the timing risk component is represented as a square root function as opposed to a squared term in a quadratic programming (QP) optimization problem. For example, portfolio construction optimization models often express risk as a variance term (risk squared) and can be directly incorporated into quadratic optimizations. Finally, there are n*m decision variables in our full formulation—one decision variable for each stock in each trading period. In portfolio construction there are m decision variables—one for each stock. Our portfolio optimization requires m times more solution variables.

Unfortunately, the time to solve these optimization algorithms increases at an exponential rate with the number of variables. For a 500-stock portfolio executed over 26 trading intervals (e.g., 15 minute intervals) this results in 13,000 decision variables and takes much more than 26 times longer to solve. Combined with the constraints above, it makes solving this optimization extremely slow.

Investors require accurate solutions within a short enough timeframe to be useful for trading. By "short enough" we mean a matter of seconds or minutes as opposed to minutes or hours.

Fortunately, there are accurate transformations and approximations that allow us to solve the trader's dilemma in a reasonable amount of time. These techniques are:

- Quadratic Optimization Approach
- Trade Schedule Exponential
- Residual Schedule Exponential
- Trade Rate Optimization

Optimization Objective Function Simplification

The full portfolio optimization objective function (Equation 9.5) includes permanent market impact cost. Because this cost is not dependent upon the specified trading strategy we can omit it from the objective function without changing the optimal solution. However, permanent impact needs to be added back into the estimated cost in order to provide investors with the full portfolio trading cost estimate.

Additionally, to simplify calculations going forward, we exclude the price appreciation term from the cost function. We only include market impact cost.

PORTFOLIO OPTIMIZATION TECHNIQUES
Quadratic Programming Approach

The trader's dilemma can be solved using a quadratic optimization (QP) by making a couple of changes to the formulation. First, formulate the problem in terms of the residual trade schedule. Second, we use the variance expression for risk, which does not include the square root function, instead of the standard deviation expression of risk, which includes the square root expression. Third, we use a variance aversion parameter in place of the traditional risk aversion parameter.

This will allow us to now solve the problem via a traditional quadratic optimization. The only outstanding issue, however, is determining the exact solution at the investor's specified level of risk aversion. This can be solved as follows. Recall that the risk aversion parameter is equal to the negative tangent of the efficient trading frontier (ETF) at the optimal trading strategy. If we solve sets of our QP optimization and plot the ETF, i.e., market impact as a function of risk using the square root function for all optimization results, we can determine the strategy on the ETF where the slope of the tangent is equal to the negative of the investor's risk aversion. This may take several iterations but it is entirely feasible.

This is an entirely valid transformation since cost-variance can be mapped to cost-risk and is consistent with Markowitz (1952) mean-variance optimization. Markowitz actually presented an optimization using return and variance but then plotted the trade-off using return and standard deviation. Markowitz's efficient investment frontier shows the trade-off between return and standard deviation but is solved using return and variance. The biggest difference here is that traders are seeking an exact point on the frontier and in an amount of time that will be useful for trading.

The QP trade cost minimization is written in terms of the residual schedule as follows:

$$\underset{r}{Min} \quad \sum_{i=1}^{m}\sum_{t=1}^{n} \frac{b_1 \cdot I_i^* \cdot (r_{it} - r_{it+1})^2}{X_i \cdot v_{it}} + \lambda^* \cdot \sum_{k=1}^{n} r_k' Cr_k \qquad (9.6)$$

Subject to:

$$
\begin{array}{ll}
r_{i1} = X_i & \text{for all } i \\
r_{in+1} = 0 & \text{for all } i \\
0 \leq r_{iJ} - r_{iJ+1} \leq x^*_{i,\max} & \text{for all } i, j \\
r_{ij} \geq 0 & \text{for all } i, j
\end{array}
$$

Notice that this formulation is written only in terms of the residual shares. This is permissible since $x_{ij} = r_{ij} - r_{ij+1}$. Additionally, $r_{i1} = X_i$ is the proper residual starting value and $r_{in+1} = 0$ is the terminal value to ensure all shares are transacted by the end of trading and satisfying our completion requirement. The last two constraints ensure the solution adheres to the shrinking portfolio constraint, and the minimum and maximum trade quantity values. Finally, λ^* is the variance aversion parameter and is different from the risk aversion parameter.

An inherent difficulty with the QP solution, however, is that there is no way to map risk aversion to variance aversion, so the actual process may need several iterations to determine the solution at the desired level of risk aversion.

Another difficulty is that the formulated problem dramatically increases in size as the number of stocks in the portfolio increases. This may diminish the efficiency benefits of the QP approach as the trade list becomes too large.

In matrix notation, the quadratic optimization is written as follows:

$$
\underset{\tilde{r}}{Min} \ \frac{1}{2} \cdot \tilde{r}' Q \tilde{r} \tag{9.7}
$$

Subject to:

$$
\begin{array}{l}
\tilde{A}_1 \tilde{r} = \tilde{b}_1 \\
\tilde{A}_2 \tilde{r} \geq \tilde{b}_2 \\
\tilde{r}_{ij} \geq 0
\end{array}
$$

Where, $\tilde{r} = m \cdot (n + 1) \times 1$, $Q = m \cdot (n + 1) \times m \cdot (n + 1)$, $\tilde{A}_1 = 2m \times m \cdot (n + 1)$, $\tilde{b}_1 = 2m \times 1$, $\tilde{A}_2 = 2m \cdot (n + 1) \times m \cdot (n + 1)$, and $\tilde{b}_2 = 2m \cdot (n + 1) \times 1$. The derivation of these matrices is provided in the appendix to this chapter.

This representation of the trader's dilemma above provides many advantages. First, there are many well known optimization algorithms suited to solve a QP minimization problem. Second, this formulation allows us to take complete advantage of diversification and hedging opportunities.

The disadvantage of this formulation is that the risk term in the objective function is expressed in terms of variance and may require several iterations to determine the trade schedule corresponding to the investor's level of risk aversion. For large trade lists this problem can be quite resource taxing. Malamut (2002) provided an adjustment to the QP model to directly convert the standard deviation risk aversion parameter to the variance risk aversion parameter which can be used to further simplify the risk aversion/variance aversion issue.

Trade Schedule Exponential

The trade schedule exponential approach parameterizes the trade schedule based on an exponential decay function with parameter θ_i. It is a non-linear optimization routine that uses the square root function for our risk expression.

The number of stocks to transact in a period is determined as follows:

$$x_{ij} = X_i \cdot \frac{e^{-j\theta_i}}{\sum\limits_{k=1}^{n} e^{-k\theta_i}} \tag{9.8}$$

The optimization formulation for the trade schedule exponential approach is a non-linear optimization formulation:

$$\underset{x}{Min} \ \sum_{i=1}^{m} \sum_{t=1}^{n} \frac{b_1 \cdot I_i^* \cdot x_{it}^2}{X_i \cdot v_{it}} + \lambda \cdot \sqrt{\sum_{k=1}^{n} r_k' Cr_k} \tag{9.9}$$

Subject to:

$$x_{ij} = X_i \cdot \frac{e^{-j\theta_i}}{\sum\limits_{k=1}^{n} e^{-k\theta_i}} \qquad \textit{for all i, j}$$

$$r_{ij} = \sum_{k=j}^{n} x_{ik} \qquad \textit{for all i, j}$$

$$LB_i \le \theta_i \le UB_i \qquad \textit{for all i}$$

Expressing the trade schedule as a parametric exponential formulation provides many benefits. First, there is only one parameter to estimate for each stock regardless of the number of specified trading periods and trading days. For example, an m-stock portfolio executed over n trading horizons has only m parameters to determine regardless of the trading horizon whereas the complete problem and QP optimization has n*m decision variables. Second, since Equation 9.9 is true for all stocks we are guaranteed of completion. Third, since $e^{-j\theta_i} > 0$ for all j we have $x_{ij} > 0$ for all periods and are ensured to adhere to the shrinking portfolio constraint.

Most essential, however, is that since the trade schedule is expressed in terms of a continuous exponential function the analytical gradient and Hessian can be easily computed. This dramatically increases the computational efficiency of a non-linear optimization algorithm. Finally, it incorporates the investor's exact risk aversion parameter.

A limitation of the exponential trade schedule, however, is that it does not allow as much freedom to take advantage of natural hedging and diversification as the exact NLP and QP approaches described above. The lower and upper bounds are included on the trading schedule parameter to ensure the order is traded within a user specified rate.

Residual Schedule Exponential

The residual schedule exponential is a technique that parameterizes the residual schedule in terms of an exponential decay function. It is a non-linear optimization routine and uses the square root function for the risk term.

The residual number of shares in each period is determined by the following:

$$r_{ij} = X_i \cdot e^{-j\omega_i} \tag{9.10}$$

This formulation is a decreasing function so it will always adhere to our decreasing portfolio constraint. But since it is always positive, i.e., $r_{ij} > 0$, we need to incorporate some terminal value to force the order to complete (within some tolerance).

The optimization formulation for the trade schedule exponential is:

$$\underset{x}{Min} \ \sum_{i=1}^{m} \sum_{t=1}^{n+1} \frac{b_1 \cdot I_i^* \cdot (r_{it} - r_{it+1})^2}{X_i \cdot v_{it}} + \lambda \cdot \sqrt{\sum_{k=1}^{n+1} r_k' Cr_k} \tag{9.11}$$

Subject to:

$r_{ij} = X_i \cdot e^{-j\omega_i}$	*for all i, j*
$r_{i1} = X_i$	*for all i*
$r_{in+1} \leq 100$	*for all i*
$x_{ij} = r_{ij} - r_{ij+1}$	*for all i, j*
$LB_i \leq \omega_i \leq UB_i$	*for all i*

Expressing the trade schedule as a parametric exponential formulation provides many benefits.

First, there is only one parameter to estimate for each stock regardless of the number of specified trading periods and trading days. For example, an m-stock portfolio executed over n trading horizons has only m parameters to determine regardless of the trading horizon whereas the complete problem and QP optimization each have decision variables. Second, since Equation 9.10 holds for all stocks we are ensured of completion of the order. Third, since $e^{-j\omega_i} > 0$ for all j we guarantee we adhere to our shrinking portfolio constraint. Most essential, however, is that since the residual schedule is expressed in terms of a continuous exponential function, the analytical gradient and Hessian can be easily computed. This dramatically increases the computational efficiency.

A limitation of the residual trade schedule exponential, similar to the exponential trade schedule, is that it does not allow as much freedom to take advantage of natural hedging and diversification as the exact NLP and QP approaches above. The lower and upper bounds are included on the trading schedule parameter to ensure the order is traded within a user specified rate.

Trading Rate Parameter

The trade strategy can also be expressed in terms of a trading rate parameter α. Here the number of shares to transact is equal to a specified percentage of market volume excluding the order shares. The process is best explained as follows. For a specified trading rate α, the expected time to complete the order (expressed as a percentage of a trading day) is:

$$t = \frac{X}{ADV} \cdot \frac{1}{\alpha}$$

If the trading day is segmented into n trading periods then the order will be completed in T periods where:

$$T = t \cdot n = \frac{X}{ADV} \cdot \frac{1}{\alpha} \cdot n \text{ (rounded up to the nearest integer)} \qquad (9.12)$$

For example, if the order size $\frac{X}{ADV} = 10\%$ and the trading rate $\alpha = 10\%$ the order will complete in a day. If the trading rate is $\alpha = 20\%$ the order will complete in one-half day, and if the trading rate is $\alpha = 5\%$ the order will complete in two days.

Market Impact Expression

For a constant trading rate the temporary market impact cost for a single stock is:

$$MI(\alpha) = b_1 \cdot I^* \alpha$$

For a basket of stock the market impact cost is:

$$MI(\alpha_i) = \sum_{i=1}^{m} b_1 \cdot I_i^* \alpha_i \qquad (9.13)$$

Timing Risk Expression

The timing risk for a portfolio cannot be expressed as a continuous function in terms of the trading rate parameter because at some time the residual shares would fall below zero. But we can overcome this problem by approximating the residual with the following continuous exponential function:

$$r_{ij} = X_i \cdot e^{-j\gamma_i} \qquad (9.14)$$

where,

$$\gamma_i = 2.74 \cdot T^{-1.22} + 0.01 \qquad (9.15)$$

This representation of residuals results in approximately the same risk that is computed using the trade schedule strategy.

The trade rate optimization problem is formulated as follows:

$$Min \sum_{i=1}^{m} b_1 \cdot I_i^* \alpha_i + \lambda \sqrt{r^t C r} \qquad (9.16)$$

Subject to:

$r_{ij} = X_i \cdot e^{-j\gamma_i}$	*for all i, j*
$\gamma_i = 2.74 \cdot T_i^{-1.22} + 0.01$	*for all i*
$T_i = \frac{X_i}{ADV_i} \frac{1}{\alpha} \cdot n$	*for all i*
$LB_i \le \alpha_i \le UB_i$	*for all i, j*

The *LB* needs to be set at a value that will ensure the order will be completed by the investor's specified end time.

Representation of the trade schedule in terms of trading rate provides many benefits. There is only one parameter per stock. The market impact cost and timing risk expressions are greatly simplified. Completion of the order and the shrinking portfolio constraint are guaranteed. And since the gradient and Hessian are easily computable, it provides efficiency and speed for the non-linear optimization.

A limitation of the trade rate formulation is that it does not provide as much freedom to take complete advantage of risk reduction opportunities as the approaches above. But it does provide guidelines to adapt to changing liquidity conditions (e.g., transact more shares in times of higher market volumes and transact less shares in times of lower market volume) throughout the trade periods which is not provided from any of the previously described techniques. And as we show below, it provides the quickest solutions even for large trade lists.

Comparison of Optimization Techniques

To compare the performance of the different optimization techniques we conducted a simulation experiment to measure solution time and accuracy. The experiment is as follows:

Sample Universe. Our sample universe was the SP500 index as of December 31, 2011.

Number of Stocks. We constructed portfolios that ranged in size from 10, 25, 50, 100, . . .450, and 500 stocks.

Order Size. We randomly defined order sizes from 0–25% ADV.

Volatility. We used actual stock volatility from the sample.

Covariance Matrix. We constructed our covariance matrix using a correlation between stocks that was equal to the average stock sector to sector correlation. For example, if the average correlation between a technology and utility stock was rho = 0.15 we used a correlation of 0.15 to compute the covariance between a technology stock and a utility stock along with their actual volatility.

Number of Simulations. We performed twenty simulations for each portfolio for each optimization technique. Ten simulations were performed for a one-sided portfolio, e.g., cash investment, five simulations were performed using a two-sided portfolio with equal weights in each side, and five simulations were performed using a 130-30 two-sided portfolio, that is, the dollar weights in one side were 130% of the total and the weights in the other side accounted for −30% of the dollar value.

Performance Measure. We recorded the time to solve each portfolio with each optimization technique and also measured the accuracy of each technique by comparing the resulting trade schedule to the true trade schedule determined by solving the portfolio using the non-linear optimization routine that solved the exact objective formulation (Equation 9.5). Advantages and disadvantages of each technique are shown in Table 9.2.

Table 9.2 Comparison of Optimization Techniques

Optimization Technique	Advantages	Disadvantages
Non-Linear Optimization	Determines exact solution to the exact problem	Takes too long to solve to be useful to traders
	Takes full advantage of diversification and hedging	Many parameters—one for each stock and period
Quadratic Optimization	Provides most accurate trade schedule	Many parameters—one for each stock and each period
	Takes full advantage of diversification and hedging	Slow solution for larger trade lists
		Could require multiple iterations
Trade Schedule Exponential	Very fast optimization solution	Does not allow full freedom in specifying trade schedule
	Few parameters—one per stock	Trade schedule is forced to follow exponential decay
	Takes very good advantage of diversification and hedging	
	Very accurate model	
Residual Schedule Exponential	Very fast optimization solution	Does not allow full freedom in specifying trade schedule
	Few parameters—one per stock	Forces a front-loaded trade schedule
	Takes very good advantage of diversification and hedging	
	Very accurate model	
Trade Rate	Quickest optimization solution	Does not take full advantage of diversification and hedging
	Adapts to changing market conditions in real-time	Requires approximation of residual risk function
	Few parameters—one per stock	Least accurate of the methods

Risk Aversion. The risk aversion parameters were randomly selected from the following values $\lambda = 0.3, 1, 2$.

Trading Days. We broke the day into thirteen intervals of equal volume.

In total, our simulation experiment took several days to run. The optimizations were run using a 64-bit pc, with an Intel i7 processor, 2.6 GHz, and with 8 Gigs of RAM. Since the actual optimization times are also dependent upon pc, processor, and memory, analysts are encouraged to setup these experiments and analyze solution time and accuracy for the approach described above incorporating the trade list characteristics most common for their fund (e.g., small cap index, global index, growth, value, momentum, one-sided, two-sided equal, 130-30, etc).

How long did it take to solve the portfolio objective problem?

Figure 9.1 plots the log of the average time in seconds for each optimization routine for each portfolio size. As expected, the non-linear optimization routine, which solved for the square root risk term and a decision variable for each stock and each period, was the slowest but did provide us with the exact trade schedule to the problem. The quadratic optimization technique, which provided exact shares to trade in each period but solved for the variance of risk, was the next slowest. This technique, however, provided reasonable solution times for portfolio sizes up to about 100 stocks (analysts need to determine what is considered a reasonable solution time for their needs). A difficulty with the QP approach is that analysts need to determine the proper variance aversion parameter from the investor's specified risk aversion parameter. So this mapping could require several runs of the problem. The trade schedule exponential and residual schedule exponential techniques provided a large improvement in solution time over the quadratic optimizer. The fastest solution was for the trade rate technique. To show the effect of the number of names in the portfolio on solution time the non-linear (NLP) optimizer took 55 minutes to solve a 500-stock portfolio. The quadratic optimizer

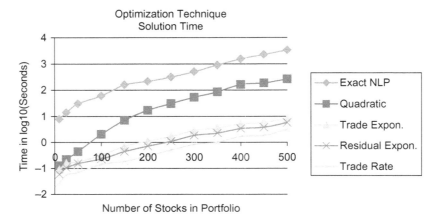

■ **Figure 9.1** Optimization Technique Solution Time.

(QP) provided dramatic improvement over the non-linear approach and only took 4.4 minutes to solve. But for a trader, even 4.4 minutes may be too long especially if they desire to perform re-optimization during the day. The fastest solutions for a 500-stock portfolio were the 14 seconds for the trade schedule exponential solution, 10 seconds for the residual schedule exponential, and only 5 seconds for the trade rate technique. These are all dramatic time improvements over the NLP and QP formulations.

How accurate was the solution for each optimization technique?

Figure 9.2 shows the accuracy of each approach. The quadratic optimizer was 98% accurate. This was followed by the trade schedule exponential 93%, residual schedule exponential 91%, and trade rate technique at 84%. Accuracy was measured as 1 minus the error between the actual trade schedule determined from the NLP solution and the trade schedule determined from each of our approaches.

This simulation experiment highlighted the inverse relationship between solution time and accuracy. The quicker we solve, the less accurate the solution. In some circles this has become known as the "developer's dilemma." Solving too fast may give an inaccurate result, but solving too slow may miss the opportunity altogether.

It appears that the exponential approaches provide the highest level of accuracy and the quickest solution times. Additionally, the exceptionally quick solution of the trade rate technique could be used in conjunction with the exponential approaches or quadratic optimization for full risk management

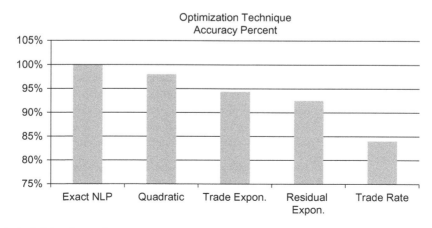

■ **Figure 9.2** Optimization Technique Accuracy Percent.

and quick solution times. Analysts need to determine the time available for the initial optimization as well as real-time re-optimizations (described below) to determine the best approach given time constraints.

PORTFOLIO ADAPTATION TACTICS

The optimization techniques provided above provide investors and algorithms with appropriate initial trading strategies. These trade schedules were determined based on expected market conditions and price movement. Unfortunately, the only thing we are sure about with regards to markets is that actual conditions will not be the same as expected conditions.

To adjust for changing conditions during the day investors can utilize the same adaptation tactics with a portfolio as they can with single stock trading (see Chapter 8). In this section we discuss the AIM and PIM tactics for portfolio trading needs.

The AIM and PIM tactics are:

$$AIM: \quad Max \ \frac{E_t[Cost] - E_0[Cost]}{E_t[Timing \ Risk]} \tag{9.17}$$

$$PIM: \quad Max \ \frac{E_0[Cost] - E_t[Cost]}{E_t[Timing \ Risk]} \tag{9.18}$$

where,

$E_0[Cost] = C^* = original \ estimated \ cost \ (including \ permanet \ impact)$

$E_t[Cost] = expected \ total \ cost \ at \ time \ t \ (includes \ realized \ and \ unrealized)$

$E_t[Timing \ Risk] = expected \ timing \ risk \ at \ time \ t \ (unexecuted \ shares \ only)$

The original cost estimate is determined from the original optimization solution. It includes temporary impact, permanent impact, and price appreciation. Even if the optimization does not include the permanent impact component (for optimization simplification) the permanent impact cost needs to be added into the estimated cost. Permanent impact cost is a true cost to investors but since it will not influence the optimization solution it is often not included in the optimization formulation.

The time expectations for cost and timing risk are computed as follows:

Then our cost equations are:

$$Realized_\$(Cost(t)) = \sum_i Side(i) \cdot Y_i \cdot (P_{ij} - P_{i0}) \tag{9.19}$$

$$Momentum_\$(Cost(t)) = \sum_i Side(i) \cdot (X_i - Y_i) \cdot (P_{it} - P_{i0}) \qquad (9.20)$$

$$E_t(MI) = \sum_{i=1}^{m} \sum_{j=t}^{n} \frac{b_1 \cdot I_i^* \cdot x_{ij}^2}{X_i \cdot v_{ij}} + \sum_{i=1}^{m} \frac{Y_{it}}{Y_{it} + X_i} \cdot (1 - b_1) \cdot I_i^* \qquad (9.21)$$

Notice that our market impact and timing risk equations only incorporate trading activity from the current period through the end of the trading horizon. Additionally, since permanent market impact will be reflected in market prices during trading we need to incorporate permanent impact cost in the re-optimization adjusted for quantity of shares traded. This is shown as the second expression on the right hand side in Equation 9.21.

Therefore we have,

$$E_0[Cost] = C^* \qquad (9.22)$$

$$E_t[Cost] = Realized + Momentum_\$(Cost(t)) + E_t[MI] \qquad (9.23)$$

Description of AIM and PIM for Portfolio Trading

Portfolio adaptation tactics are illustrated in Figure 9.3. In this scenario, the portfolio manager is rebalancing the portfolio and also investing additional cash. This results in a trade list with an initially higher buy value than sell value. Figure 9.3a illustrates how the basket will be traded under expected market conditions. Here the buy order has an initial risk of $150K and the sell order has an initial risk of $100K. The manager optimizes the trade schedule using techniques described above and this results in the buys initially being transacted at a faster rate to offset the incremental risk until the residual position is hedged. Following this optimized trade schedule, the position is traded into the hedged position at 12 p.m. After this time, the buys and sells are transacted at the same trading rate.

Figure 9.3b illustrates how the basket may be traded in a situation with favorable price movement. Suppose that by 10:15 a.m. there was a decline in market prices after the open. This makes buys cheaper but sells more expensive. But since there are more shares to buy than there are to sell investors are better off. A manager employing the AIM tactic can take advantage of the better market prices and trade into the hedged position at a faster rate. Here the manager achieves the hedged position by 11:15 a.m. After this time the portfolio is traded at the more passive rate

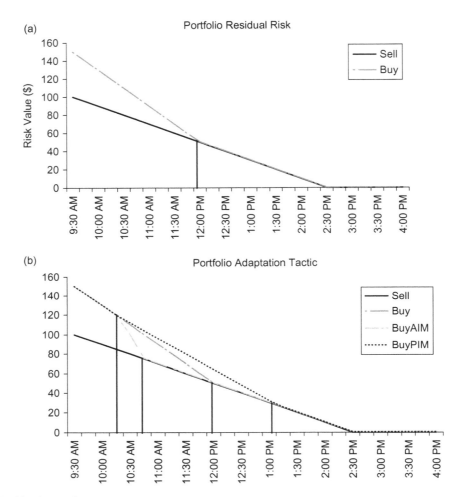

■ **Figure 9.3** Portfolio Adaptation Tactics.

until completion to reduce market impact cost. A manager employing the PIM adaptation tactic will wish to take advantage of the "good" risk and "better" prices and trade at a slower rate. Here the manager does not trade into the hedged position until 1:15 p.m. After this time the shares are traded at a passive rate to reduce market impact cost. In both situations, once the basket achieves its maximum hedged position the AIM and PIM tactics will not have any effect on the trading schedule. But there is usually something that can always be refined during trading.

How Often Should We Re-Optimize?

The next question that often arises is how often should we re-optimize the portfolio? Many of the self-proclaimed industry pundits state it should be done continuously. Others state that re-optimization should be performed at certain intervals, such as every 5, 10, or 15 minutes. We could not disagree more. Portfolio algorithms should be re-optimized if performance is projected to be dramatically different than what is expected, market conditions are different than what we planned, or if there are opportunities to take advantage of liquidity and prices. The difference between single stock and portfolio algorithms, however, is that when trading portfolios investors are interested in the overall portfolio risk and not necessarily the performance of an individual order. For example, if there is sufficient liquidity and favorable prices that would allow the trader to complete the order at great prices it may not be in their best interest to do so if the net result would adversely affect the hedge of the portfolio and increase overall portfolio risk. Portfolio analysis needs to be performed from the risk perspective of the portfolio not from the risk perspective of any individual stock.

Our recommendation for re-optimization criteria is based on the Z-score of projected performance. This is similar to the Z-score re-optimization criteria used for the single stock algorithms. This measure is:

$$Z_t = \frac{E_0(Cost) - E_t(Cost)}{E_t(Timing\ Risk)} \qquad (9.24)$$

A positive score indicates investors are performing better than projected and a negative score indicates investors are performing worse than projected. The Z-score above measures the number of standard deviations away from our original cost estimate we are projected to finish given actual market conditions. Investors could elect to re-optimize the portfolio algorithm if the Z-score at any point in time exceeds a specified range such as $|Z| > \pm 1$ or $|Z| > \pm \frac{1}{2}$. Some investors elect to only re-optimize if the Z-score is less than a specified value. In these cases re-optimization would only occur if performance is expected to be less favorable.

Investors, of course, could also re-optimize if there is opportunity to deviate from an optimally prescribed strategy to improve the overall risk characteristics of the trade list and reduce trading costs. This can also be done on an individual stock basis and is described below.

Investors should also re-optimize and change their strategy if there is reason to believe that their trading intentions have been uncovered by market participants, which would lead to higher trading costs.

MANAGING PORTFOLIO RISK

We have previously discussed adaptation techniques to manage risk from a portfolio optimization perspective following our AIM and PIM methodologies using real-time re-optimization. In this section we discuss three techniques to determine how to evaluate potential deviation tactics for individual stocks.

These are:

1. Minimize Trading Risk
2. Maximize Trading Opportunity
3. Program-Block Decomposition

From the investor's perspective, these techniques provide: improved algorithmic trading rules, better specification of market and limit orders, and appropriate utilization of non-traditional trading venues such as crossing networks and dark pools where liquidity is not transparent. These criteria have been stated in *Optimal Trading Strategies* (2003) and in *Algorithmic Trading Strategies* (2006). We expand on those findings and apply them to today's portfolio trading algorithm needs.

Residual Risk Curve

The total dollar risk in a trade period for a portfolio is:

$$Risk_\$(t) = \sqrt{r_t^t C r_t} \tag{9.25}$$

where r_t is the residual share vector at time t and C is the side adjusted covariance matrix scaled for the length of the trading interval.

The residual risk curve shows how the total portfolio risk will change as we change the number of shares of a particular stock and hold the amount of all other shares constant.

From the first and second derivatives of Equation 9.25 we find that the residual risk curve is a convex function with a single minimum value at:

$$r_{i,min} = \frac{-1}{\sigma_i^2} \sum_{j \neq i} r_j \sigma_{ij} \tag{9.26}$$

This minimum value could be either more or less than the current number of shares of the stock in the portfolio. If the minimum value is less than the current position traders could reduce portfolio risk by trading shares and reducing the holding size. If the minimum value is greater than the current position traders could reduce portfolio risk by adding shares to the portfolio and increasing the holding size. However, if a trader needs to adhere to a shrinking portfolio constraint they can only reduce portfolio risk if this minimum value is less than the current number of shares held in the basket.

Figure 9.4 depicts the scenario where the residual risk curve achieves its minimum value. In this example, an investor with a basket of stock has r_i shares of stock i. The figure shows two interesting trading values. The first is $r_{imin} = r_i - y_i$ and represents the number of shares that can be traded to achieve minimum portfolio risk. The second value is $r_{imax} = r_i - z_i$, which represents the maximum number of shares that can be transacted without adversely affecting risk. That is, the residual risk will be exactly the same after trading the shares as it was before trading the shares. These two values will be referred to as: (1) minimum trading risk quantity, and (2) maximum trading opportunity, respectively.

The minimum trading risk quantity is useful for investors continuously striving to take advantage of favorable liquidity conditions to minimize portfolio risk over time. The maximum trading opportunity is useful for investors striving to reduce trading cost without adversely affecting the overall risk of the trade basket. In both situations investors can accelerate transactions in a stock without adversely affecting risk.

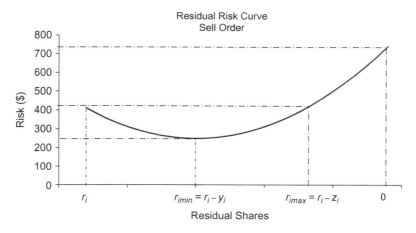

■ **Figure 9.4** Residual Risk Curve for a Sell Order.

Minimum Trading Risk Quantity

The minimum trading risk quantity is calculated as follows.

Let r be the current portfolio and let y_k indicate the number of shares to trade in stock k. Then these vectors are:

$$r = \begin{pmatrix} r_1 \\ \vdots \\ r_k \\ \vdots \\ r_m \end{pmatrix} \quad y = \begin{pmatrix} 0 \\ \vdots \\ y_k \\ \vdots \\ 0 \end{pmatrix} \quad (r-y) = \begin{pmatrix} r_1 \\ \vdots \\ r_k - y_k \\ \vdots \\ r_m \end{pmatrix}$$

where $(r-y)$ is the portfolio after trading y_k shares of stock k. Notice that the y vector only contains one value y_k for the stock that we are looking to trade. Having zeros in the other entries ensures that the other position sizes remain constant. Our goal is to determine the value of y_k that will minimize portfolio risk.

Portfolio risk after trading will be:

$$Risk = \sqrt{(r-y)'C(r-y)}$$

The number of shares to trade that will minimize total portfolio risk is determined by differentiating portfolio risk with respect to y_k, setting this derivative equal to zero, and solving for y_k. Mathematically, this is:

$$\frac{\partial Risk}{\partial y_k} \sqrt{(r-y)'C(r-y)} = 0$$

Solving, we get:

$$y_k = \frac{1}{\sigma_k^2} \sum_{j=1}^{n} r_j \sigma_{ij} \tag{9.27}$$

If an investor needs to adhere to the shrinking portfolio constraint, the feasible trading interval are values between zero and the original position size r_k. Recall that we are using a one-sided portfolio formulation and we have already adjusted the covariance matrix. This constraint is:

$$0 \leq y_k \leq r_k$$

Thus the actual number of shares that can be traded adhering to the shrinking portfolio constraint is:

$$y_k^* = \min(\max(0, y_k), r_k)$$

Maximum Trading Opportunity

Calculation of the maximum trading opportunity is as follows:

Let r be the current portfolio and z_k indicate the number of shares to trade in stock k.

Then these vectors are:

$$r = \begin{pmatrix} r_1 \\ \vdots \\ r_k \\ \vdots \\ r_m \end{pmatrix} \quad z = \begin{pmatrix} 0 \\ \vdots \\ z_k \\ \vdots \\ 0 \end{pmatrix} \quad (r-z) = \begin{pmatrix} r_1 \\ \vdots \\ r_k - z_k \\ \vdots \\ r_m \end{pmatrix}$$

where $(r-z)$ is the portfolio after trading z_k shares of stock k. Notice that the z vector only contains one value z_k for the stock that we are look-ing to trade. Having zeros in the other entries ensures that the other posi-tion sizes remain constant.

Our goal here is to determine the maximum number of shares z_k that can be traded such that the risk after trading is equal to the risk before trading.

Mathematically, this is as follows:

$$\sqrt{r'Cr} = \sqrt{(r-z)'C(r-z)}$$

Squaring both sides yields:

$$r'Cr = (r-z)'C(r-z)$$

Expanding this equation yields:

$$z'Cz - 2r'Cz = 0$$

Solving for z_k yields two solutions as is expected since this is a quadratic equation. These solutions are:

$$z_k = 0 \tag{9.28}$$

$$z_k = 2 \cdot \frac{1}{\sigma_k^2} \sum_{j=1}^{n} r_j \sigma_{ij} \tag{9.29}$$

The first solution (Equation 9.28) is the naïve solution and implies that we do not trade. If there are no transactions then of course the risk does not change. The second solution (Equation 9.29) is the value that is most interesting to traders. It signifies the most trades that can occur without adversely affecting portfolio risk. Also notice that this solution is twice the value of the minimum trade risk quantity, which makes sense since the residual risk curve is symmetric around the minimum value.

If traders need to adhere to the shrinking portfolio constraint the bounds on z_k are:

$$0 \le z_k \le r_k$$

Thus the actual number of shares that can be traded adhering to the shrinking portfolio constraint is:

$$z_k^* = \min(\max(0, z_k), r_k)$$

Figure 9.5 depicts the residual risk curve for stock II. At the initial portfolio position there are $r_2 = 7000$ shares of stock II and the total portfolio risk is \$2965. Total risk can be minimized by trading $y_2 = 2583$ shares resulting in total risk of \$2,932 and a new position size for stock II of 4417 shares. Traders can transact up to $z_2 = 5167$ shares, resulting in a new position size of stock II of 1833 shares, and still have the same overall risk of \$2965 as we had prior to trading. Trading more than 5167 shares would result in higher risk exposure.

When to Use These Values?

The minimum trading risk and maximum trading opportunity quantities provide valuable guidelines for how much an algorithm or trader can

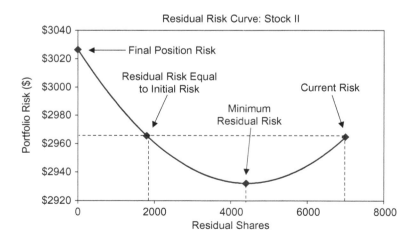

■ **Figure 9.5** Residual Risk Curve Stock II.

deviate from an optimally prescribed schedule without adversely affecting performance. The recommendations are to accelerate trading up to the maximum trading opportunity in times of favorable market prices and liquidity and to decelerate trading to the minimum risk quantity in times of high impact costs and lower liquidity.

For example, whenever faced with favorable prices algorithmic trading rules can be specified to take advantage of the displayed liquidity up to the maximum trading opportunity. Algorithmic trading rules can also be specified to enter and display limit orders up to the maximum trading opportunity. In times of illiquidity, high market impact, or short-term price drift (with expected trend reversal) algorithmic trading rules can be written to decelerate trading down to the minimum trading risk quantity.

Program-Block Decomposition

When investors enter orders into crossing venues or dark pools the executions are not guaranteed. Transactions will occur only if there is a counterparty. When entering baskets into dark pools traders are often concerned that only some of their orders will trade and the resulting residual risk will be more than the original value.

One way investors can address this problem is to decompose the basket into block and program subsets. The block subset represents those shares contributing incremental risk to the basket. These are the shares that can be entered into a dark pool and would result in less risk no matter how many shares are executed. The program subset represents those shares that are providing risk reduction through either diversification or hedging. Accelerated trading of any of these shares is not recommended because if the executions are not in the proper proportions the resulting residual risk will be higher than the starting level of risk.

For example, stock A and stock B are perfectly correlated. If we are buying $150K of A and selling $100K of B we have $50K worth of incremental risk from stock A. Thus, we could enter $50K shares of A into a dark pool without worrying about our residual risk increasing. No matter how many shares of A trade in the dark pool the resulting portfolio risk will be lower than the original value. Now suppose that we are buying $100K of stock A and selling $100K of stock B (the same value in both stocks). Since these stocks are perfectly correlated our market exposure is hedged and the total portfolio risk is equal to the stock's idiosyncratic risk values. We want to trade these names together to minimize risk and maintain our hedged position. If we enter both stocks into a dark pool but are only able to transact one of the names then the resulting residual risk will increase. These shares should be transacted as a pair to maintain risk and to minimize market impact cost. In the latter scenario it would not be advisable to submit these orders into a dark pool for execution.

The general technique to determine our program-block decomposition is through min-max optimization. That is, we seek to minimize the maximum residual risk position. This is determined as follows:

Let,

$R = (r_1, \cdots, r_k, \cdots, r_m)'$ represent the current trade portfolio

$M = \sum_{i=1}^{m} r_i$ represents the total number of shares in the portfolio

$Y = (y_1, \cdots, y_k, \cdots, y_m)'$ represents the block subset

$R - Y = (r_1 - y_1, \cdots, r_k - y_k, \cdots, r_m - y_m)'$ represents the program subset

$C =$ one-sided covariance matrix scaled for a trading period and expressed in ($/share)^2

Next, let,

$Z = (z_1, \cdots, z_k, \cdots, z_m)'$ represent the untraded shares from the block subset after submission to the dark pool. That is, Y is entered into a dark pool where some trades occur. Z represents those shares that did not transact in the dark pool. Then,

$R - Y + Z = (r_1 - y_1 + z_1, \cdots, r_k - y_k + z_k, \cdots, r_m - y_m + z_m)'$ represents the residual portfolio after submission to the dark pool.

Then, the resulting total residual risk is:

$$Risk = \sqrt{(R - Y + Z)'C(R - Y + Z)} \tag{9.30}$$

with $0 \leq z_k \leq y_k \leq r_k$ to ensure that there is no overtrading.

Mathematically, program-block decomposition can be formulated as a min-max optimization problem where we minimize the worse-case scenario. For a given number of shares S, where $S = y_1 + \cdots + y_m$, the block subset is determined as follows:

$$\underset{x}{Min}\,\underset{y}{Max} \quad (R - Y + Z)'C(R - Y + Z) \tag{9.31}$$

Subject to:

$0 \leq z_k \leq y_k \leq r_k$	*for all k*
$S = y_1 + y_2 + \cdots + y_m$	*for all k*

Notice in this case we make use of variance rather than standard deviation (square root) enabling the formulation of a quadratic optimization problem.

Figure 9.6 illustrates the program-block decomposition process. The graph shows the maximum and minimum residual risk that could arise for the corresponding number of shares that are entered into the dark pool or crossing network. The x-axis shows the number of shares from zero to M (the total number of shares). The graph shows three data series. The horizontal line is the initial risk of the portfolio. This is the residual risk that would arise if nothing is traded in the dark pool. The minimum residual risk line shows the best-case scenario for the corresponding number of shares. This is the residual risk if all shares are traded in the dark pool. This is a decreasing value. The maximum residual risk line shows the worse-case scenario that would arise if only some of the shares are entered into the dark pool. This is an increasing value. For example, suppose that an investor enters the entire hedged two-sided basket into a dark pool. If all shares are executed than the residual risk will be zero (since there are not any shares remaining). But if only one side of the portfolio is executed (such as all buy orders) the residual portfolio will have fewer shares but risk will be higher because the investor is no longer hedged to market movement.

In performing this exercise there will always be some number of shares S_1 such that the worse-case scenario will be equal to the initial risk value. This quantity S_1 is a "free block order" or "free crossing order" and represents the number of shares that can be entered into the dark pool

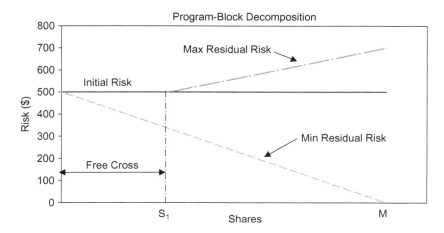

■ **Figure 9.6** Program Block Decomposition.

and ensure the residual risk will never increase, regardless of the number of shares that may or may not transact. The free block order resolves anxieties traders may have about adverse selection. For example, if an investor is buying $150K of stock RLK and selling $100K of stock LMK (with correlation = 1) the investor has market exposure of $50K due to the higher value in RLK. Hence, the investor could enter $50K of RLK into a dark pool and will always be better off if any shares are executed. This $50K represents the "free block order" or "free block cross." This number of shares should always be entered into a dark pool to reduce market risk.

APPENDIX

The matrices for the QP technique in Equation 9.7 are calculated as follows:

$$\tilde{r}_{m \cdot (n+1) \times 1} = \begin{pmatrix} \begin{bmatrix} r_{11} \\ \vdots \\ r_{1,n+1} \end{bmatrix} \\ \begin{bmatrix} r_{21} \\ \vdots \\ r_{2,n+1} \end{bmatrix} \\ \vdots \\ \begin{bmatrix} r_{m1} \\ \vdots \\ r_{m,n+1} \end{bmatrix} \end{pmatrix}$$

$$Q_{m \cdot (n+1) \times m \cdot (n+1)} = \tilde{M}_{m \cdot (n+1) \times m \cdot (n+1)} + 2\lambda^* \cdot \tilde{C}_{m \cdot (n+1) \times m \cdot (n+1)},$$

where,

$$\tilde{M} = \begin{pmatrix} M_1 & 0 & 0 & \cdots & 0 \\ 0 & M_2 & 0 & \cdots & 0 \\ 0 & 0 & \ddots & & \vdots \\ \vdots & \vdots & & M_{m-1} & 0 \\ 0 & 0 & & 0 & M_m \end{pmatrix}$$

$$M_i = \begin{pmatrix} I_i' & -I_i' & 0 & \cdots & 0 \\ -I_i' & 2I_i' & 0 & \cdots & 0 \\ 0 & 0 & \ddots & & \vdots \\ \vdots & \vdots & & 2I_i' & -I_i' \\ 0 & 0 & & -I_i' & I_i' \end{pmatrix}$$

$$I_i = \frac{n \cdot b_1 I_i}{X_i V_i}$$

$$\tilde{C} = \begin{pmatrix} C_{11} & C_{12} & \cdots & C_{1m} \\ C_{12} & C_{22} & \cdots & C_{2m} \\ \vdots & \vdots & \ddots & \vdots \\ C_{1m} & C_{2m} & \cdots & C_{mm} \end{pmatrix} \quad C_{ij} = \begin{pmatrix} \sigma_{ij} & 0 & \cdots & 0 \\ 0 & \sigma_{ij} & \cdots & 0 \\ \vdots & \vdots & \ddots & \vdots \\ 0 & 0 & \cdots & \sigma_{ij} \end{pmatrix}$$

Please note that $\tilde{C}$ is a symmetric matrix, therefore we have, $C_{ij} = C_{ji}$ for all i and j.

The equality constraint matrix $\tilde{A}_1$ and vector $\tilde{b}_1$ are:

$$\tilde{A}_1 = \begin{pmatrix} \begin{bmatrix} 1 & 0 & \cdots & 0 \\ 0 & 0 & \cdots & 0 \\ \vdots & \vdots & \ddots & \vdots \\ 0 & 0 & \cdots & 0 \\ 0 & 0 & \cdots & 1 \\ 0 & 0 & \cdots & 0 \\ \vdots & \vdots & \ddots & \vdots \\ 0 & 0 & \cdots & 0 \end{bmatrix} & \begin{bmatrix} 0 & 0 & \cdots & 0 \\ 1 & 0 & \cdots & 0 \\ \vdots & \vdots & \ddots & \vdots \\ 0 & 0 & \cdots & 0 \\ 0 & 0 & \cdots & 0 \\ 0 & 0 & \cdots & 1 \\ \vdots & \vdots & \ddots & \vdots \\ 0 & 0 & \cdots & 0 \end{bmatrix} & \cdots & \begin{bmatrix} 0 & 0 & \cdots & 0 \\ 0 & 0 & \cdots & 0 \\ \vdots & \vdots & \ddots & \vdots \\ 1 & 0 & \cdots & 0 \\ 0 & 0 & \cdots & 0 \\ 0 & 0 & \cdots & 0 \\ \vdots & \vdots & \ddots & \vdots \\ 0 & 0 & \cdots & 1 \end{bmatrix} \end{pmatrix} \quad \tilde{b}_1 = \begin{pmatrix} S_1 \\ S_2 \\ \vdots \\ S_m \\ 0 \\ 0 \\ \vdots \\ 0 \end{pmatrix}$$

The inequality constraint matrix $\tilde{A}_2$ and vector $\tilde{b}_2$ are:

$$\tilde{A}_2 = \begin{pmatrix} \begin{bmatrix} A_2' & 0 & \cdots & 0 \\ 0 & A_2' & \cdots & 0 \\ \vdots & \vdots & \ddots & \vdots \\ 0 & 0 & \cdots & A_2' \\ A_2'' & 0 & \cdots & 0 \\ 0 & A_2'' & \cdots & 0 \\ \vdots & \vdots & \ddots & \vdots \\ 0 & 0 & \cdots & A_2'' \end{bmatrix} \end{pmatrix} \quad \begin{aligned} A_2' &= \begin{pmatrix} -1 & 1 & 0 & \cdots & 0 \\ 0 & -1 & 1 & \cdots & 0 \\ \vdots & \vdots & \ddots & \ddots & \vdots \\ 0 & 0 & \cdots & -1 & 1 \end{pmatrix} \\ A''_2 &= \begin{pmatrix} 1 & -1 & 0 & \cdots & 0 \\ 0 & 1 & -1 & \cdots & 0 \\ \vdots & \vdots & \ddots & \ddots & \vdots \\ 0 & 0 & \cdots & 1 & -1 \end{pmatrix} \end{aligned}$$

$$\tilde{b}_2 = \begin{pmatrix} \begin{bmatrix} b_1' \\ b_2' \\ \vdots \\ b_m' \\ b_1'' \\ b_2'' \\ \vdots \\ b_m'' \end{bmatrix} \end{pmatrix} \quad \begin{aligned} b_i' &= \begin{bmatrix} -x_i^* \\ -x_i^* \\ \vdots \\ -x_i^* \end{bmatrix} \\ b''_i &= \begin{bmatrix} 0 \\ 0 \\ \vdots \\ 0 \end{bmatrix} \end{aligned}$$

And x_i^* is the maximum quantity that can be traded in any period for stock i defined by the trader.

Portfolio Construction

INTRODUCTION

This chapter introduces techniques to bridge the gap between portfolio construction and trading. We introduce a quantitative framework to determine the appropriate "optimal" execution strategy given the "optimal" portfolio on the efficient investment frontier.

Portfolio optimization is the process of determining an optimum mix of financial instruments. These consist of portfolios with the highest return for a specified level risk and the least risk for a specified return. These optimal portfolios are determined through advanced mathematical modeling approaches such as quadratic programming, and more recently conic optimization.

Markowitz (1952) presented a quantitative process to construct efficient portfolios through optimization. The set of all efficient portfolios constitutes what Markowitz coined the Efficient Frontier. Sharpe (1964) expanded on the efficient frontier concept by providing investors with a means to determine the most appropriate efficient portfolio on the frontier. The technique used by Sharpe was based on maximizing investor economic utility (investor happiness). Sharpe further introduced the industry to the capital asset pricing model (CAPM) which in the simplest forms is a technique to combine the market portfolio with a risk-free asset to further improve the set of risk-return above the efficient frontier. CAPM also provided the industry with metrics to quantify and manage risk, allocate investment dollars, etc. This ground breaking work by Markowitz and Sharpe paved the way for Roll and Ross (1980) with arbitrage pricing theory, Black and Litterman (1992) with alternative portfolio optimization techniques, Fama and French (1992, 1993) with their three factor model, and Michaud and Michaud (1998) with their portfolio resampling using Monte Carlo methods. Unfortunately, not as much attention has been given to portfolio construction with transaction costs. But as we show in this chapter, Markowitz and Sharpe have also paved the way to determine the most appropriate best execution strategy given the investment objectives of the fund.

The Science of Algorithmic Trading and Portfolio Management. DOI: http://dx.doi.org/10.1016/B978-0-12-401689-7.00010-6

The underlying goal of this chapter is to provide the necessary theory and mathematical framework to properly incorporate transaction costs into the portfolio optimization. This chapter builds on the findings from "Investing and Trading Consistency: Does VWAP Compromise the Stock Selection Process?" Kissell and Malamut in Journal of Trading, Fall, 2007. We expand on those concepts by providing the necessary mathematical models and quantitative framework. The process is reinforced with graphical illustrations[1].

PORTFOLIO OPTIMIZATION AND CONSTRAINTS

Many quantitative portfolio managers construct their preferred investment portfolios following the techniques introduced by Markowitz (1952) and Sharpe (1964). But often during the optimization process these managers will incorporate certain constraints into the process. These constraints are used by managers for many different reasons and in many different ways. For example, to reflect certain views or needs, to provide an additional layer of safety, or to ensure the results provide more realistic expectations. Some of the more common reasons for incorporating constraints into the portfolio optimization are:

- Fund Mandates
- Maximum Number of Names
- Reflect Future Views
- Risk Management
- Transaction Cost Management

Fund Mandates. Many funds have specified guidelines for their portfolios. Optimization constraints are used to ensure the resulting optimal mix adheres to these strategies. For example, these may specify a predetermined asset allocation process that requires certain percentages to be invested across stocks, bonds, cash, etc. These mandates may also define a predetermined max exposure to a risk factor or a sector. And some index funds may not be allowed to have their tracking error to a benchmark exceed a certain level regardless of alpha expectations.

---

[1]We would like to thank the following people for helpful comments, suggestions, insight, and especially for their helpful criticism and direction throughout several iterations of this chapter. Without their greatly appreciated help and insight, these advanced optimization techniques would not have been possible. They are: Jon Anderson, John Carillo, Sebastian Ceria, Curt Engler, Morton Glantz, Marc Gresack, Kingsley Jones, Roberto Malamut, Pierre Miasnikoff, Eldar Nigmatullin, Bojan Petrovich, Mike Rodgers, and Peter Tannenbaum.

Maximum Number of Names. Some managers will limit the number of names in the portfolio so that they can better manage the portfolio. These portfolio managers usually employ a combination of quantitative and qualitative portfolio management. These managers perform a quantitative portfolio optimization limiting the number of names to hold in the portfolio, e.g., only hold, say, fifty or fewer names at most. The portfolio manager will follow these limited number of companies in fine detail and make changes if a company has fallen out of favor or if their expectation on potential company growth or dividend stream has changed. It is much more difficult to perform fundamental analysis on a portfolio of several hundred names than on a portfolio with only fifty or fewer names.

Reflect Future Views. Managers may specify minimum weighting in a group of stocks or in a sector if they feel this group is likely to outperform the market of their benchmark index. Many times the managers may not have specific stock level alphas or stock specific views, but will apply a higher weighting to the group as a whole. Additionally, managers may specify a maximum weighting for a group of stocks if they believe a particular group will underperform the market and they do not have a view on any particular stock.

Risk Management. Portfolio managers may at times be suspicious of the estimated portfolio risk, stock volatility, or covariance across names from a particular risk model. In this case, mangers are mostly concerned about type II error—that is, the potential for the risk model to present false positive relationships. A desired property of optimizers is that the results will exploit beneficial relationships. But if these relationships are false positive relationships, the solution will actually increase, rather than decrease, portfolio risk. Thus, as means to provide an added level of safety surrounding potential false positive relationships, managers may specify maximum position sizes, maximum levels of risk exposure, or a maximum stock specific weight (e.g., hold no more than 5 or 10% of the total portfolio value in a specific stock). This constraint is intended to protect the fund from potential errors in the input data.

Manage Transaction Costs. The effect of transaction costs and their drag on performance can often be detrimental to fund performance. The larger the position size, the higher the transaction cost. Many times we observe the liquidation cost (selling the order) is much more expensive than the acquisition cost (buying the order). Managers are more likely to buy stocks in favorable market conditions and sell stocks when they have fallen out of favor, when volatility has spiked, and when liquidity dries up. Thus, the liquidation cost is often much more costly than the purchasing cost. As a means to protect the fund from these higher liquidation costs, managers may place a maximum level on the position size, such as no more than 10% of ADV.

The effect of transaction costs on portfolio returns and their drag on overall performance has been well documented. For example, Loeb (1983) found that block trading could result in an additional 1−2% of cost for large cap stocks and as much as 17−25% or more for small cap illiquid stocks. Wagner and Edwards (1993) found that implementation of trading decisions could approach almost 3% of the trade value in times of adverse market movement. Chan and Lakonishok (1995) found that the hidden trading cost components due to market impact and opportunity cost could amount to more than 1.5% of trade value. Grinold and Kahn (2000) examined the effect of transaction costs on portfolio construction. Kissell and Malamut (2006) found that inefficient executions, i.e., implementing via strategies or algorithms that are not consistent with the investment objective, can increase tracking error by 10−25 bp for passive index managers and by as much as 50 bp for actively managed funds. And more recently, studies have found that transaction costs may still account for additional slippage of up to 1% of annual performance. With such high trading costs associated with implementation it is no wonder that portfolio managers underperform their benchmarks (Treynor, 1981).

In addition to the high transaction costs and corresponding trading friction, there is often an additional drag on portfolio returns due to a misalignment between the investment objective and trading desk goals. For example, suppose a value manager enters a buy order for a stock that is undervalued in the market. This manager wants to execute the position in an aggressive manner before the market discovers the mispricing and makes a correction. If this order is executed by a trader via a full-day VWAP strategy it is very likely that the market correction will occur before the order is complete causing the manager to pay higher prices or potentially not complete the order fully, which results in high opportunity cost. In either case, the manager does not achieve the full potential of the opportunity because of the trading strategy not because of the investment decision. In this case it would be much more advantageous to trade via a more aggressive strategy, such as an arrival price or implementation shortfall algorithm, in order to transact more shares at the manager's decision price. VWAP strategy in this example is not an appropriate strategy and is misaligned with the investment objective of the fund. Even if the trader achieved or outperformed the VWAP price in this example the selection of the VWAP strategy would be an inappropriate decision.

Consider another situation where the portfolio manager constructs an optimal portfolio on the efficient investment frontier (EIF) and the decision is implemented using an optimal strategy on the efficient trading frontier (ETF). Many would argue that, since both the portfolio and underlying trading strategy are optimal, the fund is positioned to achieve its maximum performance, and hence, the trading decision is best execution. But this is not necessarily true. Suppose the trader executes the trade list via a passive full-day VWAP strategy. If there is adverse price movement over the day the manager would realize less favorable prices and incur a higher trading cost. If there is favorable price movement over the day the manager would realize better prices and a lower trading cost. Regardless of the actual prices incurred and resulting trading cost, this fund may have been exposed to unnecessary incremental market exposure. And in Markowitz's terminology, this results in lower investor utility. The same situation would hold true for a trader who executes more aggressively than necessary. In this case the fund will incur an unnecessarily high trading cost and again lower investor utility. It is imperative that both investor objectives and trading goals be aligned in order for investors to achieve the targeted level of investor utility.

Portfolio managers and traders are often at odds with each other regarding what constitutes best execution and how a portfolio decision should be implemented. Managers often wish to use the benchmark price that was used in their optimization process. Traders often seek to achieve the price that is being used to measure their performance such as the VWAP price. This results in an inconsistency between the investment objectives and trading goals, and often leads to suboptimal portfolios and lower levels of investor utility. Investors and traders need to partner across all phases of the investment cycle to capture maximum levels of return.

The true magnitude of underperformance is probably understated in the industry even after accounting for market impact and opportunity cost. This is primarily due to the inconsistency across portfolio manager objectives and trader goals. This inconsistency often leads to higher cost and/ or higher risk and ultimately lower ex-post investor utility. While alpha decay and transaction costs are often discussed in the literature, the reduction in investor utility is seldom if ever discussed, and this is even more difficult to observe than market impact.

TRANSACTION COSTS IN PORTFOLIO OPTIMIZATION

Transaction costs as part of the portfolio optimization process are not a new concept. There have been many attempts to account for these costs during stock selection. A brief history of these approaches is described below.

First Wave: The first wave of portfolio optimization with transaction costs focused on incorporating the bid-ask spread cost into the optimization process. The belief was that, since trading costs are generally lower for large cap liquid stocks and trading costs are generally higher for small cap illiquid stocks, spreads would be a good proxy for costs since spreads are generally lower for large cap stocks and higher for small cap stocks. By decreasing expected returns by the round-trip spread cost, managers felt that the optimizer would determine a more appropriate mix of stock and a more accurate expected return. The optimized solution would apply larger weights to stocks with smaller spreads and lower weights to stocks with higher spreads. While this process was a good first step in the process it still did not account for the possibility that a large number of shares of a large cap liquid stock could, in fact, be more expensive than a small number of shares of a small cap illiquid stock. The first wave of portfolio optimization with transaction costs did not account for the cost associated with the size of the order.

Second Wave: The second wave of portfolio optimization with transaction costs focused on incorporating a market impact estimate that was dependent upon order size. These types of models have been previously formulated by Balduzzi and Lynch (1999) and by Lobo, Fazel, and Boyd (2006). In this process, larger orders will have higher market impact cost than smaller orders in the same names. The expectation is that the optimization process would determine sizes that could be easily absorbed into the market without incurring inappropriate levels of impact and the resulting optimal solution would provide a more efficient allocation of dollars across the different stocks. In this approach, however, the market impact formula used was based on a "static" cost-size relationship and does not provide any cost reduction benefits from the underlying execution strategy.

This means that estimated cost will be exactly the same for the number of shares transacted regardless of whether those shares were to be transacted with a high level of urgency or passively through the day. Furthermore, the optimization process, even though it considered the risk term to determine the optimal mix of stocks and portfolio weightings, does not consider the risk composition of the other names in the trade list to determine corresponding impact cost. For example, suppose the trade list consists of only a single buy order for 500,000 shares of RLK. The manager is exposed to both market risk and company specific (idiosyncratic) risk from the order and will more often trade in an aggressive manner and incur higher cost. Next suppose that the trade list consists of the same 500,000 share buy order for RLK and an additional 100 buy orders. Here the manager achieves diversification of company specific risk due to the large number of names in the trade basket and hence is only exposed primarily to market risk. The manager could trade this list at a more moderate rate since there is less risk exposure (only one source of risk as opposed to two sources of risk). This results in a lower market impact cost for the 500,000 shares of RLK. Finally, suppose the manager performs a rebalance of the portfolio and the trade list is comprised of the 500,000 buy order for RLK plus an additional 100 stocks to buy and an additional 100 stocks to sell (with equal dollars across both the buy list and sell list). This trade list will now achieve risk reduction from diversification of company specific risk just like it did in the previous scenario and will also achieve market risk reduction from having a two-sided portfolio (due to the buy and sell orders). Now RLK can be traded in a passive manner and the corresponding trading cost will be even lower.

The second wave of portfolio optimizers did not take into account the corresponding trading cost resulting from the actual implementation strategy—which as we demonstrated above can vary dramatically. The second wave of optimizers would assess the same exact cost to the order regardless of the other names in the trade list and regardless of how those shares would be transacted. Aggressive, moderate, and passive strategies would all be assumed to incur the same trading cost. Finally, the second wave of portfolio optimizers did not provide managers or traders with any insight at all into how the targeted portfolio should be best implemented. It is left to the managers and traders to determine. But as we have discussed above, the goals of these parties are often conflicting.

Third Wave: The third wave of portfolio optimization with transaction costs consists of incorporating a market impact function that is dependent upon the size of the order, the overall risk composition of the trade list, and the underlying trade schedule. This portfolio optimization problem has been studied by Engle and Ferstenberg (2006, 2007) and Kissell and Malamut (2006). The advantages of this type of optimization are (1) it will properly account for the trading cost based on the underlying trading strategy, and (2) it will provide as output from the process the exact trading schedule to achieve the targeted portfolio so there will be perfect alignment between portfolio manager and trader. For example, these optimization processes will have different costs for the 500,000 share order of RLK (from above) based on the composition of the optimal trade list. This will also result in a more appropriate allocation of dollars across assets and across stocks. The portfolio will be more efficient. Furthermore, since the byproduct of the optimization will include both the new targeted portfolio and the underlying instructions and trading schedule to achieve that portfolio there will be no ambiguities between the investment objectives and trading goals. Traders will be provided with the underlying execution strategy (directly from the optimizer) to be used to transact those shares. Portfolio manager and trader goals will finally be aligned!

In the remainder of the chapter, we provide the necessary background and quantitative framework to assist portfolio managers and traders properly align investment objectives and trading goals. This results in a single best execution trading strategy for the specific investment decision. We expand the Markowitz efficient trading frontier to include transaction costs and show there are various cost-adjusted frontiers but only one efficient trading frontier. The chapter concludes with an introduction to the necessary mathematics and optimization process to develop multi-period portfolio optimizers incorporating transaction cost analysis.

Some of the highlights of the chapter include:

- Unification of the investment and trading decisions resulting in consistency across all phases of the investment cycle and providing a true best execution process.
- There exist multiple cost-adjusted efficient investment frontiers but a single optimal trading strategy.
- The Sharpe ratio determines the appropriate level of risk aversion for trade schedule optimization.
- Evidence that a naïve VWAP strategy is often an inefficient execution strategy because it may lead to lower levels of investor utility and suboptimal ex-post portfolios.
- Evidence that a passive VWAP strategy and an aggressive execution strategy may result in identical levels of investor utility.
- Portfolio optimization framework that properly incorporates market impact and timing risk estimates. This leads to an improved best execution frontier and optimal ex-post portfolios.
- An approach to determine whether a suboptimal Markowitz portfolio exists resulting in a more efficient and pareto optimal portfolio after trading costs. For example, it may be possible for an ex-ante suboptimal portfolio to have higher risk-return characteristics (and be more optimal) than the originally optimal portfolio ex-post.

PORTFOLIO MANAGEMENT PROCESS

Quantitative portfolio managers pride themselves on making rational investment decisions, constructing efficient investment portfolios, and maximizing investor utility. In fact, this is the central theme of modern portfolio theory (MPT). Portfolio managers following this course of action will seek to maximize return for a specified quantity of risk (variance).

This optimization is formulated as follows:

$$
\begin{aligned}
Max \quad & w'r \\
s.t. \quad & w'Cw \le \sigma_p^{2*} \\
& \sum w = 1
\end{aligned}
\tag{10.1}
$$

where w is the vector of weights, r is the vector of expected returns, C is the covariance matrix, and σ_p^{2*} is the targeted or maximum level of risk.

This optimization can also be formulated as the dual of Equation 10.1 where the goal is to minimize risk for a targeted return r^*. This optimization is formulated as:

$$
\begin{aligned}
Min \quad & w'Cw \\
s.t. \quad & w'r \le r^* \\
& \sum w = 1
\end{aligned}
\tag{10.2}
$$

The set of all solutions to the portfolio optimization problem above results in the set of all efficient portfolios and comprises the efficient frontier. This is the set of all portfolios with highest return for a given level of risk or the lowest risk for a specified return. Proceeding, to avoid confusion, we use the term efficient investment frontier (EIF) to denote the set of optimal investment portfolios (Markowitz and Sharpe) and the efficient trading frontier (ETF) to denote the set of optimal trading strategies (Almgren and Chriss).

The set of all optimal portfolios can also be found using Lagrange multipliers as follows:

$$
\begin{aligned}
Max \quad & w'r - \lambda \cdot w'Cw \\
s.t. \quad & \sum w = 1
\end{aligned}
\tag{10.3}
$$

where λ denotes the investor's risk appetite (e.g., level of risk aversion). Solving for all values of $\lambda \ge 0$ provides us with the set of all efficient portfolios.

Example: Efficient Trading Frontier w/ and w/o Short Positions

In this example, we construct the efficient frontier using the techniques above for a subset of 100 large cap stocks from the SP500 index. We construct these portfolios with and without a short sales constraint. That is, one portfolio is a long-only portfolio where all weights have to be positive, and the other portfolio contains both long and short positions and the weights could be positive or negative. This is shown in Figure 10.1. Notice how the efficient frontier with short sales allowed provides higher returns for the same risk than the no short sales case. This is because investors can use short positions to better manage risk in the portfolio. Additionally, this example demonstrates how the usage of constraints (in this example no short sales allowed) may result in reduced portfolio performance.

Example: Maximizing Investor Utility

In this example, we show how investors determine their investment portfolio based on their utility preferences. Utility preferences, expressed in terms of indifference curves, are the set of all return-risk portfolios that provide the investor with equal quantities of "happiness." Investors, therefore, are indifferent to which portfolio on the indifference curve they actually own since all of these portfolios provide the same quantity of economic utility. Investors, of course, will always prefer higher returns for the same level of risk than lower returns. Thus, the goal of the investor is always to be on the highest indifference curve possible.

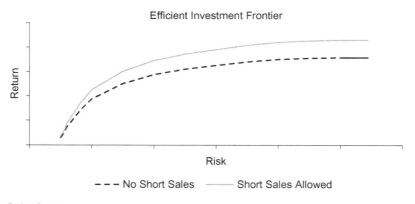

■ **Figure 10.1** Efficient Trading Frontier.

In Figure 10.2 we show the efficient frontier with three different utility curves. These indifference curves are ordered such that $U_1 > U_2 > U_3$. Therefore, investors will prefer portfolios on U_1 to portfolios on U_2, and will prefer portfolios on U_2 to portfolios on U_3. Notice that these utility curves are shaped in such a way that investors will only accept more risk if they receive higher returns. For utility curve U_3 investors are equally happy with either portfolio A2 or A3 since they are on the same indifference curve. But investors prefer portfolio A1 to either A2 or A3 since utility curve U_2 is higher than utility curve U_3 ($U_2 > U_3$). Unfortunately, utility curve U_1 does not intersect with the efficient trading frontier and does not contain any optimal portfolios—it is an unattainable level. The best that investors can achieve is A1 on curve U_2.

This utility maximization proves to be an invaluable exercise for not only determining the preferred optimal portfolio, but also for determining the optimal trade schedule to achieve that optimal portfolio. We make further use of utility optimization below.

TRADING DECISION PROCESS

Once the optimal portfolio has been constructed traders are tasked with determining the appropriate implementation plan to acquire that new position. As discussed through the text, when implementing these decisions investors encounter the by now all too well known trader's dilemma—trading too quickly results in too much market impact cost but trading too slowly results in too much timing risk.

To determine the best way to implement the portfolio manager's decision, Almgren and Chriss (1999, 2000) provided a framework similar

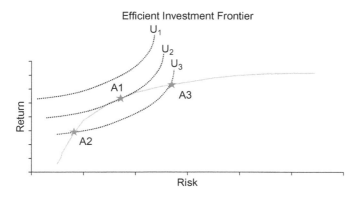

■ **Figure 10.2** Maximizing Investor Utility.

to Markowitz (1952) to solve this trader's dilemma by balancing the trade-off between market impact cost (*MI*) and timing risk (*TR*). Mathematically, these optimal trading strategies are computed as follows:

$$Min \quad MI(x) + \lambda \cdot TR(x) \tag{10.4}$$

where x denotes the optimal trading schedule (e.g., how shares are to be transacted over the trading horizon) and λ denotes the investor's level of risk aversion. The appropriate formulation of the market impact function and portfolio optimization techniques have been the focus of earlier chapters.

If we solve Equation 10.4 for all values of $\lambda \geq 0$ we obtain the set of all optimal strategies. When plotted, these strategies constitute the Almgren and Chriss efficient trading frontier (ETF). This is illustrated in Figure 10.3. In this example, we have highlighted three different strategies. Strategy x denotes a moderately paced trade schedule with market impact 25 bp and timing risk 50 bp, strategy z is an aggressive strategy with high market impact 200 bp but low timing risk 25 bp, and strategy y is a passive strategy, e.g., VWAP, with low market impact 10 bp but much higher timing risk 300 bp. The efficient trading frontier is illustrated in Figure 10.3.

What is the appropriate optimal strategy to use?

There has been quite a bit of research and industry debate focusing on how to determine the appropriate optimal strategy. Bertsimas and Lo (1998) propose minimizing the combination of market impact and price appreciation (price drift) without regards to corresponding trading risk.

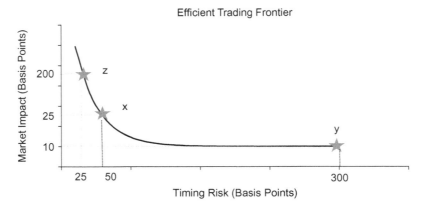

■ Figure 10.3 Efficient Trading Frontier.

Investors in this case need to specify their alpha component. In a situation where there is not any directional view of natural price appreciation, the underlying strategy is a VWAP that will minimize market impact cost. Almgren and Chriss (1999, 2000) propose two solutions. First, balance the trade-off between market impact and timing risk at the investor's level of risk aversion. Second, minimize the value-at-risk at the investor's alpha level (e.g., 95%). Kissell, Glantz, and Malamut (2004) provide a macro-level decision making framework (also, see Chapter 8) to determine the most appropriate strategy based on the investment objective of the fund.

In order to determine the most appropriate trade schedule for the trade list we do need further information regarding the underlying investment objective. An investor who has uncovered a market mispricing may choose to execute more aggressively and take advantage of the temporary market inefficiency. A manager performing a portfolio rebalance may elect to trade via a strategy that best manages the cost and risk trade-off. And an index manager who is purchasing shares and quantities in order to replicate the underlying benchmark index may not have any momentum expectations and may trade passively over the day following a VWAP strategy to minimize impact.

Fundamental and active managers who do not construct portfolios based on mean-variance optimization will often achieve better performance utilizing pre-trade analysis and following trade schedule optimization techniques. And portfolio managers who do utilize mean-variance optimization can achieve even better results by combining the investment and trading decisions.

UNIFYING THE INVESTMENT AND TRADING THEORIES

In this section we provide techniques to bridge the gap between the investment and trading theories. We follow the approach outlined by Engle and Ferstenberg (2007) and Kissell and Malamut (2007) below.

Let us first start by re-examining our optimal trading strategies, this time from the context of portfolio theory.

A portfolio manager constructs the efficient investment frontier (EIF) utilizing quadratic optimization (Equation 10.3) and then determines their preferred optimal portfolio utilizing investor utility maximization following Sharpe (Figure 10.2).

Suppose that this preferred portfolio has expected return $u^* = 10\%$ and risk $\sigma^* = 20\%$. The trader then performs trade schedule optimization (Equation 10.4) and constructs the efficient trading frontier (ETF) using values of $0 \leq \lambda \leq 10$ to determine the best way to implement the portfolio.

Rather than analyze our trade schedules in the traditional cost-risk space, let us examine our trading cost consequences by overlaying the efficient trading frontier on the efficient investment frontier (Figure 10.4). Notice that the efficient trading frontier is now inverted from its more traditional appearance and shows the cost-adjusted potential risk-return profile for optimal portfolio A1. The efficient portfolio A^1 is no longer associated with a single expected return and risk. There are multiple sets of potential return and risk depending upon the trading strategy.

The adjusted return for the portfolio will be reduced by the estimated impact cost. That is:

$$Adjusted\ Return = Portfolio\ Return - Strategy\ (Cost) \qquad (10.5)$$

The new timing risk, however, will increase due to the market exposure incurred while acquiring the position. The actual increase in risk is additive in variance (risk value squared). We add the one day timing risk to the annualized portfolio risk value. This is calculated as follows:

$$Adjusted\ Risk = \sqrt{(Portfolio\ Risk)^2 + (Strategy\ Risk)^2} \qquad (10.6)$$

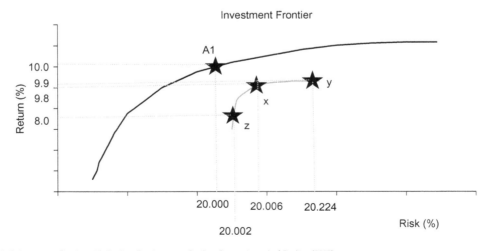

■ **Figure 10.4** Investment Frontier with Trading Frontier as an Overlay. *Source: Journal of Trading (2007).*

Let us now examine three potential trading strategies represented by *x*, *y*, and *z*. The expected market impact and corresponding timing risk for these strategies and their consequence on the overall risk-return profile for optimal portfolio A1 is shown in Table 10.1.

Notice the reduction in return is equal to the trading cost corresponding to the strategy but the increase in risk is actually less than the timing risk of the strategy. This is because risk is sub-additive (the variance expression is additive making the square root of this term less than additive). For example, to implement portfolio A1 using strategy *y* returns are expected to decline by 0.10% (10 bp) and risk will increase by 0.224% (22.4 bp) due to the corresponding timing risk of the transaction. Notice that the overall risk consequence from the strategy is much less than the risk incurred on the day of 3% (300 bp) due to the sub-additive nature of risk. Variance is additive and risk is sub-additive (Equation 10.6). On the other extreme, let us evaluate implementation via strategy *z*. Here return will be reduced by 2% due to the market impact of strategy *z* and the increase in portfolio risk will be negligible at 0.002% or 0.2 bp. Notice that for the investment portfolio the underlying market impact cost of the strategy has a much more dramatic effect on the ex-ante portfolio than the timing risk of the strategy. This is an important observation for the portfolio manager when devising the appropriate strategy to execute the trade.

Therefore, even a strategy with a large quantity of timing risk will have a much smaller effect on overall portfolio risk. But a strategy with a large quantity of market impact cost will have a large effect on overall portfolio returns.

Following the example in Table 10.1, the best the manager can expect to do after trading costs is to realize an ex-post portfolio return from 8 to 9.90%, with corresponding portfolio risk of 20.002 to 20.224%,

Table 10.1 Cost-Adjusted Risk-Return Values

	Trading Costs		Adjusted Risk-Return	
Scenario	Impact	Timing Risk	Return	Risk
Portfolio A1			10%	20%
Strategy *y*	0.10%	3.00%	9.90%	20.224%
Strategy *x*	0.25%	0.50%	9.75%	20.006%
Strategy *z*	2.00%	0.25%	8.00%	20.002%

respectively. Notice the effect of trading costs on portfolio performance. This cost dominates the effect of the timing risk of the strategy.

Important Note: The investor will not be able to achieve the expected portfolio return even via a very passive strategy such as VWAP because even a very passive VWAP strategy will incur permanent impact cost. The portfolio risk calculation is sub-additive whereas portfolio variance is an additive relationship. This results in trading costs primarily due to market impact cost having a much more dramatic effect on portfolio performance than corresponding timing risk of the strategy.

Which execution strategy should the trader use?

As stated above, some investors may wish to trade passively, such as with strategy y, to minimize market impact cost and some may wish to trade aggressively, such as with strategy z, to minimize timing risk or possibly lock in a market mispricing or realized profit. Further, there are other investors who prefer a strategy somewhere in the middle, such as strategy x. But which strategy is most appropriate? Since each of these strategies lie on the efficient trading frontier can they all be considered a best execution strategy?

The answer is no. There is only a single best execution strategy. (This answer may surprise some readers.)

Our conclusion is described following the same investor utility maximization that was used to determine the preferred optimal portfolio A1 on the efficient trading frontier.

Combining the efficient investment and efficient trading frontiers onto one chart provides investors with the ability to determine the proper optimal strategy for a specified portfolio. To show this, first recall that investors determine their preferred optimal portfolio through maximizing their utility function. But now let us maximize investor utility for both frontiers. This is illustrated in Figure 10.5.

The portfolio manager selected portfolio A1 as the preferred strategy because it was the portfolio that maximized the investor's utility function (shown as U_2 in Figure 10.2). Now let us apply the same technique used to determine portfolio A1 to determine the optimal trading strategy. First, utility U_4 passes through two strategies on the efficient trading frontier: aggressive strategy z and passive strategy y. Since both of these strategies lie on the same indifference curve they provide the investor with equal utility. This may be surprising to many. This analysis shows that two seemingly opposite strategies (aggressive and passive) can have the same

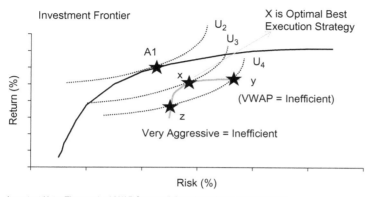

Important Note: The passive VWAP Strategy "y" and Very Aggressive Strategy "z" have identical investor utility and are both are inefficient strategies!

■ **Figure 10.5** Using Indifference Curves to Determine the Most Appropriate Execution Strategy. *Source: Journal of Trading (2007).*

effect on investor utility and thus the investor is indifferent as to whether they trade aggressively or passively.

But investors can achieve higher results. Indifference curve U_3 is higher than U_4 and so provides the investor with a higher level of utility. Additionally, curve U_3 intersects with a single trading strategy x. Therefore, strategy x is the single strategy that maximizes investor utility and represents the single best execution strategy. It is also the strategy that is most consistent with the investment objective.

There are two additional insights that need to be highlighted. First, the VWAP strategy represented by strategy y is not necessarily consistent with the investment object because it results in a lower level of investor utility. The VWAP strategy corresponds to U_4 which is below the optimal utility U_3 corresponding to strategy x. Investors wishing to hold portfolio A1 and trade via VWAP are not aligning their trading decisions with their investment goals. Investor utility is not being maximized to its fullest extent.

Second, the VWAP strategy is also associated with the same level of investor utility as an aggressive strategy denoted as z on the ETF. Since both strategies lie on the same investor utility curve U_4 they are providing the same level of happiness and investor utility. Here we have what appears to be two conflicting strategies but with the same level of utility, which means that investors are completely indifferent as to which strategy they use to acquire portfolio A1. But neither strategy is the preferred strategy since they are not associated with the highest value of utility.

The optimization process used here is the same process utilized by investors to determine their optimal preferred efficient portfolio. In this depiction, the maximum level of achievable utility is U_3 and corresponds to trade strategy x. Therefore, investors with preferred portfolio A1 need to implement their investment decision utilizing strategy x to ensure consistency between investment and trading goals. Notice that in the case of a VWAP trading strategy y investors incur too much risk, resulting in a lower utility than associated with strategy x. Also, there additionally corresponds an aggressive strategy z with an equivalent level of utility as associated with the VWAP strategy. The last important point here is that there is a single "optimal" trading strategy corresponding to each efficient portfolio.

The importance of this representation in Figure 10.5 is that it clearly illustrates there is a "single" optimal trading strategy consistent with the underlying investment portfolio. This is the strategy that maximizes investor utility.

COST-ADJUSTED FRONTIER

The cost-adjusted frontier is the efficient investment frontier after adjusting for trading costs (e.g., the ex-post frontier). An example of the derivation of the cost-adjusted frontier is as follows:

First, start with three efficient portfolios on the efficient investment frontier. These portfolios represent the Markowitz (ex-ante) efficient portfolios.

Second, perform trade schedule optimization for each portfolio. This will result in the set of all optimal trading strategies for each of the portfolios. It provides three different efficient trading frontiers.

Third, overlay the efficient trading frontier for each portfolio onto the efficient investment frontier. These adjusted portfolios portray the set of risk-return profiles for each of the efficient portfolios after adjusting for trading costs.

The cost-adjusted frontier is then the highest envelope of all cost-adjusted portfolios. This process is illustrated in Figure 10.6 and shows multiple cost-adjusted frontiers. For each portfolio on the frontier there is a corresponding ETF. We can draw the cost-adjusted frontier as the curve through all corresponding points of the same strategy. For example, the VWAP cost-adjusted frontier is drawn by connecting all VWAP strategies on the ETF, and the same process is carried out for the aggressive

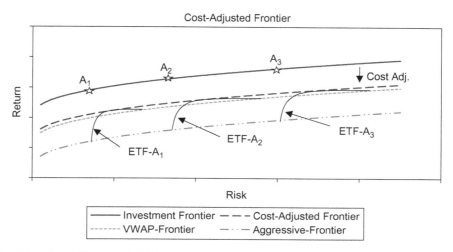

■ **Figure 10.6** Cost Adjusted Frontier. *Source: Journal of Trading (2007).*

and normal strategies. Then, the best the investor can achieve is the envelope of the highest cost-adjusted points. This frontier is now referred to as the cost-adjusted frontier.

The VWAP frontier shows the expected risk-return profile for the optimal portfolios if VWAP was used to implement the decision. The aggressive frontier shows the expected risk-return profile for the optimal portfolios if an aggressive strategy was used to implement the decision. The optimal cost-adjusted frontier is the upper envelope of all the cost-adjusted portfolios.

It is interesting to point out here that in our example, the VWAP strategy is an inefficient ex-post frontier because the VWAP frontier lies below the cost-adjusted frontier and it is associated with a lower level of investor utility.

Another interesting aspect is that the VWAP frontier is equivalent to a cost-adjusted frontier constructed from an aggressive strategy (aggressive frontier). Notice that the VWAP frontier passes through the most passive strategies on the efficient trading frontier as well as a more aggressive strategy on the efficient trading frontier. Neither the passive VWAP frontier nor the aggressive frontier is a preferred strategy since they do not maximize investor utility. Therefore, execution via a VWAP or overly aggressive strategy leads to decreased utility. To maximize utility it is essential that the underlying trading strategy not incur too much cost (primarily market impact) or too much risk.

Investors will always seek out the highest efficient investment portfolio. Engle and Ferstenberg (2006, 2007) provide an alternative discussion of the cost-adjusted frontier. In their article, the authors present a framework to incorporate transaction costs directly into the investment process in order to determine a more efficient ex-post portfolio.

DETERMINING THE APPROPRIATE LEVEL OF RISK AVERSION

Suppose a manager constructs a portfolio by maximizing investor utility and then submits the list to the trader for execution. In most situations the trader does not have sufficient time or tools to perform a detailed cost analysis to determine the appropriate cost-adjusted frontier and corresponding execution strategy. However, the trader usually does have sufficient time to perform a single trade cost optimization as defined in Equation 10.4. But how should the trader specify the level of risk aversion to ensure the trading decision is consistent with the investment decision?

A joint examination of the efficient investment frontier and the cost-adjusted frontier provides some insight into our question (Figure 10.7). In the figure A represents the selected optimal portfolio and X represents the single best execution strategy. The question now shifts to finding this strategy. If we assume that all investors are indeed rational investors then the tangent to the efficient investment frontier at the optimal portfolio A is equal to the Sharpe ratio S of the portfolio (Sharpe, 1966),

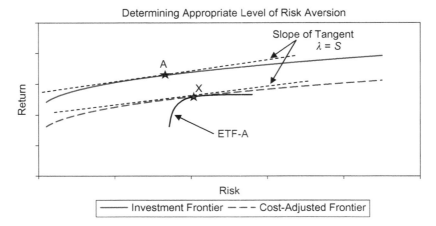

■ **Figure 10.7** Determining the Appropriate Level of Risk Aversion. *Source: Journal of Trading (2007).*

e.g., $S = dReturn/dRisk$. The corresponding level of risk aversion on the efficient trading frontier at the best execution strategy X is $\lambda = dCost/dRisk$. This is also equal to the slope of the tangent at the point of the intersection of the cost-adjusted frontier and the ETF overlay. Notice that the slopes of the two tangents are approximately equal. Therefore, the corresponding level of risk aversion to ensure consistency across investment and trading decisions can be determined from the Sharpe ratio of the trade, e.g., $\lambda \cong S$. While this may not be the exact value it does at least ensure a large amount of consistency between investment and trading decisions. And it provides the trader with an appropriate input into the trade schedule optimization process which is extremely useful at times when they do not have enough time to perform a detailed analysis.

$$S = dReturn/dRisk \cong dCost/dRisk = \lambda$$
$$S \cong \lambda$$

(10.7)

BEST EXECUTION FRONTIER

The next step in the portfolio construction process is for managers to consider the possibility that there may be a suboptimal Markowitzian portfolio, but after adjusting for trading costs and trading risk, this portfolio may in fact be pareto efficient with improved risk-return characteristics over the set of optimal Markowitzian portfolios. For example, is it possible that a portfolio that does not lie on the theoretical Markowitz efficient investment frontier, but after accounting for variable market impact cost and timing risk, the resulting cost-adjusted risk-return profile lies above the cost-adjusted frontier?

This problem is illustrated in Figure 10.8. First, consider the three efficient portfolios on the efficient investment frontier $(A_1, A_2, \text{and } A_3)$. After accounting for trading costs, we arrive at the corresponding portfolios $(x_1, x_2, \text{and } x_3)$. The set of all cost-adjusted optimal portfolios results in the cost-adjusted frontier (described above). Next, consider the possibility that there exist a set of suboptimal portfolios $(B_1, B_2, \text{and } B_3)$. These are portfolios that do not initially lie on the efficient investment frontier but due to more favorable trading statistics (e.g., higher liquidity, lower impact sensitivity, lower volatility, etc.) they result in cost-adjusted portfolios $(y_1, y_2, \text{and } y_3)$ with higher risk-return characteristics. The resulting cost-adjusted portfolios correspond to a higher level of investor utility.

This frontier is defined as the best execution frontier. If these portfolios do in fact exist, investors would greatly benefit by investing in portfolios

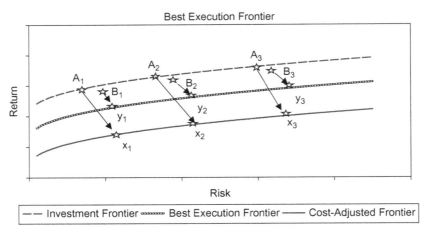

■ Figure 10.8 Best Execution Frontier. *Source: Journal of Trading (2007).*

that appear suboptimal prior to trading but after trading they result in portfolios that achieve a higher risk-return trade-off and investor utility than starting with the optimal portfolio. Notice that the utility associated with the best execution frontier is higher than the utility associated with the cost-adjusted frontier.

While it is not guaranteed that a suboptimal portfolio will always result in a higher cost-adjusted frontier, it is entirely possible. The best execution frontier can only be uncovered via incorporation of trading costs and risks directly in the portfolio optimization decision. For many portfolio managers, the quest for the best execution frontier has become the next generation of portfolio research. This is exactly what Wayne Wagner was referring to when he defined best execution as the process of maximizing the investment idea.

PORTFOLIO CONSTRUCTION WITH TRANSACTION COSTS

The integration of transaction costs into the investment decision process has been previously addressed in the academic literature. For example, Leland (1996) studied the appropriate time to rebalance a portfolio consisting of stocks and bonds in the presence of transaction costs. Michaud (1998) introduces a portfolio optimization technique based on Monte Carlo methods to construct optimal portfolios in the presence of risk and return uncertainty. Ginold and Kahn (2000) examined various techniques to incorporate transaction costs into the investment decision. Tutuncu and Koenig (2003) address the optimal asset allocation problem under the scenario where the estimated returns are unreliable. Balduzzi and Lynch (1999) study a multi-period optimization problem in the presence of costs that are either fixed or proportional to trade value. Malamut and Kissell (2002) study efficient implementation of a multi-period trade cost optimization from the perspective of the trader. Mitchell and Braun (2004) also consider portfolio rebalancing in the presence of convex transaction costs where costs are dependent solely on the quantity of shares transacted. Engle and Ferstenberg (2007) discuss a cost-adjusted frontier.

Most of these research works fall into what we coined the second wave of portfolio optimizers earlier in the chapter. In this section, we introduce the necessary techniques to solve the portfolio optimization problem (third wave of optimizers) and determine the best execution frontier. We differentiate from the above works in many ways (see Kissell & Malamut, 2007). For example:

1. We examine the portfolio optimization in terms of both market impact cost and trading risk.
2. We define market impact to be dependent upon the size of the order and the underlying execution strategy. In this case, investors have the opportunity to achieve further cost reduction through trading a diversified and/or well-hedged portfolio. Thus, depending upon the underlying trade list, the number of shares in an order could vary.
3. Our solution is based on a multi-period optimization problem that separates the total investment horizon into a trading period where shares are transacted and a holding period starting after the acquisition of the targeted portfolio and where no other shares are transacted. These problems are linked by the total shares to trade S and the trade schedule used to acquire those shares, that is, $S = \sum x$.
4. Our ultimate goal in this section is the quest for the best execution frontier.

An interesting aspect of the current portfolio construction environment is that the process appears to be backwards. For example, the results of the current portfolio optimizer provide us with a targeted future portfolio. The next question is to determine how we should best get to that end point. But what if the road is too bumpy or if no efficient road exists? Then what? We would only find this out after setting out on our journey.

A better process is a forward looking view of portfolio construction. Rather than start with the future targeted portfolio and work our way backwards through the unknown as we do now, we begin to move forward from the current portfolio and buy and sell shares efficiently until we arrive at the optimal end point. Proceeding in this manner will ensure that we only take an efficient implementation path or only a difficult path if the end return more than offsets the incremental cost. The end product in this case is determined directly from the trade schedule and ensures the most frictionless path was taken.

Quest for best execution frontier

The quest for the best execution frontier is centered on proper integration of trading costs into the portfolio optimization. The trick is to incorporate a variable market impact function dependent upon the number of shares transacted (size), volatility, trade strategy, and the overall risk composition of the trade list (covariance) to take advantage of potential diversification and hedging opportunities.

For consistency of notation and uniformity across trading horizon and holding periods we express decisions in terms of dollars and shares to trade rather than the traditional investment units of weights and returns. This is also more important because it is the dollar value and shares traded that affect market impact cost and not the weight of the stock in our portfolio.

Using these new units, the original portfolio construction optimization problem (Equation 10.3) can be written in terms of a cash investment and shares to trade as follows:

$$\underset{S}{Max} \quad S'(P_t - P_0) + \lambda \cdot S'CS$$
$$s.t. \quad S'P_0 = V_\$ \tag{10.8}$$

where S is the vector of shares to hold in the portfolio (decision variable), P_t is the vector of expected prices at time t, P_0 is the vector of current prices, and C is the covariance matrix expressed in ($/shares)^2.

To properly incorporate trading costs into Equation 10.8 it is necessary to introduce a new decision variable x_k to denote how shares are to be transacted over time (e.g., the underlying trading strategy). This solution is best accomplished via a multi-period optimization formulation that considers both a trading horizon where investors acquire shares from $t = 1$ to $t = n$, and an investment or holding horizon where no other shares are transacted from $t = n$ to $t = T$.

Let us now consider the effect of trading costs on portfolio return and risk in terms of the multi-period context.

Return

Expected stock return is the difference between expected future price and expected average execution price multiplied by the number of shares in the position. Here, the expected average execution price needs to incorporate market impact cost.

For a general market impact function in the absence of natural price appreciation during the trading horizon, the expected execution value for a stock can be computed as follows:

$$S_i \overline{P}_i = \sum_{j=1}^{n} x_{ij}(P_{i0} + MI(x_{ij}))$$

$$S_i = \sum_{j=1}^{n} x_{ij} \tag{10.9}$$

where P_{i0} is the current price, x_{ij} is the number of shares of stock i to trade in period j, and $MI(x_{ij})$ is the market impact cost expressed in \$/share for transacting x_{ij} shares, S_i is the total number of shares, $S_i, x_{ij}, MI(x_{ij}) > 0$ for buys or long positions, and $S_i, x_{ij}, MI(x_{ij}) < 0$ for sells or short positions. Notice that this representation of the market impact cost $MI(x_{ij})$ does indeed provide costs that are dependent upon the underlying trading strategy x.

The total expected dollar return for the stock is:

$$\begin{aligned} \mu_i &= S_i P_{it} - S_i \overline{P}_i \\ &= S_i P_{it} - \sum_{j=1}^{n} x_{ij}(P_{i0} + MI(x_{ij})) \\ &= S_i P_{it} - S_i P_{i0} - \sum_{j=1}^{n} x_{ij} MI(x_{ij}) \end{aligned} \tag{10.10}$$

For an m-stock portfolio, the total expected dollar return accounting for market impact is:

$$\mu_p = \sum_{i=1}^{m} \left(S_i P_{it} - S_i P_{i0} - \sum_{j=1}^{n} x_{ij} MI(x_{ij}) \right) \tag{10.11}$$

Further insight and formulation of market impact models customized for the portfolio construction process can be found at www.KissellResearch.com.

Risk

The risk (variance) of a portfolio over a specified trading period is determined from the number of shares held in the portfolio and the corresponding covariance matrix. The total portfolio risk for either a held portfolio or a portfolio that is changing over time is computed by summing the portfolio variance over each period. The total risk borne by a portfolio manager over a multi-period horizon is thus determined as follows:

Let r_k be the vector of shares held in the portfolio at time k, e.g.,

$$r_k = \begin{pmatrix} r_{1k} \\ r_{2k} \\ \vdots \\ r_{mk} \end{pmatrix} \tag{10.12}$$

where r_{ij} is the number of shares of stock i held in the portfolio at the beginning of period j. That is,

$$r_{ij} = \sum_{k=1}^{j} x_{ik} \tag{10.13}$$

Notice that this is the reverse notation used for residual shares in the trade schedule optimization. Then the total risk borne by the portfolio manager over the entire T-period horizon is:

$$\sigma_p^2 = \underbrace{r_1' C^* r_1 + \cdots + r_n' C^* r_n}_{\text{Trading Horizon}} + \underbrace{r_{n+1}' C^* r_{n+1} + \cdots + r_t' C^* r_t}_{\text{Holding Period}} \tag{10.14}$$

where C^* is the covariance matrix expressed in $(\$/\text{shares})^2$ scaled to the length of the trading period. For example, if each trading interval is fifteen minutes and C is the annualized covariance matrix, we have $C^* = \dfrac{1}{250} \cdot \dfrac{1}{26} \cdot C$ since there are approximately 250 trading days in a year and 26 fifteen minute intervals in a day. Now, since there are no additional transactions in the portfolio after the end of the trading horizon (e.g., $k = n + 1, \ldots, t$) we have $r_{in+1} = r_{in+2} = \cdots = r_{it} = S_i$ for all stocks. In compressed form, Equation 10.14 is written as:

$$\sigma_p^2 = \underbrace{\sum_{j=1}^{n} r_j' C^* r_j}_{\text{Trading Horizon}} + \underbrace{(t-n) S' C^* S}_{\text{Holding Period}} \tag{10.15}$$

since there are n trading periods and $(t - n)$ periods where the portfolio is held and unchanged.

The full investment optimization incorporating market impact and timing risk can now be expressed as follows:

$$\underset{x}{Max} \quad \sum_{i=1}^{m} \left(S_i P_{it} - S_i P_{i0} - \sum_{j=1}^{n} x_{ij} MI(x_{ij}) \right) - \lambda \cdot \left(\sum_{j=1}^{n} r_j' C^* r_j + (t-n) S' C^* S \right)$$

$$\tag{10.16}$$

Subject to:

i. $S_i P_{i0} + \sum_{j=1}^{n} x_{ij}(P_{i0} + MI(x_{ij})) = V_\$$

ii. $S_i = \sum_{j=1}^{n} x_{ij}$

iii. $r_{ij} = \sum_{k=1}^{j} x_{ij}$

iv. $x_{ij} \geq 0$

The new objective function in Equation 10.16 correctly incorporates a variable trading cost function along with estimated return and portfolio risk. The decision variables are S and x_{ij} although S is computed from x_{ij}. We distinguish between these two variables for notation purposes only—the only true decision variable for the optimizer is x_{ij}. The important decision variable for portfolio managers is the number of shares to hold in the portfolio S, while the important decision variable for traders is how those shares need to be transacted over time.

The first constraint above ensures that the entire cash value $V_\$$ will be invested into the portfolio and follows from the definition of expected transaction value in Equation 10.9. The second constraint defines the number of shares that will be held in the portfolio. The third constraint defines the cumulative number of shares transacted in period j. The fourth constraint is an optional constraint and can be specified for a cash investment only $x_{ij} \geq 0$, liquidation only $x_{ij} \leq 0$, no constraint to incorporate both buys and sells.

Following Equation 10.16 the new portfolio optimization problem can be separated into the traditional investment and trading horizons as follows[2]:

$$\underset{x}{Max} \left(\sum_{i=1}^{m}(S_i P_{it} - S_i P_{i0}) - \lambda \cdot (t-n)S'C^*S \right) - \left(\sum_{i=1}^{m}\sum_{j=1}^{n} x_{ij} MI(x_{ij}) + \lambda r_j' C^* r_j \right)$$

(10.17)

This new formulation brings four interesting aspects to light. They are:

1. Risk aversion λ is the same for both portfolio manager and trader.
2. Trader's dilemma is not dependent upon any benchmark price.
3. Portfolio optimization with trading costs requires a multi-period process.
4. The portfolio manager's and trader's decisions are not separable—they are linked by $S = \sum x$.

---

[2]This separation of the portfolio optimization problem into corresponding trading and investment horizons was first presented publicly in December, 2003 expanding on the work of Kissell and Malamut presented in *Optimal Trading Strategies*.

These are explained as follows:

First, as shown in Equation 10.16 the complete portfolio optimization is only based on a single risk aversion parameter. This ensures consistency across the investment and trading decisions. A consequence of this formulation is that any trading strategy derived using a risk aversion parameter that is different than that used during portfolio construction will result in lower investor utility since it would not correctly quantify trading risk with investment risk. This is most notable for a VWAP strategy where risk aversion is set to be $\lambda = 0$ and results in higher risk and lower utility.

Second, notice that the expression in the trading horizon section of Equation 10.17 is not dependent upon any benchmark price. Indirectly it is based on the current price since we are starting with the current portfolio value. Thus, any post-trade analysis based solely on a specified benchmark price or computed as the difference between average execution price and some benchmark is not the ideal approach to evaluate a trader's performance or skill because it does not consider the underlying goal of the manager or trader. Since the newly formulated portfolio optimization is now based on an expected market impact cost and corresponding trading risk estimates, subsequent post-trade performance attributions need to incorporate these values in order to be able to provide any meaningful benefits. It is, however, essential that post-trade analysis be performed to assess the accuracy of the market impact and trading risk estimates to ensure appropriate future investment decisions. Furthermore, with the advent of algorithmic trading, algorithms based on achieving a specified benchmark price rather than a specified cost will surely hinder overall portfolio performance. For example, a VWAP strategy will likely increase risk exposure and reduce overall utility.

Third, portfolio construction with trading costs needs to be formulated as a multi-period optimization problem. This requires both a trading period that will accommodate a market impact estimate based on size, volatility, and composition of the trade list (e.g., diversified market impact effect), and an investment holding period where there will not be additional changes to the portfolio. As shown above, to achieve the maximum benefit it is essential that the market impact cost be based on both size and strategy. Thus allowing managers to implement their decisions in an appropriate manner—aggressive, passive, or normal—depending upon the risk composition of the trade list.

Fourth, the portfolio manager's and trader's decisions are not separable. A decision making framework that first maximizes the risk-return profile in the investment problem then minimizes trading costs is not guaranteed to maximize the entire objective function. This type of decision making timeline is also exactly opposite to what happens in practice where shares are transacted first in order to arrive at the optimal portfolio. The process formulated in Equation 10.17 is based on transacting shares in the market first to be consistent with practice. And addressing these issues in reverse order—e.g., determine first shares S and then trade schedule x—is only part of the whole picture and is likely to make overall performance even worse.

CONCLUSION

In this chapter we presented a process to unify the investment and trading theories. We presented a framework that overlays the efficient trading frontier (ETF) onto the efficient investment frontier to determine a set of cost-adjusted frontiers. The analysis showed that while a traditional Almgren-Chriss trade cost optimization will result in numerous efficient strategies, there is only a single "optimal" execution strategy consistent with the underlying investment objective.

The analysis also shows that a traditional VWAP strategy is not consistent with the investment objective and may compromise the portfolio manager's stock selection ability by resulting in lower levels of investor utility. The reason is that the corresponding VWAP frontier is inferior (lies below) to the cost-adjusted frontier. Furthermore, an overly aggressive execution strategy is also an inappropriate strategy because its cost-adjusted frontier lies below the optimal cost-adjusted frontier. To maximize investor utility it is essential that the trading strategy not incur too much impact (aggressive strategy) or too much risk (VWAP strategy). Doing so is likely to result in sub-par performance.

In the last part of this chapter we presented a methodology to incorporate variable trading cost estimates (market impact and timing risk) directly into the investment optimization process. Recent attempts in this arena have been insufficient since resulting cost estimates have only been dependent upon the number of shares transacted not on the overall list composition. Managers could achieve performance improvement by incorporating market impact cost estimates directly into the investment process such that costs will be dependent upon shares transacted and trading strategy, taking advantage of overall risk reduction. This in turn could dramatically reduce the overall cost of the list.

The resulting procedure, however, is a relatively difficult non-linear multi-period optimization problem but recent advancements in optimization routines and computational power allow the required formulation to be solved quickly and efficiently. For example, see Malamut (2002) and www.KissellResearch.com.

The appropriate optimization technique is based on a multi-period process that segments the time horizon into a trading period where shares are transacted and an investment holding period where there are no further changes to the portfolio. With this new multi-period formulation it is possible that a suboptimal Markowitzian portfolio (e.g., below the efficient investment frontier) will result in better performance and higher utility due to more favorable trading statistics (liquidity and volatility). We refer to this set of ex-ante optimal portfolios as the best execution frontier.

To summarize, our main findings in this chapter are:

- There exist multiple sets of cost-adjusted frontiers for every efficient portfolio on the efficient investment frontier.
- There is a single "optimal trading strategy" that is consistent with the investment objective resulting in a single optimal cost-adjusted frontier. This is the best execution strategy for the investment portfolio.
- The proper level of risk aversion for a trade cost optimization to be consistent with the investment objective of the fund is the Sharpe ratio of the portfolio, or the forecasted Sharpe ratio of the investment decision.
- Evidence that a VWAP strategy is seldom consistent with the investment objectives and may lead to a suboptimal portfolio and lower levels of investor utility.
- A formulated multi-period investment portfolio optimization problem that considers both market impact cost and trading risk with the investment decision and leads to the best execution frontier.
- The formulated model provides opportunity to achieve cost reduction for a diversified trade list.
- The formulated model provides the preferred portfolio and corresponding road map (trade strategy) to build into those holdings.
- Market impact dominates ex-post performance much more than timing risk. Market impact results in a direct reduction in cost whereas timing risk is a sub-additive function and does not have the same linear relationship with portfolio risk.
- Post-trade analysis needs to incorporate the estimated costs of the trade (e.g., market impact and trading risk), and not solely rely on a benchmark price.

Quantitative Portfolio Management Techniques

INTRODUCTION

Transaction cost analysis (TCA) has become an important decision-making tool for portfolio managers. It allows managers to uncover hidden opportunities that may otherwise not be as transparent, especially given the vast array of data propagating the marketplace. Portfolio managers who once treated transaction costs as an unavoidable cost of business have turned to TCA as a valuable source of incremental alpha. TCA has finally made it to mainstream portfolio management.

Below are just a few ways that TCA is being incorporated into the stock selection phase of the investment cycle.

Quantitative Overlays. Managers select the universe of stocks for potential inclusion into the portfolio. They then determine a subset of stocks from that universe that meet specified investment criteria such as market cap, price to earnings, book value, forecasted profit, etc. As a final filter, managers further reduce the potential investment list based on the expected trading cost. Stocks that are too expensive to transact are eliminated from potential inclusion into the portfolio.

MI Factor Scores. Incorporates both liquidity and volatility to determine a market impact factor score to rank stocks based on trading cost. The higher the score the more expensive it is to transact the stock. The MI factor score provides an equal and fair comparison across all stocks. We show below that MI factor scores provide a large benefit over other techniques that simply rely solely on liquidity or volatility (such as a maximum % ADV to hold in the portfolio).

Cost Curves. Cost curves provide managers with the expected market impact cost for various share quantities and execution strategies. Share quantities are usually expressed in terms of percentage of ADV and the

The Science of Algorithmic Trading and Portfolio Management. DOI: http://dx.doi.org/10.1016/B978-0-12-401689-7.00011-8

strategies are usually expressed in terms of percentage of market volume (POV rate) or in terms of trading time.

Alpha Capture. Managers determine the expected profit level of an investment idea based on the stocks projected alpha and the corresponding trading cost. This helps to maximize expected ex-post stock return (e.g., returns after incurring trading cost). Managers then select stocks based on ex-post return.

Investment Capacity. Determine how many shares of a stock can be transacted before the trading cost erodes the expected stock return beyond a specified level. For example, suppose that a manager has determined that an investment strategy is expected to achieve an incremental return of 3% over its benchmark. The manager turns to TCA to determine how many shares can be purchased with a trading cost equal to the incremental return of 3%. After this point, the manager is better off investing in their next most attractive investment idea.

Portfolio Optimization. Portfolio optimization techniques are being developed to provide managers with the "optimal" weightings and the underlying transaction strategy to achieve those positions. These optimizations incorporate expected returns, volatility, correlation, and market impact to determine the optimal mix of stock. Market impact cost is determined from the underlying market impact model parameters. And, the resulting execution strategy is the "best execution" strategy that provides exact consistency between investing and trading decisions. The optimization technique will take advantage of any synergies resulting from diversification or market hedging opportunities. Portfolio optimization with transaction cost analysis has become one of leading areas of research for portfolio managers and is discussed in Chapter 10, Portfolio Construction.

Back Testing. Managers use market impact back-testing series to test investment ideas and determine if those ideas will be profitable in different market conditions. All too often, however, managers find a strategy works well in the back-testing environment but once the strategy goes live it does not provide the expected level of return due to implementation costs. Some of the more forward thinking managers have begun incorporating historical trading costs into their back-testing scenarios. The biggest issue we have encountered here is that while there are participants providing historical costs these are based on the actual market structure and actual cost of trading at that specific point in time. This could result in dramatically overstating the true cost of trading (such as in the early to mid-1990s) when stocks were quoted in 1/8ths (well really odd-eights or quarters, see Christie and Stoll 1994). Overstating the true costs in a back-testing environment could have the opposite results and cause managers to eliminate an investment idea on the basis of it being cost prohibitive when in fact its costs may be much lower now given market structure improvements and increased efficiency from competition. It is imperative that any cost index developed for back testing is based on today's market structure, regulations, and competition, and the trading characteristics at the historical point in time (liquidity, volatility, size, etc.). Only then can a manager determine the strategy's true feasibility and the realistic return expectations of the investment idea. We discuss techniques to develop back-testing cost series in Chapter 12, Cost Index and Multi-Asset Trading Costs.

Liquidation Cost. The cost of trading, unfortunately, is not symmetric. The cost to enter (buy) the position is usually less expensive that the cost to exit (sell) the position. This cost, however, is not due to any structural difference between buying and selling stock, but it is rather due to a difference in the underlying investment decision at the time of the stock purchase and stock sale. Managers will buy stocks under the most advantageous market conditions and sell stock under more dire circumstances. For example, managers tend to buy stocks with attractive company fundamentals, low volatility, and at times when there is liquidity. But managers tend to sell stocks when they fall out of favor, when company fundamentals tank, volatility spikes, and liquidity dries up. All of which increase trading cost of favor.

Sensitivity Analysis. Managers are beginning to incorporate their own market views in investment planning phases. Managers are performing sensitivity analysis to better determine trading cost under various scenarios such as increased and decreased volatility scenarios such as were present during the financial crisis of 2008−2009. Portfolio managers who are able to incorporate their views of the market conditions will improve the portfolio construction process, which will result in portfolios that are more consistent with their underlying investment objective.

ARE THE EXISTING MODELS USEFUL ENOUGH FOR PORTFOLIO CONSTRUCTION?

The needs of traders and portfolio managers are very different when it comes to market impact analysis. Traders use market impact models to estimate trading costs, and to evaluate and select trading algorithms. Portfolio managers use market impact models for cost estimates that can be incorporated directly into the stock selection process. Portfolio managers, however, need to be able to run these models independently of brokers and vendors so that these parties will not have any opportunity to reverse engineer the manager's decision making process. Managers also need to be able to perform sensitivity and "what-if" analyses to determine the cost of trading under various market conditions. Managers need to be able to incorporate their own proprietary view of markets including their volatility and liquidity estimates, as well as their proprietary alpha estimates. Finally, managers do not want to be reliant upon what other brokers and vendors feel are appropriate values for the input variables, especially if these views differ from their own.

The current state of market impact models falls well short of the needs of portfolio managers. Broker models are often black box models, and most do not provide managers with sufficient transparency into the approach to allow the managers to evaluate or critique the results. Much of the reason why brokers keep these models so secretive and hidden is that they do not want the investment community to judge their models. To test this point, simply ask the broker salesperson to write the formulation of their market impact model, the definition of their input variables, and the model parameters. Then sit back and observe their responses. And if these parties do provide this information, try to duplicate their results for a few different samples of stocks.

Brokers will usually state numerous reasons why they are unable to provide their model to the client. They often claim that the model needs to be connected to a tic database, that the model is specific for their algorithms, and that the model uses a proprietary approach, or that the data cannot be redistributed. Regardless of the reason stated, investors should be extremely cautious of using any model or approach that is not amply described or transparent. These models have to be analyzed, tested, and verified.

To be fair, there may be some truth to why these models cannot be provided to the client. But it is still likely that the vendor is hesitant to provide the functional form because it may reveal that the model is not nearly as complex or sophisticated as claimed. Keeping the functional form of market impact models hidden from potential users makes it difficult for users to properly evaluate the model.

Suffice to say, that current industry market impact models have not evolved to a level needed by portfolio managers. A summary of these reasons, as stated in Journal of Trading (see Kissell, 2012), is as follows:

Current State of Vendor Market Impact Models

- Vendor models are black box approaches with no transparency. They do not provide the underlying formulas and often do not provide the complete set of input variables and explanatory factors used to estimate costs. And while many of these models do provide accurate pre-trade estimates, their lack of transparency does not allow portfolio managers to perform "what-if" analysis under various scenarios, or incorporate their own market expectations and alpha views into the process.

- Pre-trade cost calculations are often performed on the vendor's server. Managers need to pass their portfolio from their site to the vendor's server to obtain cost estimates. When portfolio holdings, data, or information leaves the manager's site there is always the potential for information leakage allowing the outside party to reverse engineer the manager's decision process. This could be detrimental to the fund's competitive edge.

- Pre-trade impact models incorporate the vendor's market expectations. These systems do not easily allow managers to revise factor expectations. For example, these models do not allow managers to change volatility, average daily volume, or expected liquidity over the trading horizon. If managers have better forecasts of explanatory variables there is no easy way for them to incorporate these values into the pre-trade estimates. And even if vendors make necessary provisions, there is still no way to do so without alerting these vendors of their own proprietary forecasts.

- Portfolio managers are very sensitive to alpha erosion. In other words, how much of their alpha will they capture given trading costs. But since managers are reluctant to pass these alpha estimates to any vendor's system these models are not able to structure strategies to minimize alpha erosion. Furthermore, managers are suspicious of any party providing alpha estimates for free and to a large array of customers.

- Constructing in-house market impact models is resource intensive and time consuming. Firms developing in-house models using their own trade data have the advantage of knowing the full order size, including shares cancelled and the decision price, and they can also incorporate their own proprietary market views into the cost estimate. This allows the market impact model to be customized for the fund's specific investment behavior. But this still does not allow the fund to perform thorough sensitivity analysis for a situation where they want to analyze an order that may be traded differently than they have in the past because they do not have any historical observations. These models could potentially suffer from in-sample bias.

The current approach being used by managers to incorporate TCA estimates into the stock selection process is to utilize systems such as vendor/broker web-sites, APIs, etc. Managers are asked to send their portfolios or potential portfolios to the vendor so that they can perform the analysis. The vendor will then analyze the basket, estimate costs, and send the results back to the manager.

This approach, however, requires that the information be passed from the manager's site to the vendor's server where the data will be computed, and possibly even stored, before being sent back to the manager. Investors who are using the process need to ensure that their data queries are not being saved or stored at the vendor's or broker's site without their prior approval. If the data is stored on the vendor's site it could potentially allow parties outside the manager's firm to reverse engineer the investment decision.

To alleviate this fear, some vendors provide results of their models to managers in the form of cost curves and include a specified universe of stocks with various sizes and execution strategies. PMs can query and filter these data points for the stocks they are interested in analyzing, but this is a very inefficient process and requires an iterative approach to determine optimal solutions.

Portfolio managers continuously state that they are leery of anything that could potentially result in any kind of information leakage or reverse engineering of the investment decision process. And rightly so! It is the stock selection and portfolio construction process that is the true value of the manager. Even if managers use a verified secure FTP or API protocol that is not viewable by the vendor, the process still does not allow investors to incorporate their own proprietary variables into the analysis. For example, they still cannot integrate proprietary volatility estimates or expected liquidity conditions into the model to perform "what-if" analysis. And we have yet to meet a portfolio manager willing to share their proprietary alpha estimates with any outside party for improved pre-trade analysis.

If the vendor or broker will not provide the model to managers what are they to do? Managers could develop their own market impact model using tic data or by incorporating their own inventory of orders and trades to calibrate the model. But this is very often very time consuming, resource intensive, and could suffer from in-sample error if they rely only on their own trade data. An alternative approach is for managers to develop and build their own models but incorporate broker and vendor pre-trade cost estimates to calibrate the model. This will also allow the managers to incorporate their proprietary views of liquidity, volatility, and even their own alpha estimates.

This latter approach is referred to as the pre-trade of pre-trade approach and it has become very popular among portfolio managers. An abstract of the approach was published in Journal of Trading (Kissell, 2011), and subsequently presented at the Northfield Risk Conference (August 2012). We follow the process described in the Journal of Trading below.

PRE-TRADE OF PRE-TRADES

The pre-trade of pre-trade modeling approach consists of using broker-dealer and/or vendor cost estimates as input into the market impact model. Managers then calibrate their preferred market impact model with these cost estimates. This allows managers to focus on stock selection and analysis rather than spending valuable resource time and dollars managing data, corporate actions, and building system infrastructure.

But why can a portfolio manager not request cost estimates for various stocks and trading strategies across different brokers and vendors, and then take the average cost as the estimate rather than calibrating their own model? While this type of approach is being used in the industry, it does have some limitations.

- First, it does not provide managers with the ability to determine how costs will vary by company characteristics such as volatility, market cap, or liquidity states. For example, if volatility in the stock increased what would be its effect on cost?
- Second, the modeling approach used by vendors is still a black box approach and without a functional form managers are not able to integrate trading cost estimates with their proprietary stock selection models.
- Third, managers cannot express their views of market conditions (volatility and liquidity) or incorporate their own proprietary alpha estimates. There is always the potential that the vendor's view of the market conditions will be dramatically different from the view of the portfolio manager. This creates another level of inconsistency between trading and investing.
- Finally, these approaches do not allow managers to perform sensitivity analysis. For example, managers need to be able to investigate the cost of buying stock in the current market environment and be able to investigate the cost of selling stock at a future point in time and under an entirely different set of market conditions.

Estimation Process

The pre-trade of pre-trade cost estimation process is described through the following example. For illustrative purposes, we use two stocks and three brokers. In practice, investors are encouraged to use a much larger sample of stocks with various order sizes and transaction strategies, and solicit cost estimates from more than three brokers.

Analysts can also implement this approach without revealing their hand to their brokers simply by using a financial data provider such as Bloomberg since many brokers have embedded their pre-trade models into these financial systems. Managers can generate a large enough sample of trades through these systems to calibrate the pre-trade of pre-trade model without their brokers becoming any the wiser or learning their true intentions (also see Kyle, 1985).

This approach will also allow analysts to test and critique broker models from various perspectives. Our pre-trade of pre-trade process is as follows:

Step I: Select Preferred Market Impact Model

The first step in the process is to select the preferred market impact model. In this example, we use a simplified version of the I-Star model that does not separate temporary and permanent impact. Analysts are encouraged to experiment with different cost models to determine the formulation that works best for their needs. This model is:

$$MI_{bp} = a_1 \cdot Size^{a_2} \cdot \sigma^{a_3} \cdot POV^{a_4} \tag{11.1}$$

Step II: Solicit Broker-Dealer Cost Estimates

The second step in the process is to collect cost estimates for a universe of stocks from different brokers. Be sure to use a large enough range of stock characteristics, sizes, and trading strategies.

It is important to note that the vendors who are providing the cost estimates for the estimation process are likely to have constructed their models from their own trade data. In these cases, the model may be well suited their client's trading styles but it may not be as accurate for different trading styles. There is the potential that this model would have a very good in-sample fit but may not be as accurate out-of-sample (suffer from in-sample bias). An example of this is as follows. Suppose broker A's clients consist only of traditional long-term buy and hold indexer managers who usually incur low impact cost, and broker B's clients consist only of short-term trading funds who typically transact in a high cost environment due to price momentum. Both broker models may be accurate for their particular client base, but if the model is applied to the other broker's clients the model would provide inaccurate estimates.

Investors also need to ensure that the brokers they solicit cost estimates from have properly corrected for multicollinearity and heteroscedasticity. Analysts are encouraged to explore these correction methods with their brokers and vendors.

Table 11.1 shows cost estimates for two stocks from three brokers. Notice how these estimates can vary across brokers. For example, the cost estimates for trading 1% ADV of RLK using a POV = 20% strategy is 6.0 bp from broker I, 11.4 bp from broker II, and 4.7 bp from broker III. For an order of 30% ADV using a POV = 5% strategy the estimates are 31.0 bp from broker I, 17.4 bp from broker II, and 21.0 bp from broker III. Broker II had the highest cost estimate in the first situation and the lowest cost estimate in the second situation. It is not uncommon for broker cost estimates to vary to this extent.

Step III: Estimate Model Parameters

The third step of the process is to estimate the model parameters. Since vendors will have positive cost estimates we are able to log-transform our model and solve for the parameters using ordinary least squares (OLS). Since OLS regression analysis is well understood, analysts can easily critique and evaluate the performance of these models.

If any broker or vendor provides a negative cost estimate, meaning you will profit on the trade immediately, it may be time to find a different broker!

The log-transformed simplified I-Star model is:

$$\ln(MI_{bp}) = \ln(a_1) + a_2 \cdot \ln(Size) + a_3 \cdot \ln(\sigma) + a_4 \cdot \ln(POV) \qquad (11.2)$$

Estimates for the model parameters obtained using data from Table 11.1 are shown in Table 11.2.

The best fit log-transformed model is:

$$\ln(MI_{bp}) = 6.66 + 0.57 \cdot \ln(Size) + 0.78 \cdot \ln(\sigma) + 0.52 \cdot \ln(POV) \qquad (11.3)$$

The model has a high R^2 ($R^2 = 0.93$), significant t-stats ($t \gg 0$), and a high F-stat ($F = 513.47$). This indicates that this is a very reasonable model. If the regression results had low R^2 and/or insignificant t-stats or F-stat, then the proposed analytical form of the preferred model (see step I) should not be used. The good thing is that since most pre-trade models include size,

Table 11.1 Broker-Dealer Cost Estimates

Stock	Size	Volt.	POV	Broker I	Broker II	Broker III
RLK	1%	20%	5%	3.6	4.7	2.2
RLK	1%	20%	10%	5.9	6.1	2.4
RLK	1%	20%	20%	6.0	11.4	4.7
RLK	5%	20%	5%	10.1	8.0	7.8
RLK	5%	20%	10%	14.8	14.0	12.2
RLK	5%	20%	20%	23.9	20.7	12.2
RLK	10%	20%	5%	18.2	12.7	10.9
RLK	10%	20%	10%	18.5	18.3	11.3
RLK	10%	20%	20%	24.4	29.1	19.9
RLK	15%	20%	5%	16.3	13.9	14.8
RLK	15%	20%	10%	22.6	18.9	17.8
RLK	15%	20%	20%	36.6	35.1	30.4
RLK	20%	20%	5%	19.6	21.8	16.6
RLK	20%	20%	10%	27.1	26.0	24.2
RLK	20%	20%	20%	47.3	37.2	36.4
RLK	25%	20%	5%	23.8	23.2	21.2
RLK	25%	20%	10%	29.1	33.0	34.0
RLK	25%	20%	20%	40.6	39.4	37.7
RLK	30%	20%	5%	31.0	17.4	21.0
RLK	30%	20%	10%	33.1	32.0	27.9
RLK	30%	20%	20%	62.3	52.9	38.0
ABC	1%	30%	5%	3.7	6.9	3.1
ABC	1%	30%	10%	7.8	6.9	4.3
ABC	1%	30%	20%	11.6	13.1	7.2
ABC	5%	30%	5%	12.5	13.0	10.4
ABC	5%	30%	10%	16.5	19.1	14.0
ABC	5%	30%	20%	29.5	27.9	14.5
ABC	10%	30%	5%	21.8	17.9	16.7
ABC	10%	30%	10%	32.0	24.3	26.2
ABC	10%	30%	20%	33.5	39.1	31.7
ABC	15%	30%	5%	26.6	18.3	15.2
ABC	15%	30%	10%	34.8	28.0	25.0
ABC	15%	30%	20%	55.9	48.3	41.7
ABC	20%	30%	5%	33.4	23.2	26.7
ABC	20%	30%	10%	41.1	40.3	31.7
ABC	20%	30%	20%	52.0	49.8	56.0
ABC	25%	30%	5%	36.3	22.0	27.8
ABC	25%	30%	10%	49.4	37.7	32.3
ABC	25%	30%	20%	57.4	50.7	65.1
ABC	30%	30%	5%	30.7	31.9	27.3
ABC	30%	30%	10%	43.5	33.5	56.2
ABC	30%	30%	20%	65.3	59.1	72.3

Table 11.2 Regression Results

	ln(a_1)	**a_2**	**a_3**	**a_4**
Estimate	6.66	0.57	0.78	0.52
Std Error	0.15	0.02	0.09	0.03
t-stat	43.88	34.59	8.76	16.34
R^2:	0.93			
SeY:	0.20			
F-Stat.	513.47			
Calculation				
ln(a_1)	6.66			
seY	0.020697			
a_1*	6.68			
a_1	793.3635			

volatility, and strategy (in terms of percentage of volume, time, or trading rate), users will find that the simplified I-Star model works extremely well in the pre-trade of pre-trade process.

Therefore, the simplified I-Star model is:

$$MI_{bp} = 793 \cdot Size^{0.57} \cdot \sigma^{0.78} \cdot POV^{0.52} \qquad (11.4)$$

Recall that $E[\exp\{\ln(a_1)\}] = a_1 + \frac{1}{2} \cdot \sigma^2$.

Investors can also estimate the parameters of the full I-Star model including the b_1 parameter to differentiate between permanent and temporary impact cost if desired. This technique is also fairly straight-forward but requires more sophisticated estimation process such as non-linear least squares or maximum likelihood estimation since the equation cannot be linearized via log-transformation.

This pre-trade of pre-trade technique (also referred to as the aggregated model) allows investors to construct models that are a general consensus of market participants. It is a way to survey the market and draw a general conclusion of expectations. The most appealing aspect of using a pre-trade of pre-trade model is that it provides investors and analysts with a quick and easy process to generate the necessary cost data to fit these models. Investors and buy-side analysts often have difficultly constructing the input data set for their in-house model since the process is extremely data intensive and resource draining.

Applications

We are now at a point where we can begin to incorporate our pre-trade market impact cost estimates into the investment decision process for stock selection, portfolio optimization, alpha capture, and "what-if" analysis.

The model we will use for our analysis is:

$$MI_{bp} = 793 \cdot Size^{0.57} \cdot \sigma^{0.78} \cdot POV^{0.52} \tag{11.5}$$

We next illustrate how our pre-trade of pre-trade model can be used to address various portfolio manager needs. This builds on Journal of Trading, "Creating Dynamic Pretrade Models: Beyond the Black Box," Fall 2011.

Example 1

An investor wishes to determine expected impact cost for RLK for an order of 10% ADV utilizing a POV = 20% strategy. The volatility of RLK is 20%. Cost estimates were provided by all three brokers (Table 11.1) and are shown below in Table 11.3.

The pre-trade of pre-trade (aggregated model) cost estimate is computed from the following:

$$MI_{bp} = 793 \cdot (0.10)^{0.57} \cdot (0.20)^{0.78} \cdot (0.20)^{0.52} = 26.3 \text{ bp}$$

The average cost from the three brokers is 24.5 bp and the estimated cost using our pre-trade of pre-trade model is 26.3 bp. Notice that the estimate from the pre-trade of pre-trades is consistent with the average of the broker models.

The advantage now is that the average estimated cost across brokers can be computed directly from Equation 11.5 without having to access broker models or shift through broker data. The pre-trade of pre-trades is an important tool for those parties wishing to minimize information leakage.

Table 11.3 Estimated Costs	
Broker	**Est. Cost**
I	24.4
II	29.1
III	19.9
Avg:	24.5
I-Star Model:	26.3

Example 2

Next suppose the portfolio manager expects volatility in RLK to jump from 20 to 40%. What is the expected market impact cost for the same order with the new volatility estimate?

The only way the PM can obtain estimates from the three brokers is to provide the brokers with their proprietary volatility forecast 40% and ask the broker to re-run the scenario with this volatility estimate. But it is likely that the PM will not be willing to provide any broker with their market view and proprietary expectations for any stock.

But in the case of the aggregated pre-trade of pre-trade cost model the portfolio manager can easily re-compute the cost estimate with the new volatility estimate directly. This is as follows:

$$MI_{bp} = 793 \cdot (0.10)^{0.57} \cdot (0.40)^{0.78} \cdot (0.20)^{0.52} = 45.2 \ bp$$

Portfolio managers can generate these estimates incorporating their expectations without providing proprietary information to brokers.

Example 3

The portfolio manager is interested in the cost of transacting an order of 7.5% ADV of stock ABC using a full day VWAP strategy. But unfortunately, the manager did not request cost estimates for this size order from their brokers. The options for the PM are to request cost estimates for ABC from their brokers but then the brokers would now know that the manager is interested in stock ABC, or interpolate cost estimates for this scenario. Unfortunately, linear interpolation is not a direct process because neither the order size nor POV rate was provided by any of the brokers. But it still can be done in three steps.

The PM can easily utilize the aggregate pre-trade model to determine the expected cost under this scenario. This calculation is:

$$MI_{bp} = 793 \cdot (0.075)^{0.57} \cdot (0.30)^{0.78} \cdot (0.0698)^{0.52} = 17.7 \ bp$$

Example 4

A PM is evaluating a worse case scenario to liquidate a 10% ADV position of RLK under extreme situations using a full day VWAP strategy. The PM is interested in the cost to liquidate the position if volatility spikes to 40% and volume on the trade day is only half of its normal volume. Here, the POV rate for the full day VWAP strategy is $0.10/(0.5 + 0.10) = 0.167$.

Once again, the broker models are not flexible enough to provide cost estimates for this situation without the PM providing the brokers with proprietary volatility and liquidity. But we can utilize the aggregate pre-trade model to determine the expected cost under this scenario. This calculation is:

$$\mathrm{MI}_{bp} = 793 \cdot (0.10)^{0.57} \cdot (0.40)^{0.78} \cdot (0.167)^{0.52} = 41.1 \text{ bp}$$

Notice once again that this cost is significantly higher than what we would find from any of the vendor pre-trade models under current market conditions.

The more important concepts of the pre-trade of pre-trade modeling approach are:

■ Simplified I-Star provides a valuable starting point and serves as an appropriate workhorse model.
■ This allows investors to infer essential information from broker black box models.
■ Vendor pre-trade models incorporate the current point in time variables such as current volatility, current liquidity conditions. But we often want to understand the exit costs that will occur under an entirely different set of market conditions.
■ Managers can incorporate their own market views into the analysis (e.g., volatility, liquidity, as well as proprietary alpha signals).
■ Managers can perform analyses independent of other brokers and vendors (minimizes information leakage).
■ A transparent model allows stress testing, "what-if," and sensitivity analyses.

HOW EXPENSIVE IS IT TO TRADE?

All too often we hear portfolio managers complain that their incremental alpha was lost during trading and the fund underperformed their benchmark due to transaction costs. But how true is this statement? Is the underperformance really due to the transaction drag on the fund or is it due to inferior stock selection? As we show below, the corresponding trading cost of an investment idea is often much more expensive than originally anticipated. And this is especially true when managers liquidate a position (e.g., sell the holding).

To begin, let us compare trading costs across large cap (SP500) and small cap (R2000) stocks. Table 11.4 provides the average trading characteristics

Table 11.4 Comparison of Trading Characteristics: June 2012

Index	Avg Dollar Turnover*	Avg Daily Volume*	Avg Price	Avg Volatility	Avg Rho	Median Spread (cps)	Median Spread (bp)
SP500	202,511,240	5,666,180	$54.28	30%	0.57	1.80	3.32
R2000	6,674,599	503,553	$20.34	43%	0.39	10.01	49.16
Net Diff	195,836,641	5,162,627	$33.93	−13%	0.18	−8.21	−45.84
Ratio	30.34	11.25	2.67	0.70	1.45	0.18	0.07

* = Stock level averages, e.g., the avg ADV for an SP500 stock was 5,666,180 shares per day in June 2012.

for these samples (as of June 2012). For example, the average daily trading volume for an SP500 stock is 5,666,180 shares per day, and the average daily trading volume for an R2000 stock is 503,553 shares per day. On average, SP500 stocks trade 11.25 times more daily share volume than R2000 stocks. The average daily dollar turnover value in these names is even more exaggerated. An SP500 stock trades $202,511,240 per day and an R2000 stock trades only $6,674,599 per day. This is more than 30 times more traded dollars per day per stock in SP500 names than in R2000 names. Additionally, R2000 stocks have higher volatility and larger spreads than SP500, thus also increasing trading costs.

What does this have to do with trading costs? Well everything. Trading costs are usually stated for order sizes or share quantities expressed in terms of % ADV. In these cases, when we compare the actual cost of trading across SP500 and R2000, the difference between the stock categories of stock is large but not outrageously large. And the difference is mostly due to volatility, spreads, company specific risk, and higher perceived information-based trading in small cap stocks compared to large cap stocks.

For example, the average cost of trading an order of 10% ADV via a VWAP strategy is 19.8 bp for large cap and 27.8 bp for small cap. Small cap stocks are 70% more expensive. Utilizing a POV = 20% strategy the cost is 37.4 bp for large cap and 55.7 bp for small cap. Small caps are 149% more expensive. Much of this difference, as mentioned, is explained by small cap volatility (43%) being higher than large cap volatility (30%), and small cap spreads (49 bp) being higher than large cap spreads (3.3 bp). Figure 11.1 shows this difference in trading cost across various order sizes for large and small cap stocks for a POV = 20% strategy.

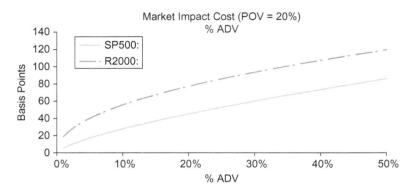

■ **Figure 11.1** Market Impact Costs Based on Order Size as Percentage of ADV.

The trading cost difference between large and small cap stocks becomes even more dramatic when we analyze the dollar amount of a trade. For example, if we invest $5 million in a large cap stock it results in an average order size of 5.8% ADV and a cost of 17.6 bp (POV = 20%). But the same amount invested in a small cap stock results in an average size of 394% ADV and a cost of 266.6 bp (POV = 20%). This is now more than 15 times more expensive to trade small caps compared to large caps for the same dollar investment. This is an outrageous difference! Figure 11.2 shows the comparison of trading costs across large and small cap stocks across equivalent dollar amounts. Notice how dramatically more expensive small cap stocks are to transact compared to large cap stocks even when holding the execution strategy constant (POV = 20%).

Another useful way to compare trading costs is by total dollar allocation. For example, index funds allocate their dollar investment across stocks based on the stocks' weightings in the index. Stocks with higher weightings receive a large dollar investment and stocks with smaller weightings receive a smaller dollar investment. A $3 billion investment allocated to each index based on market capitalization weightings results in an average order size of 2.8% ADV for SP500 and a corresponding cost of 10.4 bp. The same investment in the R2000 index results in an average order size of 42.8% and corresponding trading cost of 101.4 bp. Small cap stocks are 9.7 times more expensive to trade than large cap—even for a passive index fund. This differential is especially dramatic considering that the dollars are allocated across a much larger number of stocks for the R2000 index (1992 stocks in June 2012) compared to the SP500 index (500 stocks in June 2012). Figure 11.3 compares the difference in trading costs for various investment amounts across market cap weighted replication of the indexes.

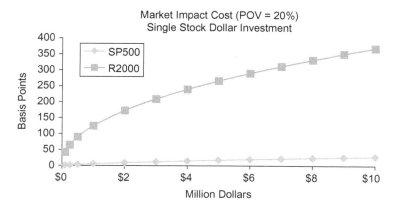

■ **Figure 11.2** Market Impact Cost Based on Dollar Investment Value.

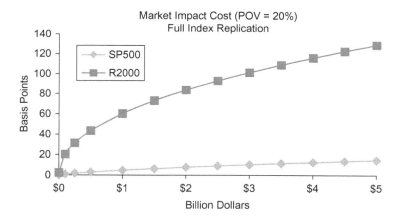

■ **Figure 11.3** Market Impact Cost Based on Full Index Replication.

Table 11.5 shows the estimated market impact parameters for large cap and small cap stocks using data from 2011.

Acquisition and Liquidation Costs

An important issue that needs to be fully understood in the portfolio management process is that the cost to acquire a position is often lower than the cost to liquidate that same position. Earlier we mentioned that we have not found any true statistical difference between the cost to buy shares and the cost to sell shares. So how can this be true? First, the cost to buy and sell shares is the same when everything else is the same, such as volatility, market volumes, and incremental buying and selling

Table 11.5 Estimated Market Impact Parameters

Scenario	a_1	a_2	a_3	a_4	b_1
All Data	708	0.55	0.71	0.50	0.98
SP500	687	0.70	0.72	0.45	0.98
R2000	702	0.47	0.69	0.55	0.97

* = Estimation Period: 1H2011.

pressure from other investors. Quite often, however, market participants observe, and data confirms, that buy orders are less expensive to transact than sell orders. But we have found that this is due to managers selling stocks more aggressively, as well as survivorship bias where managers have a complementary stock to buy when prices become too expensive but do not have a complementary stock to sell when the company has fallen out of favor.

But what also occurs at this point in time is that the stock volatility has spiked and liquidity has decreased. Even in cases where there is more trading in the name (such as during the financial crisis), the amount of transactable liquidity is often much lower because there are several investors on the same side of the order as the manager, thus everyone is competing for the smaller liquidity pool. All of which increase the cost to trade. Portfolio managers buy in times of favorable conditions and sell in times of adverse conditions and market stress. For example, an index manager may hold 5% of the ADV of a stock in the portfolio. If this stock is suddenly deleted from the index the expected trading cost to liquidate the position will likely be greater than the expected cost of trading an order of 5% ADV. This is because all index managers who own the stock will have to sell the shares and liquidate the position from their portfolio and thus the aggregated market selling pressure will be equal to the aggregated number of shares that need to be sold. In many situations, the number of shares that need to be transacted due to an index reconstitution could be much greater than 100% of the stock's ADV. Thus, the trading cost for the index event trade is dramatically higher than the costs that would occur from a non-index event trade.

To further highlight this point we examined a $100 million small cap portfolio comprised of 100 stocks. The average order size corresponding to this investment amount is 35% ADV and the volatility 42%. If the portfolio is purchased via a full day VWAP strategy the expected cost is 106 bp. This is a very realistic cost estimate for a small cap strategy under normal market conditions. But now suppose that liquidity has dried

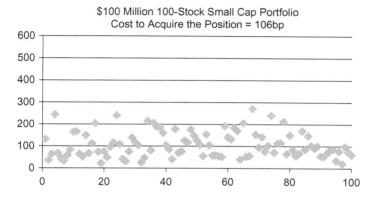

■ Figure 11.4 Cost to Acquire Each Position in the 100 Stock Portfolio.

up and volatility has spiked. If transactable liquidity is now only half of normal levels and volatility has doubled, the cost to liquidate the position will jump from 106 to 215 bp. Costs are more than 2 times more expensive to sell than to buy. This was caused by market conditions at the time of the sale and not due to any difference in buying or selling sensitivity.

The overall round-trip trading cost of this strategy is approximately 321 bp. If the manager was expecting and planned for a round-trip transaction cost of 200–220 bp they would quickly realize that the liquidation cost of the position eroded their entire incremental alpha and caused them to incur a loss. And the end result is likely that they underperformed their benchmark.

Figure 11.4 shows the market impact cost to acquire each of the 100 stocks in our example. Notice the extent that this cost can vary across all names in the portfolio. The range of cost is from 21 bp (least expensive) to 271 bp (most expensive). The actual cost to trade is determined from the dollar allocation to each name as well as the stock's liquidity (ADV) and volatility. And as this analysis shows—actual trading cost by stock can vary tremendously.

Figure 11.5 shows the market impact cost to liquidate each of the 100 stocks in our example. The stock by stock cost in this example varies from 45 to 527 bp with an average of 215 bp. Again, notice how much higher the liquidation cost is compared to the acquisition cost shown in Figure 11.4.

Formulaically, the liquidation cost in a stressed trading environment computed directly from the simplified I-Star market impact model is as follows.

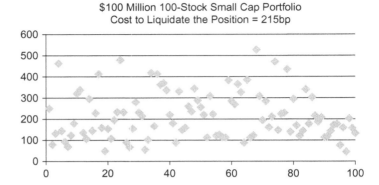

$100 Million 100-Stock Small Cap Portfolio
Cost to Liquidate the Position = 215bp

■ **Figure 11.5** Cost to Liquidate Each Position in the 100 Stock Portfolio.

In a normal environment the market impact cost is:

$$MI_{Normal} = a_1 \cdot Size^{a_2} \cdot \sigma^{a_3} \cdot \left(\frac{Size}{1+Size}\right)^{a_4} \tag{11.6}$$

In a stressed environment where volatility doubles and liquidity is halved the portfolio manager incorporates these expectations into the cost estimation model as follows:

$$MI_{Stressed} = a_1 \cdot Size^{a_2} \cdot (2 \cdot \sigma)^{a_3} \cdot \left(\frac{Size}{0.5+Size}\right)^{a_4} \tag{11.7}$$

Therefore, in a stressed environment the cost premium is equal to:

$$Cost_{Premium} = 2^{a_3} \cdot \left(\frac{1}{0.5}\right)^{a_4} = 2^{a_3} \cdot 2^{a_4} = 2^{a_3+a_4} \tag{11.8}$$

This cost premium for the full I-Star model can also be approximated from the simplified I-Star equation since the value of b_1 is often small. In our example, the increase in cost can be approximated with the above cost premium equation and our small cap market impact parameters as follows:

$$Cost_{Premium} = 2^{a_3+a_4} = 2^{0.69+0.55} = 2^{1.24} = 2.36$$

Portfolio Management—Screening Techniques

Given the large potential variation in trading costs across stocks, managers have begun using many different techniques to screen and filter for those names that could potentially be too expensive to buy and/or sell. One common technique is to limit the position size based on a percentage of the stock's average daily volume, for example, set a maximum size of 10% ADV. The belief in this case is that the market would always be able to absorb an order of this size without the investor inflicting too much impact into the stock price. But the maximum % ADV value is often an arbitrary value and even at this level there are names that are still potentially very expensive to transact.

To better show this point, we analyzed the trading cost corresponding to 10% ADV position size for large (SP500) and small (R2000) cap portfolios. The trading cost for a 10% ADV order size in each of the stocks in the SP500 index is shown in Figure 11.6a. The average cost is 20 bp but these costs vary greatly from a low of 3 bp to a high of 48 bp. For small cap stocks, the average cost of a 10% ADV order is 37 bp with a range of 6 to 160 bp (Figure 11.7a). There are also a very large number of names with costs greater than 60 bp.

Two questions arise when performing this type of PM screening process. First, why should the portfolio managers limit the order size to only 10% ADV for those stocks with very low trading costs? And second, should the maximum size not be set lower for those very expensive to trade stocks? The answer to both of these questions is: Yes. If a stock has low trading costs and is a very appealing investment opportunity the manager should not limit the holding size to some arbitrary value. And if a stock is very expensive to trade, that stock should be held in a much lower quantity in the portfolio—or possibly not held at all.

To help determine the maximum % ADV size to hold in a portfolio, managers can reverse engineer the filtering process. Rather than starting with an arbitrary size constraint, managers can specify the cost level that they feel is appropriate for the portfolio. For example, managers may deem a reasonable cost to trade large cap stocks is 20 bp. This cost level is often tied to the expected alpha of the fund and will be further discussed below. Then managers compute for the investable universe the size that can be traded resulting in the cost of 20 bp. The maximum size for the stock is then determined for all stocks by solving the MI equation by setting the LHS equal to 20 bp and solving for size. Mathematically, this is found by solving for size in the following:

$$20 = b_1 \cdot (a_1 \cdot Size^{a_2} \cdot \sigma^{a_3}) \cdot POV^{a_4} + (1 - b_1) \cdot (a_1 \cdot Size^{a_2} \cdot \sigma^{a_3}) \qquad (11.9)$$

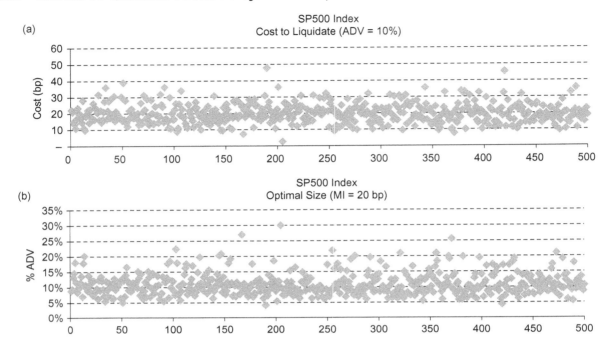

Figure 11.6 SP500 Index Cost Analysis.

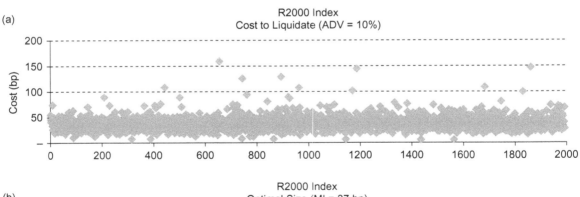

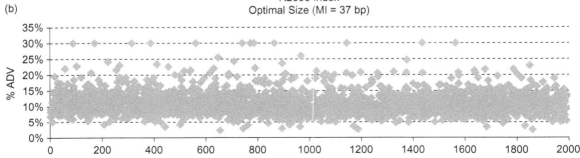

Figure 11.7 R2000 Index Cost Analysis.

The one caveat here is that the manager needs to further specify the underlying execution strategy. For example, if the strategy is a VWAP strategy, then the corresponding POV rate is:

$$POV = \left(\frac{Size}{1 + Size} \right)$$

Therefore, we are not able to solve Equation 11.9 in terms of size directly. But we can determine the size via fairly straightforward numerical methods but we still need to solve for each stock individually.

The size corresponding to a cost of 20 bp using a full day VWAP strategy is found from the following equation where we substitute POV with $\frac{Size}{1 = Size}$.

$$20 = b_1 \cdot (a_1 \cdot Size^{a_2} \cdot \sigma^{a_3}) \cdot \left(\frac{Size}{1 + Size} \right)^{a_4} + (1 - b_1) \cdot (a_1 \cdot Size^{a_2} \cdot \sigma^{a_3}) \quad (11.10)$$

Notice that in this formulation the portfolio manager can also incorporate stress testing to determine the position size that will result in a cost of 20 bp under adverse market conditions by changing the volatility and liquidity.

Figure 11.6b shows the holding sizes for SP500 stocks resulting in a trading cost of 20 bp (the average cost for an order of 10% ADV). The average position size is about 10% as expected and the range is from 4 to 76% ADV. (The figure truncates the scale at 35%). Figure 11.7b shows the position sizes for R2000 stocks resulting in a trading cost of 37 bp (the average cost for an order of 10% ADV). The average position size is again about 10% and the range in order sizes is from 2 to 67% ADV.

This analysis shows that there are many stocks the manager can hold in the portfolio at more than 10% ADV and not worry about incurring unnecessary transaction costs, even under stressed market conditions. And if these stocks do have appealing alpha expectations the manager can greatly enhance portfolio performance by determining appropriate order sizes to maximize profits. Additionally, there are other stocks that should not be held even at the 10% ADV level because the corresponding costs will be too high and will erode too much of the uncovered alpha. These are the stocks that need to be held in much lower quantities or possibly not held in the portfolio at all. Unfortunately, the computational process above does not have a direct analytical form but could be solved via optimization or other non-linear solution techniques.

Portfolio managers could improve fund performance by analyzing appropriate holding sizes based on both alpha expectations and trading costs.

MI FACTOR SCORES

The preceding section showed how portfolio managers could back into the order size that could be transacted for a specified cost. But this often requires complicated numerical procedures to find the solution and this must be solved for each stock individually. This analyses can be done, but they are often quite time consuming.

The market impact factor score (MI Factor Score) is an alternative method to efficiently screen stocks based on trading costs. The MI Factor Score incorporates the I-Star market impact model, corresponding parameters, and stock specific trading characteristics (liquidity, volatility, and market price) to determine a trading cost score. The higher the score the more expensive the stock is to trade and the lower the score the less expensive the stock is to trade.

Manager's use the MI Factor Score to improve their stock screening process by filtering for the more expensive and difficult names to trade. In addition to being a direct calculation, this provides an improvement over methodologies that only consider liquidity or only consider volatility when filtering for trading costs.

The advantage to our MI factor score is that it does not require complicated or sophisticated numerical procedures, and it is a more accurate representation of the trading cost environment. The derivation of the MI factor score is as follows:

Derivation of the MI Factor Score for Shares

Step 1. Start with the I-Star model:

$$I^*(Shares) = a_1 \cdot \left(\frac{Shares}{ADV}\right)^{a_2} \cdot \sigma^{a_3} \tag{11.11}$$

Step 2. Rearrange the terms in the expression by factoring out shares:

$$I^*(Shares) = \left\{ a_1 \cdot \left(\frac{1}{ADV}\right)^{a_2} \cdot \sigma^{a_3} \right\} \cdot Shares^{a_2} \tag{11.12}$$

Step 3. The MI factor score is then:

$$K(Shares) = \underbrace{a_1 \cdot \left(\frac{1}{ADV}\right)^{a_2} \cdot \sigma^{a_3}}_{MI\ Factor\ Score(Shares)} \tag{11.13}$$

The MI factor score provides managers with the stock's market impact sensitivity K(*Shares*). For the same number of shares to trade and the same strategy, the MI factor score will provide a ranking value of each stock's trading cost. This score allows a fair and consistent comparison of cost of trading across stocks and includes both liquidity and volatility terms. If the MI factor score is twice as high for one stock compared to another stock, the cost to trade the first stock will be twice as high as for the second stock.

Portfolio managers finally have a trading cost factor score that will alleviate the need to utilize broker-dealer pre-trade models, and perform time consuming numerical procedures. The factor score can be easily computed on the manager's desktop or integrated into a proprietary in-house model. The only requirement to compute the score is to have the market impact parameters and stock trading characteristics (see Kissell, 2013).

In many situations, portfolio managers are not setting out to invest in a specified number of shares or in a specified order size (% ADV). They are more often setting out to invest a specified dollar value into a stock or basket of stocks. To accommodate these needs we can reformulate the MI factor score in terms of dollars.

The number of shares that can be purchased for a fixed dollar amount is:

$$Shares = \frac{Dollars\$}{Price} \tag{11.14}$$

Then we can compute the MI factor score in terms of dollars as follows:

Step 1a. Start with the I-Star model:

$$I^*(Shares) = a_1 \cdot \left(\frac{Shares}{ADV}\right)^{a_2} \cdot \sigma^{a_3} \tag{11.15}$$

Step 1b. Convert shares to dollars:

$$I^*(Dollars\$) = a_1 \cdot \left(\frac{Dollars\$}{Price} \cdot \frac{1}{ADV}\right)^{a_2} \cdot \sigma^{a_3} \tag{11.16}$$

Step 2. Rearrange the terms in the expressions and factor out dollars:

$$I^*(Dollars\$) = \left\{ a_1 \cdot \left(\frac{1}{Price} \cdot \frac{1}{ADV} \right)^{a_2} \cdot \sigma^{a_3} \right\} \cdot Dollars\$^2 \qquad (11.17)$$

Step 3. The MI factor score is then:

$$K(Dollars\$) = \underbrace{a_1 \cdot \left(\frac{1}{Price} \cdot \frac{1}{ADV} \right)^{a_2} \cdot \sigma^{a_3}}_{MI \; Factor \; Score(Dollars\$)} \qquad (11.18)$$

Current State of MI Factor Scores

Portfolio managers are using the MI factor scores as an additional layer of quantitative screening, and as part of asset allocation and stock selection. Since the factor score incorporates both liquidity and volatility, and provides a consistent comparison across all stocks, they are quickly becoming the preferred TCA screening tool for funds. Results incorporating MI factor scores have been found to adhere to best execution practices by better ensuring consistency between trading goals and investing needs.

Kissell Research Group has begun providing MI factor scores to portfolio managers for global equities and various financial instruments across the multi-assets classes[1].

MI Factor Score Analysis

We compared the MI factor scores expressed in terms of dollars for large and small cap stocks. Scores are sorted from smallest (cheapest to trade) to largest (most expensive to trade). Large cap stocks had an average MI factor score of $K = 0.001$ (Figure 11.8a). Visual inspection of the scores finds approximately three distinct categories. The first grouping consists of the 100 least expensive stocks to trade, the middle grouping consists of the 300 stocks with an average trading cost, and the last grouping consists of the 100 most expensive stocks to trade.

The first grouping of stocks represents, the cheapest trading cost stocks. Managers could transact these names with the least amount of worry of adversely affecting prices. Managers could also select to hold larger quantities of these stocks without incurring abnormally high costs. The third grouping of stocks represents the expensive trading cost stocks. These are the names that will result in the highest market impact cost of the group.

---

[1]Kissell Research Group, www.KissellResearch.com.

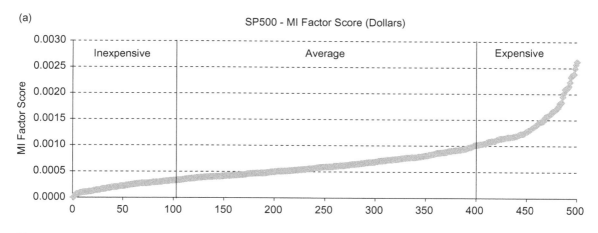

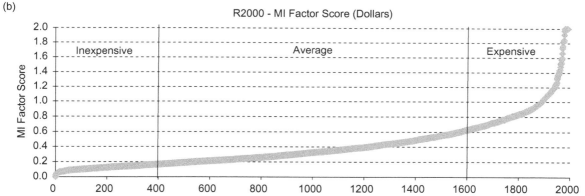

■ **Figure 11.8** Market Impact Factor Scores.

Managers should analyze these names to fully understand their trading characteristics and they may be best served by holding few shares of these stocks in the portfolio unless, of course, the incremental alpha will more than offset the incremental trading cost.

Stocks in the R2000 index also have a very similar MI factor score (dollar) shape. This is shown in Figure 11.8b. The average MI factor score for R2000 stocks is K = 0.42. Notice that this is much larger than the average score for SP500 (K = 0.001). This difference is primarily due to SP500 stocks having higher prices and much higher liquidity. The market is also more sensitive to trading small cap stocks in general. Our market impact analysis found higher values for small cap stocks compared to large cap stocks. Visual inspection finds that there are approximately 400 stocks that are relatively inexpensive to trade (in relation to other small cap stocks) and approximately 400 stocks that are relatively very expensive to trade (in relation to other small cap stocks). Managers could take advantage of these MI scores by increasing holdings in the stocks with low factor scores without the worry of adversely inflicting abnormal levels of market impact cost above acceptable levels. The 400 very expensive to trade stocks could result in dramatically higher levels of market impact cost much above and beyond what is expected, especially in times of a stressed market environment. These are the trades that often turn a great investment opportunity into one that is just moderately profitable and possibly incurs a loss.

Notice in Figure 11.8b how quickly the MI factor score increases for the tail end of the small cap universe. The last 100 stocks are extremely costly and very sensitive to investment dollars. Managers should think about excluding these stocks from their portfolio or at least holding smaller investment dollars in these names, unless, of course, the expected stock alpha will more than offset these incremental transaction costs.

Please note that these break points are found by visual inspection and analysts need to determine actual break points based on their investment needs and alpha expectations. In the end, these MI factor scores provide a vast improvement over other screening techniques that only use liquidity or only use volatility.

ALPHA CAPTURE PROGRAM

An alpha capture analysis provides the portfolio manager with the quantity of forecasted alpha that can be achieved via an appropriately structured trading strategy. Forecasted alpha in this manner has alternatively been referred to as price return, price appreciation, price trend, price evolution, and drift (Kissell, 2003).

The quantity of the alpha that can be captured by managers is dependent upon the size of the order, the alpha forecast, the cost of the trade, and the underlying strategy. For example, if a strategy is expected to provide a return of 10% over a period the alpha capture analysis will provide information about how much of the expected return the manager will be able to achieve for various order sizes. An order of 5% ADV may be able to capture 9.8%, an order of 10% ADV may be able to capture 9.5%, an order of 25% ADV may only be able to capture 9.0% of the return, etc. The larger the order size, the lower the expected alpha the manager will realize due to trading costs. In other words, the alpha capture strategy will estimate the profitability of a strategy.

Alpha capture programs provide managers with answers to many of their investment related questions: How much alpha will my investment achieve? How much should I invest? And most importantly, how is transaction cost analysis being utilized to analyze profitability concerns?

To accurately compute expected alpha capture, managers need to specify their alpha estimates and have accurate market impact modeling capabilities. And this is yet another reason why TCA has historically gained so little traction in the industry. Portfolio managers are unwilling to provide brokers with their alpha estimates, and brokers have been unwilling to provide managers with the underlying market impact models.

Market impact and alpha cost are conflicting terms. Trading too fast will incur too much impact but trading too slow will miss too much alpha (or missed profit opportunity). Subsequently, we refer to this conflicting expression as the portfolio manager's dilemma.

In Chapter 5 we provided techniques to develop and test a market impact model. In this chapter we provided a pre-trade of pre-trade model to decipher broker models, to calibrate preferred market impact parameters, and allow those models to function as a stand-alone application on the investor's own desktop and as part of their own in-house proprietary systems.

The pre-trade of pre-trades approach is the easiest way for analysts and managers to solve the issue of not having a market impact model on their desktop. The process provides managers with a functional form of a model and also allows them to run that model from their own desktop. They can then incorporate their own liquidity and volatility views, and perform sensitivity analysis with different alpha estimates and various different market conditions. This last piece is the most important since it allows managers to keep their alpha expectations proprietary. Imagine if fund managers did provide their brokers with their alpha views!

Example 5

Portfolio managers develop alpha capture programs by incorporating the simplified I-Star impact model and the manager's alpha forecast. For a continuous trading strategy and an alpha estimate following a linear trend, the manager will incur an alpha cost equal to one-half the total alpha movement over the trading period t. This mathematical representation is:

$$Alpha\ Cost_{bp} = \frac{1}{2} \cdot \frac{\mu_{bp}}{d} \cdot t \qquad (11.19)$$

Where μ_{bp} is the alpha forecast, d is the time horizon of the alpha forecast, and t is the time to complete the order with the condition $0 \leq t \leq 2$.

For example, if the alpha forecast is that the stock will increase 5% over the next three days we have $\mu_{bp} = 500$ bp and $d = 3$.

We can further express our trading time t in terms of our trading strategy α as follows:

$$t = \frac{Shares}{ADV} \cdot \frac{1}{\alpha} \tag{11.20}$$

Then our alpha cost is:

$$Alpha\ Cost_{bp} = \frac{1}{2} \cdot \frac{\mu_{bp}}{d} \cdot \frac{Shares}{ADV} \cdot \frac{1}{\alpha} \tag{11.21}$$

Our simplified I-Star market impact model is:

$$MI_{bp} = a_1 \cdot \left(\frac{Shares}{ADV}\right)^{a_2} \cdot \sigma^{a_3} \cdot \alpha \tag{11.22}$$

In a properly structured alpha capture program the manager will seek to maximize the expected profit from this opportunity. This is determined as:

$$Max \quad \pi = \mu_{bp} - \left(\frac{\mu_{bp}}{2d} \cdot \frac{Shares}{ADV} \cdot \frac{1}{\alpha} + a_1 \cdot \left(\frac{Shares}{ADV}\right)^{a_2} \cdot \sigma^{a_3} \cdot \alpha\right) \tag{11.23}$$

The solution to the problem is:

$$\alpha^* = \sqrt{\frac{\mu_{bp}}{2d} \cdot \frac{Shares}{ADV} \cdot \left(a_1 \cdot \left(\frac{Shares}{ADV}\right)^{a_2} \cdot \sigma^{a_3}\right)^{-1}} \tag{11.24}$$

The maximum alpha capture opportunity is:

$$\pi^* = \mu_{bp} - \left(\frac{\mu_{bp}}{2d} \cdot \frac{Shares}{ADV} \cdot \frac{1}{\alpha^*} + a_1 \cdot \left(\frac{Shares}{ADV}\right)^{a_2} \cdot \sigma^{a_3} \cdot \alpha^*\right) \tag{11.25}$$

Example 6

A small cap stock is expected to increase 3% in the next 3 days. The next most attractive investment will increase 2% in the next three days. The manager wants to answer three questions:

1. How much alpha can a manager capture if the order size is 10% ADV?
2. How much can be invested in this stock before we begin to incur a loss?

3. How much should the manager invest in the stock?

Part 1. How much alpha can a manager capture if the order size is 10% ADV?

The solution to how much alpha can a manager capture if the order size is 10% ADV is found by maximizing our expected profit (Equation 11.23). In this example, the alpha forecast is 3% over three days. For simplicity, we illustrate this concept using a linear appreciation model. In practice, managers can incorporate any trend preference they have, such as a compounded model, non-linear, exponential, or even a step function where the return only occurs overnight and is constant during the day.

The alpha capture optimization is illustrated in Figure 11.9 for a small cap stock with annualized volatility = 43%. If we execute an order of 10% ADV ultra-aggressively the market impact cost will be 132 bp (the instantaneous impact cost) but we will not incur any alpha cost. Since the stock will return 3% or 300 bp over the period this urgent execution strategy will earn us a net return of 168 bp. If we trade passively over the entire three day period our impact cost will be 23 bp but our alpha cost will be 150 bp. The total cost will be 173 bp and out net profit will be 127 bp which is less than trading the entire position at once. A naïve analyst may elect to trade the entire order at once to earn a higher expected return. But this would not be an appropriate option. Take a look at the graph in Figure 11.9. Notice how the market impact cost is decreasing over the period and alpha cost is increasing over the period as expected. But most importantly, take a look at the total cost curve. This cost starts high when market impact is dominating the total cost, decreases, and then begins to increase again when the alpha cost starts to dominate the total cost. The total cost function will always be a convex function unless the manager is buying stocks that are decreasing in value or selling stocks that are increasing in value.

The minimum cost occurs at a trading time of 0.45 days (just slightly less than half of a day in volume time). The corresponding market impact cost is 54 bp and corresponding alpha cost is 23 bp for a total cost of 77 bp. The net profit the manager can earn is 223 bp (300 bp − 77 bp = 223 bp). Notice that this is much larger than the profit of 168 bp for the ultra-aggressive instantaneous strategy and the 127 bp for the passive strategy. The maximum alpha capture for this order is 223 bp.

Part 2. How much can be invested in this stock before we begin to incur a loss is found by maximizing the number of shares that be transacted at

Trade Characteristics		Analysis Results (Basis Points)		Profit Analysis (bp)	
Size:	10%	Size:	10%	Size	Net Profit
Volatility:	43%	Volatility:	28%	1%	282
Alpha/Day (bp):	100			5%	250
Alpha/Total (bp):	300	Min Total Cost:	77	10%	223
		Market Impact:	54	15%	201
		Alpha Cost:	23	20%	182
		Time:	0.45	25%	164
				30%	148
		Alpha 3 Days (bp):	300		
		Net Profit (bp)	223		

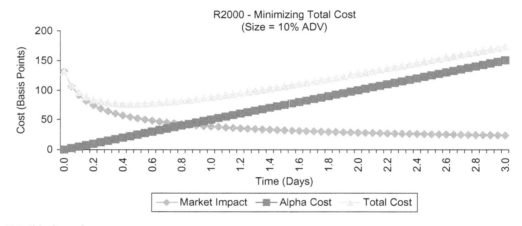

a cost equal to the projected alpha of 3% or 300 bp. This is determined by solving the following optimization:

$$Max \quad Shares$$

$$s.t. \quad a_1 \cdot \left(\frac{Shares}{ADV}\right)^{a_2} \cdot \sigma^{a_3} \cdot \alpha = 300 \text{ bp} \quad (11.26)$$

In this example, the manager could purchase up to 91% of the stock's ADV over 1.5 days at a cost of 300 bp. This is provided in Figure 11.9.

Part 3. How much should the manager invest in the stock is determined by performing economic opportunity cost analysis. The portfolio manager will invest dollars in the stock until the net profit is equal to the expected return of the next most attractive vehicle (economic opportunity cost). In this example, the manager can purchase an order equal to 15% ADV resulting in a net profit of 201 bp. Purchasing any more than 15% ADV will cause the manager's profit to fall below 200 bp, and then the manager would be better by investing the investing amount in the next most attractive opportunity and earning the profit of 200 bp. This is shown in Figure 11.9.

In practice, many portfolio managers may elect to investigate a profit maximizing strategy that consists of the appropriate allocation of dollars across both stocks simultaneously.

Alpha Capture Curves

The alpha capture curve is the portfolio manager's answer to the trader cost curve. Alpha capture curves provide the maximum quantity of total alpha that can be achieved (captured) for a given order size. These calculations follow Equation 11.23 and an example of an alpha capture curve is shown in Table 11.6 using stock data from Figure 11.9. The left hand column shows the order size as a percentage of average daily volume. The columns show the maximum alpha that can be achieved for different alpha forecasts and time horizons. Similar to the trading cost curves, the alpha capture curves are specific for the stock and alpha forecast. Alpha capture curves provide managers with invaluable reference data to determine the size or dollar value that can be invested into a stock.

Table 11.6 Alpha Capture Curves

Portfolio Manager Profit Curves

Maximum Trading Profit

Alpha over 3 days

% ADV	1%	2%	3%	4%
1%	87	184	282	380
5%	65	156	250	346
10%	45	132	223	317
15%	29	112	201	293
20%	15	95	182	272
25%	2	79	164	253
30%	−10	64	148	236
35%	−22	51	133	219
40%	−32	38	118	204
45%	−43	25	105	190
50%	−52	14	92	176

Alpha capture curves provide managers with the following (Table 11.6):

1. Expected alpha capture (profit level) for a specified order size and alpha forecast. For example, if the manager wants to transact an order for 30% ADV in a stock where the forecasted alpha is 3% over three days, the maximum profit (excluding trading costs) the manager can expect to achieve is 148 bp.

2. It provides the manager with the the maximum order size that can be traded for a profit. For example, if the alpha forecast is 1% over three days the manager can trade up to 25% ADV. At 25% ADV the manager will net a profit of 2 bp. Trading more shares will cause the manager to incur a loss on the trade (e.g., the trading cost was higher than the alpha forecast). If the manager sets out to trade 30% ADV in this stock they would expect to incur a cost of −10 bp.

3. It provides the manager with the means to determine the appropriate order size while evaluating the economic opportunity cost of the trade. For example, following the scenario in Example 6, a manager whose most attractive investment opportunity is a stock with an alpha forecast expectation of 3% over three days and second most attractive opportunity is a stock with an alpha forecast of 2% over three days could trade up to 15% ADV in the first stock before having to allocate dollars to the second stock. This is shown in Table 11.6. Notice that for an order of 15% ADV, the maximum trading profit for the investment vehicle at 3% over three days is 201 bp. Investing any more than 15% ADV will cause the net profit level to be less than 200 bp (2%). Hence, the manager would be better off investing the incremental dollars in the next most attractive investment.

Important Note: Alpha capture curves provide managers with a quick reference for profitability. In our examples above, we only included the one-way implementation cost of the trade. The expectation in this analysis is that the manager would hold the acquired position over a longer period of time. Investors who are looking to take advantage of a short-term trend and trade in and out of these positions in shorter time horizons will also need to include the liquidation cost (sell cost) of the trade. If the expectation is that the market conditions will be exactly the same during liquidation of the order investors can simply double the implementation cost (buy cost). But if the market conditions are expected to be different during liquidation than acquisition managers could use techniques provided above to determine realistic liquidation costs for the order based on their expectation for market conditions during these times. Yet another reason why managers need their own market impact models. These expectations can be incorporated into the optimization process described above.

Cost Index & Multi-Asset Trading Costs

INTRODUCTION

This chapter is separated into two sections. In the first section we introduce readers to the cost index methodology. In the second section we introduce techniques for investing and trading across multi-asset classes.

First, cost indexes provide investors with trading cost estimates based on actual market conditions and aggregated buying and selling pressure across all market participants. We derive cost indexes for historical and real-time data sets. These assist investors to improve stock selection, back-test investment ideas, perform portfolio attribution analysis, critique algorithms, and hold their providers accountable for performance in real-time. We also present a market impact simulation experiment that can be used to evaluate and construct stock specific market impact models. This simulation technique shows the difficulty in formulating these models and estimating corresponding parameters at the stock level even when the true parameters are known. The cost index has become an essential (if not the most essential) trading cost metric for investors.

We next turn our attention to investing methodologies across different asset classes. We discuss the trading caveats of trading similar instruments such as equities, exchange traded funds, and futures. For example, many times a portfolio manager's investment objective is to acquire a specified beta exposure to the market or SP500 index. Other times a manager's objective is to acquire exposure to a different factor or macro economic indicator. This is commonly referred to as "Investing in Beta" or "Investing in Factor Exposure." The manager determines what the end goal is but is indifferent as to which underlying instruments are used to acquire that exposure. We provide techniques to determine how to best acquire the specified exposure given market conditions and trading costs. Managers could improve portfolio performance through more efficient implementation. As we show, many times the optimal portfolio mix will be an allocation across the three asset types: equities, exchange traded funds, and futures. The chapter concludes with the introduction of an

The Science of Algorithmic Trading and Portfolio Management. DOI: http://dx.doi.org/10.1016/B978-0-12-401689-7.00012-X

impact model that can be applied across the different markets, regions, and asset classes. We provide readers with the estimated coefficients by the different asset classes, and corresponding cost estimates for various position sizes.

Highlights of the chapter include the following:

- Cost Index
  - Historical
  - Real-Time
  - Back-Testing
  - MI Simulation Exercise
- Multi-Asset Class Investing
  - Investing in Beta
  - Multi-Asset Market Impact Model

COST INDEX

The cost index was developed to assist investors gauge the trading cost environment and the market's overall sensitivity to order flow.

The cost index helps portfolio managers understand changing trading cost dynamics, explain portfolio slippage and tracking error, and also provides necessary input into portfolio construction models. These indexes also serve as an important input factor when back-testing invest-ment ideas. The cost index assists traders evaluate broker performance and algorithms, and also provides a real-time cost metric based on actual market conditions and overall buying and selling pressure in the stock. This provides traders with a valuable reference point to hold their provi-ders honest during the chaotic trading day.

The cost index can be defined across market capitalization categories such as large cap and small cap stocks, and also in different global regions, markets, and countries, as well as for individual stocks and other financial instruments. The cost index can also be customized for any individual benchmark, market index, or other type of investor need. Cost indexes have been previously studied by Kissell and Tannenbaum (2009) and Kissell (2006).

In the industry, there are numerous performance based indexes such as the SP500 and Russell 2000 that provide insight into market movement and value, and numerous volatility indexes such as the VIX® (SP500) and RVX (R2000) providing insight into market uncertainty and risk. Our cost index is intended to provide investors with an indication of the trading cost environment. But unlike performance and volatility indexes, the cost index is dependent upon order size and trading strategy, in addition to actual market conditions. But it is entirely possible to report trading costs on a normalized basis.

How is the cost index constructed?

The cost index is constructed based on actual market conditions (volumes and volatility) and corresponding market impact parameters. This differs slightly from forward looking pre-trade estimates based on expected market conditions. In situations where the cost index is requested for a historical time period such as the previous five years the calculation is based on the actual market conditions over those five years and the corresponding point-in-time market impact parameters. If the market impact parameters are updated weekly the construction of the index will require five years of weekly market impact parameters. Notice that this is similar to the calculation of an index such as the SP500 index which requires the adjustment factor for changing index membership. Here the adjustment factor changes whenever there is a change to the index membership as well as when there are corporate actions affecting the valuation of the index. Many managers have begun maintaining a history of market impact parameters on a daily basis.

Cost Basis

The underlying basis for the cost index can be constructed in various ways depending upon the investment needs of the fund. The three most common are size, dollar value, and share quantity. In addition to the cost basis the cost index is dependent upon the corresponding implementation strategy. These are described below.

Size. Trading costs are computed for a specified trading rate such as 10% ADV. This is the most common representation of the cost index. A size metric is often preferred because it serves as a nice comparison point across stocks and across time. But if the stock has exhibited an increase or decrease in volumes, the size metric would no longer be a consistent basis because these share quantities would not be constant over time.

Dollar Value. Portfolio managers often request the cost indexes to be constructed for a specified dollar value such as $10 million for a single order or possibly $100 million or more if investing in an index or portfolio. Many portfolio managers have set dollar amounts to invest each month and are interested in the cost to transact that dollar quantity and how those costs have evolved over time. One limitation of using a dollar value, however, is that the actual shares to transact are dependent on price levels. For example, investing $10 million when prices are high will results in fewer shares to purchase and lower cost and investing $10 million when prices are low will result in more shares to purchase and higher costs. The price level will cause the trading costs to be higher or lower even if the investment value and market conditions remain unchanged.

Shares. The share quantity cost basis is most often used when computing cost indexes for individual stocks. This works well for the individual stock and trading over time since it is not dependent upon overall volume levels, but it makes it difficult to compare the cost of trading across different stocks with different prices. For example, a one million share order of IBM represents a fairly large percentage of overall daily trading volume but a one million share order of MSFT represents a much smaller percentage of daily trading volume. Therefore, these costs will likely be much different for the same share quantity. This is helpful for managers evaluating the acquisition cost of a position and a future liquidation cost for the same number of shares.

Cost Strategy

When constructing the cost index we must also specify the underlying trading strategy. For example, a passive strategy (e.g., VWAP) will incur lower costs than an aggressive (e.g., POV = 20%). It is important when we calculate the cost index that we use a constant strategy for consistency.

The cost index for 2012 is illustrated for two different scenarios. Figure 12.1 illustrates the trading cost index in 2012 for an order size of 10% ADV

■ **Figure 12.1** Cost Index 2012.

executed via a full day VWAP strategy for large and small cap stocks. The index was computed based on average stock volatility in each category in each month and the corresponding market impact parameters. This index shows how actual costs for this order size change from month to month and also shows the difference in trading costs between large and small cap stocks. Our large cap index was in the $20-25$ bp range for the majority of the year. The small cap index was in the $40-50$ bp range over the year but spiked as high as 50 bp in November 2012. The spike in the index in November 2012 was much more dramatic for small cap stocks than for large cap stocks. A short-term cost index such as this helps portfolio managers perform portfolio attribution to understand tracking error and the trading cost environment. It also allows portfolio managers to analyze whether the missed alpha was due to poor stock selection or inferior trading, or whether it was due to unavoidable market conditions.

Some portfolio managers have requested the cost index to be computed based on the I-Star expression without incorporating the strategy or based on an optimized cost index incorporating market impact and timing risk. These strategies are:

I-Star Index (without strategy):

$$I^* = a_1 \cdot Size^{a_2} \cdot \sigma^{a_3} \tag{12.1}$$

Optimized Cost Index:

$$Min \quad \mathcal{L} = MI + \lambda \cdot TR \tag{12.2}$$

The advantage of using the I-Star calculation (Equation 12.1) is that it is not strategy dependent and so provides managers with a fair comparison of trading costs over time which then can be mapped to any investor strategy. The difficulty is that this value is a proxy for the instantaneous trading cost (ultra-aggressive) and will thus provide cost estimates much higher than will be realized in the market. It works well, however, if used to normalize the change in cost across time (see below).

The advantage of the optimized cost metric (Equation 12.2) is that the cost reflects both market impact cost and timing risk so it also provides investors with a measure of cost and trading difficulty (uncertainty). The difficulty is that this metric will tend to be higher than the expected market impact cost of the trade since it does add the timing risk expression to the total value. Investors further need to specify the risk aversion parameter λ. In this case, investors should use the corresponding MI estimate at the optimized solution.

Normalization Process

As indicated above, there are some difficulties with the I-Star and optimized cost index derivations. Mainly, these metrics will overstate the cost to trade because they either indicate the ultra-aggressive trading cost or incorporate an extra term (timing risk) into the estimate. For example, if the expected cost of a trade is 20 ± 50 bp the expected cost is 20 bp. If we quote a cost of 70 bp we would be overstating the true expected cost. In these cases, these derivations are often normalized based on a base time period with a value of 100. Costs are then scaled up or down based on the actual cost metric.

For example, if we set the index in Jan. 2012 = 100, then a cost index of 105 would indicate that costs are 5% higher than in Jan. 2012 and a cost index of 95 would indicate that costs are 5% lower than Jan. 2012. As we show below, the normalized cost index is extremely useful for funds to construct and customize their own historical cost index series.

The cost index normalization process is as follows:

- Start with the trading cost calculation such as that in Equation 12.1 or Equation 12.2 and compute this cost metric over the required period.
- Select the base time period. For illustrative purposes we set Jan. 2012 to be our base month (e.g., Jan. 2012 = 100).
- Compute the normalized cost in each period following:

$$Normalized\ Cost(t) = \frac{I^*(t)}{I^*(Jan\ 2012)} \tag{12.3}$$

Figure 12.2 illustrates the normalization process using the I-Star cost calculation for large cap stocks. We compute the I-Star cost in each month based using an order size of 10% ADV and then normalize based on Jan. 2012 as the base year. That is, Jan. 2012 = 100. The normalized cost index can be constructed for different sizes or by dollar value or share quantity.

Portfolio managers are often interested in the cost of trading different size orders such as 5% or 25% ADV or trading via a different strategy than was used in the historical index calculation. These costs could vary greatly from the cost computed for 10% ADV. In these situations, managers have a few options available. They could recreate the cost index for the specified order size and market conditions. But computing the cost index for the desired order size over the required time period could become a very time consuming exercise, especially if the portfolio manager wishes to evaluate several different order sizes and various different strategies, as well as a large number of potential stocks.

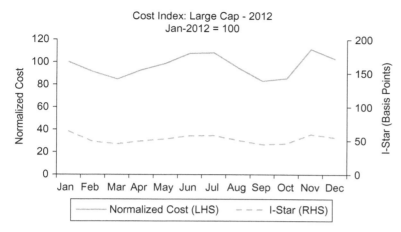

■ **Figure 12.2** Normalized Large Cap Cost Index.

An alternative approach is to construct the cost index for the desired time period using the current trading cost for the specified order size and adjusting for the normalized index value. For example, suppose the manager is interested in the cost of trading an order of 20% ADV with a POV = 25% strategy. The current trading cost for this order is 51 bp, and the current and previous month's normalized index are 105 and 95 respectively. From the normalized cost index we estimate that the trading cost in the previous month was 95/105 = 0.90 or 90% less. Therefore, the expected cost for trading 20% ADV via POV = 25% is 95/105*51 bp = 46 bp.

The normalized cost index provided the necessary information to recreate historical trading costs without the need to reconstruct the index for this size and strategy. Furthermore, we did not need access to historical market conditions or market impact parameters since this information was already embedded into the normalized cost index time series.

The backwards-propagation process can be generalized as follows:

$$Trade\ Cost(t-k, Size, POV) = Current\ Trade\ Cost(t, Size, POV)$$
$$\cdot \frac{Normalized\ Index(t-k)}{Normalized\ Index(t)} \qquad (12.4)$$

Where *Trade Cost*$(t-k, Size, POV)$ is the historical cost at time $t-k$ for the specified *Size* and *POV*, and *Current Trade Cost*$(t, Size, POV)$ is the current trading cost at time t for the identical *Size* and *POV* rate. The current trading cost is estimated via the preferred market impact model (e.g., see Chapter 5 or Chapter 11).

This approach has been found to be extremely useful for investors looking to evaluate market trends and to evaluate the performance of their brokers and algorithms. Furthermore, it has proven extremely valuable for portfolio managers performing portfolio attribution and evaluating potential investment opportunities.

Customized Indexes

There are many times when portfolio managers need customized cost indexes for specified indexes, such as market capitalization, growth/value, sector, country, or specific stocks, etc. These cost indexes could be constructed using size (% ADV), dollar value, or share quantity as discussed above, as well as for a specified implementation strategy defined by the portfolio manager.

Each of these customized cost indexes can be constructed providing we have the required data, such as the normalized cost index described above, or historical market impact parameters and corresponding market conditions. We discuss three customization approaches for developing cost indexes below.

Actual Trading Costs. Many times portfolio managers are interested in historical costs for their specific trading style and investment strategies. As mentioned, these costs can be used for portfolio attribution as well as to assist managers improve future stock selection. In this case, portfolio managers can follow the approach described above and back-propagate their current trading costs using the normalized cost index. This is a very efficient and accurate estimate of the historical trading cost environment for portfolio managers whose investment styles and trading strategies will be similar over these periods.

Reconstruction. Portfolio managers at times will need to understand historical trading costs for investment styles that may be different from the investment styles and/or trading strategies they are currently using. This could be used for portfolio attribution or peer group comparison, or more likely, to investigate the profiting opportunity from a different portfolio construction approach via back-testing and simulation. In these situations, portfolio managers will recreate the historical cost index for the specific market conditions and specified orders. This requires managers to have access to not only the historical cost index but also the underlying historical market impact parameters.

Stock Specific Indexes. There are times when managers wish to analyze or back-test trading ideas on a per stock basis. To do so historically we would simply apply the historical market impact parameters and stock specific trading characteristics to compute the historical trading cost. But many active managers require a finer market impact estimate in order to differentiate cost across companies more than is possible from stock volatility and market cap. These are the active managers who are seeking to uncover superior investment opportunities through fundamental analysis using the company's balance sheet and accounting data. In these cases, it would be appropriate to correlate or define the market impact sensitivity parameter a_1 with company fundamentals to try to establish a relationship. Then, as long as we have historical company specific fundamentals, we can determine a stock specific trading cost sensitivity parameter. In Chapter 5, Estimating I-Star Model Parameters, we analyzed the error of the estimation model and showed that the error term is positively correlated with spreads, beta, and tracking error, and negatively correlated with log of market cap, and log of price. Thus these data relationships could be a very good starting point to estimate a stock specific sensitivity parameter. It is very important to point out here that this is a very time consuming and very data intensive exercise.

The question that often arises is, "Would it be better to construct stock specific market impact model?" Unfortunately, stock specific models are extremely difficult to uncover due to the nature of the data. For many stocks, price movement is dominated by the general market movement, volatility, and company news. Market impact for many stocks only has the dominating price effect if the order size is significant enough to dominate market movement and volatility. Additionally, many of the studies that uncover a stock specific relationship are not necessarily incorrect, but analysts will find it extremely difficult to establish a stock specific model across all stocks in all markets.

We encourage suspicious readers to test the validity of these findings. And to assist in this process, we outline a single stock market impact simulation exercise that can be used as the foundation for testing market impact modeling approaches later in the chapter (Market Impact Simulation).

REAL-TIME COST INDEX

A real-time cost index provides investors with the expected trading cost given actual market conditions and aggregated market imbalance. This metric combines actual volume patterns, volatility, and buying and selling pressure across all investors, and determines the fair value market price for these conditions. It is important to point out here that the derivation of the real-time cost index is not based on a specified order size. It is derived from aggregated (estimated) overall buying and selling pressure. The real-time cost index provides investors with the expected cost they should have incurred by trading over the specified trading horizon.

For example, if the customer buy order size is 5% ADV but there are additional incremental buyers in the market comprising an additional 30% ADV in aggregate, market impact will be representative of an order of $+35\%$ ADV not $+5\%$ (e.g., $+30\% + 5\% = 35\%$). If the customer buy order is 5% ADV but there are incremental sellers in the market comprising an additional -30% ADV in aggregate then market impact will be representative of an order of -25% ADV and not $+5\%$ ADV (e.g., $-30\% + 5\% = -25\%$). Actual impact cost is a function of aggregated market imbalance and actual market conditions.

Traders can use real-time cost indexes to measure and evaluate trading performance in real-time and hold their broker's accountable for trading decisions before the end of the day and while they still have time to revise the trading strategy or change algorithmic parameters. Portfolio managers can use these metrics in an attempt to uncover a mispricing or other potential profiting opportunity.

Let us examine how a real-time cost index can provide insight to hold the broker accountable. Suppose the broker provides a pre-trade cost estimate of 30 bp for the order. But by 1:30 p.m. the cost is already at 80 bp and is 50 bp more than quoted. So the typical scenario is for the buy-side trader to call the broker and ask why the cost is so much higher than the pre-trade estimate. The typical conversation that often follows is that the sell-side trader quotes to the buy-side trader that volatility is up, there is less volume and the market moved away or from us. While all of this may be true, it still does not address whether the $+80$ bp was justified for these market conditions. The question that should be asked here is: what should the cost be given higher volatility, less volume, and adverse market movement? Should the actual cost given these conditions only be 60 bp and not 80 bp, which would imply that the broker is providing sub-par performance by 20 bp, or should the actual cost in these conditions be 120 bp, in which case the broker would be adding value by 40 bp?

The real-time cost index will provide the expected trading cost given these market conditions and actual aggregate buying and selling pressure. Buy-side firms can easily quote these metrics and push their brokers to do a better job in explaining the issues surrounding actual trade performance. "Volatility is up and the market moved away" simply won't cut it anymore!

Another example for the usefulness of the real-time cost index is its application to critique the VWAP price. For example, suppose the fund achieved the VWAP price over the trading period but the VWAP price was 50 bp higher than the arrival price. The cost index will tell us if the 50 bp cost was the appropriate cost given market conditions and aggregated buying and selling pressure, or if the broker achieved the VWAP price by trading aggressively and pushing the price up.

Portfolio managers, on the other hand, can use real-time cost indexes to evaluate buying and selling opportunities. By analyzing pressure in the stock and market conditions, managers can determine whether the price is too higher or too low and whether a short-term reversal is likely to occur. These are just some of the high frequency strategies being tested by many managers.

Figure 12.3 illustrates the real-time cost index for AAPL on April 4, 2012 as of 1:30 p.m. The top graph shows the real-time trading cost

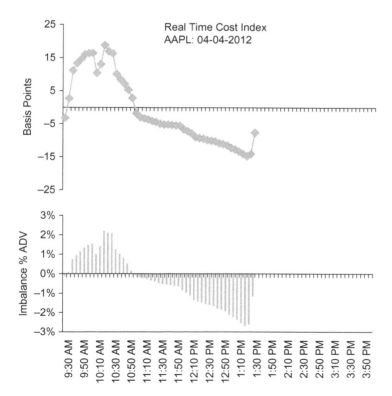

■ **Figure 12.3** Real Time Cost Index. *Source: NYSE Taq Data, I-Star Impact Model, Yahoo Finance, Google Finance.*

based on actual market conditions and aggregated buying and selling pressure from the open through 1:30 p.m. The bottom graph shows the corresponding (estimated) cumulative buy-sell imbalance in the stock over the same period of time. We used the I-Star model and corresponding parameters from Chapter 5 to compute these costs. The buy-sell imbalance was computed from tick data following the modified Lee & Ready tick rule. Then based on market conditions, volumes, volatility, and the buy-sell imbalance we are able to compute the trading cost. The graph shows that there was early buying pressure in the stock that pushed the trading cost up but the buying pressure was short lived. There was a subsequent sell-off in the stock beginning at around 10:15 a.m. continuing through about 1:15 p.m. causing prices to also decline.

The real-time cost index provides investors with a metric that can be used to measure the performance of their broker in real-time. Furthermore, this metric is transparent and can be verified by both parties, and computed over any specified time period used for the trading horizon. Brokers can no longer hide in the shadows and provide vague and elusive responses when questioned about their trading performance in real-time. The real-time cost index has leveled the playing field for buy-side and sell-side firms.

The most important aspect here is that it allows both broker and investor to better partner and determine when it is most appropriate to change intraday strategies and revise algorithmic trading parameters.

Figure 12.4 illustrates the end of day real-time cost index for AAPL during April 2012. The top graph shows the end of day real-time cost index given actual market conditions and buying-selling pressure. This is the cost an investor trading over the full day was expected to incur based on what actually occurred in the market conditions over that horizon. In actuality, the real-time cost index can be computed for any intraday time period. The bottom graph shows the estimated buy-sell imbalance on the day expressed as % ADV.

This metric can be used to further evaluate trader, broker, and algorithmic performance since it incorporates actual market activity. This measure can assess whether the VWAP price on the day was reasonable given market activity. Money managers have been using these metrics to determine whether there is any persistence of order imbalance in any stocks over time, which could signal a mispricing due to the market impact cost (here we have temporary impact disguised as permanent) with likely trend reversal. Money managers who uncover these signals could potentially earn a short-term profit.

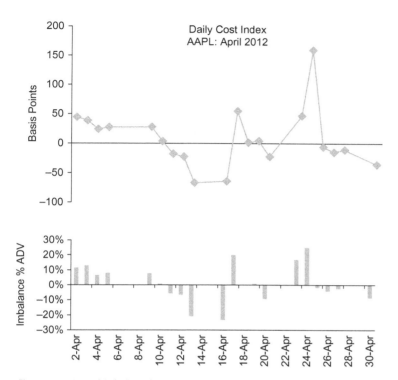

■ **Figure 12.4** Historical Daily Cost Index. *Source: NYSE Taq Data, I-Star Impact Model, Yahoo Finance, Google Finance*

Another way that daily imbalance data is being used by traders is in conjunction with index changes. Often the expected index change is known in advance, although the exact number of shares in the index and weighting may not be known until the announcement date. The annual Russell Index Reconstitution is a primary example of when the index changes could be pre-traded. The daily cost index is a key tool to help determine if the change has been overtraded on the actual reconstitution day. Changes that have been overtraded are likely to be those trades that have "gone the wrong way" on reconstitution day. That is, the adds to the index fall in price and the deletes from the index increase in price.

The error that we have seen most often with regards to the cost index and index changes is that brokers and vendors use the same market impact parameters as with normal non-event trading days. For index reconstitution, investors are best served with a set of impact parameters specific to the index event.

Back-Testing

Money managers will often construct portfolios via optimization processes. To help in the construction process these same managers will often run numerous studies testing different ideas, how prices react to different sets of factors, and how these portfolios perform under various market conditions. Interest lies not only in which companies will outperform or achieve excess returns, but rather how stocks will perform compared to different factors.

Quantitative managers spend much time and energy testing, re-testing, and verifying ideas and ensuring the uncovered relationships are statistically sound before settling on the preferred portfolio. Often this requires the investment ideas to be tested over a very long horizon, such as twenty to thirty years, or possibly more if data is available.

All too often, unfortunately, portfolio managers find a strategy works well in the back-testing analysis but once it is put into production the strategy does not achieve the expected level of return or worse case loses money. Why does this happen you ask. Well the primary reason is due to the implementation cost of the strategy. Costs are often much higher in reality than expected or planned. Of course, some of this is driven by the quant managers themselves. Quantitative managers do a great job at uncovering opportunities, but unfortunately all quants seem to find the same opportunities at the same time. This increases the buying and selling pressure in the stocks, leading to a higher cost. For example, if the manager's order is 5% ADV it is more likely that the group of quantitative managers will exert buying pressure in the stock close to possibly 50−75% ADV or more. This leads to higher trading costs in reality than would be reflective of the 5% order and could lead to reduced profits or losses. But it is also due to managers often using an unrealistic transaction cost estimate such as 20 bp or possibly just one half of the spread.

Another issue that could potentially arise is one where the manager deems the trading cost is too expensive and offsets the incremental alpha. Thus, the manager would abandon this strategy because they believe it would be unprofitable. However, many times managers are using historical costs that are too high or do not properly reflect today's trading environment. For example, it is possible that if we traded the same list historically based on today's market structure (decimals, algorithms, etc.) the transaction cost would be much lower than we expect thus making the strategy profitable. If the manager does not include a realistic trading cost expectation in their back-testing environment it may cause them to abandon a potentially profitable investment idea.

To combat these situations where managers incur a loss due to unexpected high trading costs or abandon a strategy that seems unprofitable due again to inaccurate transaction costs, managers are back-testing investment ideas using historical cost index series.

The important part of the cost index is to provide the historical trading cost based on today's market structure conditions not on the historical structure. For example, in the early 1990s the market was trading in 1/8ths or really 1/4s according to the odd-eighths paper of Christie and Stoll (1994). Let's take a look at what has happened since then. The market changed from the 1/8th quoting system to one of 1/16ths (teenies), the SEC order handling rules came into play, decimalization, algorithms, Reg-NMS, growth of electronic trading venues and the proliferation of dark pools, and now having ten, twelve, or more displayed venues compared to only two mutually exclusive exchanges, NYSE and NASDAQ.

All of these regulatory changes have dramatically improved the efficiency of our financial markets and reduced trading costs. Therefore, when we back-test using trading costs our goal is to construct a back-testing cost index series based on the costs that would have occurred historically based on today's market structure and trading environment, not what actually occurred back then. We use today's market structure because we do not want to miss out on an opportunity that would not have been profitable during yesterday's market structure, but given today's market environment the strategy is greatly profitable.

To assist managers resolve these issues and potentially uncover additional investment opportunities, we constructed a cost index based on today's market structure and the historical market conditions. This is shown in Figure 12.5 for US large cap stocks. Our cost index covers the periods from 1991−2012 (22 years). The cost index shows costs gradually decreasing over the 1990s but spiking in the late 1990s (Latam Crisis) and beginning of the tech boom with increased volatility. Costs remained at higher levels with large fluctuations until the tech bubble crash in March 2003. Cost then remained low through the quant crisis and dramatically spiked during the financial crisis, flash crash, and again during the US debt crisis. Quantitative portfolio managers could suffer large losses if they use a constant trading cost such as 20 bp to enter and exit positions. Many times a strategy appears profitable due to lower modeled costs, but these are the strategies that suffer the largest losses when implemented. This cost index could also result in portfolio managers finding that a strategy that would not have been profitable due to the

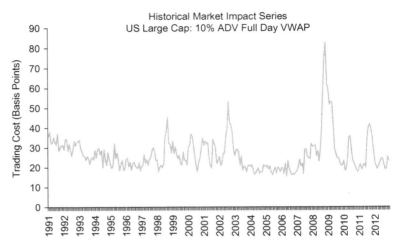

Historical Market Impact Series
US Large Cap: 10% ADV Full Day VWAP

■ **Figure 12.5** Historical Back-Testing.

spread size of 1/8th or 1/4 in the 1990s many now only incur a trading cost of a few cents and result in a profitable strategy.

These historical cost index series could be used in conjunction with a portfolio manager's current costs to determine an appropriate customized series via our back-propagation technique (Equation 12.4).

Market Impact Simulation

In this section we present a market impact simulation experiment to highlight the difficulty in developing stock specific market impact models. The approach also presents techniques that can be used by investors to test and evaluate different types of financial models, not just market impact models. The techniques are centered around simulating market conditions and prices following a defined model and model parameters. The exercise then sets out to estimate the model parameters based on the simulated data. If the process can uncover the "true" parameter values then the modeling and estimation approach is reasonable. Otherwise, even if the model is perfect, the difficulty in estimating model parameters may cause the approach to be unusable. Remember in this case that, since we are simulating data based on a specified model, the "true" parameters are those parameters that were used to simulate the data.

How difficult is it to derive a stock specific market impact model? In Chapter 5 we discussed techniques to derive a general equation market impact formula and a universal set of parameters across all stocks $(a_1, a_2, a_3, a_4, \text{and } b_1)$. Since this model incorporates stock volatility and stock ADV into the formulation it in a way provides different estimates across stocks. But recall the last part of the chapter where we uncovered a relationship between stock errors and company fundamental data. We found strong evidence that market impact is negatively correlated with log of market cap and log of price, and positively correlated with beta, spreads, and stock specific risk (tracking error). This finding suggests that we might still be able to improve our results at the stock level by including these data. But then why not simply calibrate a stock specific model similar to how we have stock specific volatility and beta estimates. The short answer is that we simply do not have enough data available. Price movement is often dominated by market movement, market noise (volatility), and buying and selling pressure from all other investors. This makes it very difficult to uncover stable statistical relationships between price movement and customer orders.

To illustrate the difficulty associated with estimating stock specific parameters let us examine the estimation process using simulated data[1]. We know that one of the biggest drawbacks with market impact estimation is the "Heisenberg uncertainty principle of trading"; that is, we can only observe the price trajectory with the order or the price trajectory without the order—not both. Therefore, we are not able to accurately determine price movement caused solely by the order. Well this is certainty true in reality. But let's take a step back for a moment. Suppose that we do know the exact relationship between price movement and buying and selling pressure of a trade. Then we can simulate trade data and test our market impact estimation approach on the simulated data to determine if we have an accurate estimation technique. The important point to keep in mind is that stock price is driven by many factors such as stock specific alpha, general market movement, impact from the order, buying and selling pressure from other participants, and price volatility. If we know exactly the relationship between price impact and buying-selling pressure we can simulate a market impact data series, and then test our model against that simulated series to determine if our modeling approach is able to uncover the true relationship.

---

[1]A variation of this exercise was previously given to my Cornell University Financial Engineering graduate students as a final project, and it does a great job to highlight the difficulty in constructing stock specific parameters.

Analysts are encouraged to duplicate these simulation tests with different order sizes, volatility, liquidity, and market impact sensitivity to observe the difficulty with calibrating a stock specific model. Then repeat the same experiment for a group of stocks, say 100 plus stocks, and then again with say 500 stocks. A pattern will start to emerge.

Simulation Scenario

Stock RLK. Current price $P_0 = \$50$, annualized volatility $\sigma = 30\%$, the stock trades 1,000,000 shares per day. The customer order will consist of order sizes from -25% to $+25\%$ ADV. The average trade size is 200 shares per trade and the distribution of trade size is shown below as $Shares(t)$.

Simulate market trades for the customer and all other investors. Let the total volume traded from all other market participants be equal to one million shares where the side of the order is randomly assigned (50% chance of a buy and 50% chance of a sell), and let the customer order size and side be specified in advance. Then the simulated data is as follows. Let,

$$P_0 = 50$$

$$Side(t) = \begin{cases} +1 & 0.50 \\ -1 & 0.50 \end{cases}$$

$$Customer\ Side(t) = +1 \text{ (buy order)}$$

The customer order side is specified in advance:

$$Shares(t) = \begin{cases} 0.90 & 100 \\ 0.06 & 500 \\ 0.03 & 1000 \\ 0.01 & 5000 \end{cases}$$

$$Avg\ Share\ Size = 200$$

$$Number\ trades\ during\ the\ day = 5000$$

$$ADV = 1,000,000$$

$$MI(Shares(t)) = 0.0000025 \cdot Shares(t)$$

For a 20% order size on a stock that trades one million shares per day, this market impact cost will be equivalent to 40 bp (and is consistent with our findings in Chapter 5). For simplicity, we assume a linear impact relationship but analysts are encouraged to try various formulations of the impact model.

$$Beta = 1$$

$$\sigma_{market} = 0.20$$

$$\sigma_{stock} = 0.30$$

$$\sigma_\varepsilon = \sqrt{0.30^2 - 0.20^2}$$

R_m = simulated market returns following random walk with 5000 trades in the day with volatility scaled for one day and one trading period, i.e.,

$$\sigma_{market} \ Per \ Trade \ Period = 0.20 \cdot \frac{1}{\sqrt{250}} \cdot \frac{1}{\sqrt{5000}}$$

The simulation process is as follows:

Step 1. Specify the customer order size and side. For example, in iteration 1 specify a buy order for 200,000 shares of RLK. This represents 20% ADV and will consist of approximately 1000 customer trades since stock RLK has an average trade size of 200 shares. Analysts performing this simulation exercise will change the order size and side in each iteration.

Step 2. Simulate 5000 trades from other market participants and the number of trades from the simulated order. Sequence the customer trades throughout the day following any preferred methodology. For example, sequence customer trades over the day following a VWAP strategy or an aggressive front-loaded strategy. Customer order trades should be alternated with market participant trades in a random fashion but following the specified strategy. We encourage analysts to experiment with various sequencing schemes to simulate different trading algorithms. For the first iteration there will be 6000 trades in total: 5000 from other market participants and 1000 from the customer order.

Step 3. Simulate market prices for these orders:

$$P_t = P_{t-1} + (MI(Shares(t)) \cdot Side(t)) - 0.95 \cdot (MI(Shares(t-1)) \\ \cdot Side(t-1)) + Beta \cdot R_m + \xi_t$$

where $\xi_t \sim N(0, \sigma_\varepsilon)$ and $t = 1$ to 6000. Notice that in the formulation above we add in the full market impact of the trade and subtract 95% of the impact from the previous trade. This accounts for the dissipation of temporary impact—we assume an immediate dissipation of 95%. The 95% factor is consistent with our findings in Chapter 5. Readers are encouraged to experiment with various different temporary percentages and market impact sensitivities. Also notice that the side of the order for the customer will be either $+1$ or -1 in the full iteration depending if the side was specified to be a buy order or a sell order respectively. The side of the order from all other market participants will be randomized (50% chance the trade was initiated from a buy order and 50% chance the trade was initiated from a sell order). The side of the order in our example only affects the market impact cost of the trade.

Step 4. Compute the average execution price of the customer's trade using the simulated price data above.

Step 5. Compute the customer trade cost as the difference between the average execution price and the starting price (adjusted for the side of the order).

Step 6. Repeat this experiment 22 times (one month of data) changing the customer's order size and side designations in each iteration.

Step 7. Plot the customer's trading cost on each day as a function of order size.

Step 8. Estimate the market impact sensitivity for one month of data and observe how close the estimated value is to the true value.

Step 9. Repeat this experiment several times to mimic several months of data. For each month estimate the market impact sensitivity.

Analysts are encouraged to perform these simulation exercises changing the variables above such as the temporary impact dissipation rate, market participant side parameter, total volume on the day, beta, volatility, etc. This will further highlight the difficulty in uncovering the true relationship between market impact and customer order size.

Figure 12.6 illustrates the above simulation exercise for one month (22 trading days) of data. The order sizes range from 0% to 25% ADV. But the relationship we uncovered from the data indicates that cost and size are negatively related which would suggest that larger order sizes have lower costs. The reason we have difficulty in uncovering a relationship between customer order and impact is, as mentioned previously, actual price movement is often dominated by stock alpha, market movement, buying and selling pressure from other market participants, and volatility.

Readers who carry out this simulation exercise are sure to have a difficult time estimating accurate market impact sensitivities using the customer order. In addition, this exercise also shows how unstable these parameters could be month to month (readers are encouraged to simulate several months worth of data and estimate parameters in each month). Notice that we used a very simple market impact model above and very basic assumptions and we

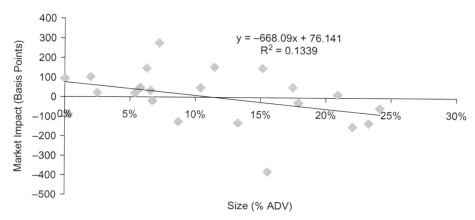

■ Figure 12.6 Single Stock Market Impact Simulation.

still encountered difficulty. With a more sophisticated model it would be even more difficult to obtain accurate results. But if we simulate data for a universe of stocks (e.g., for 500 stocks) over a month we will see a pattern begin to emerge, even when using stocks with different volatilities and different average daily volumes. Suspicious readers are encouraged to duplicate the analysis outlined above. This should convince our doubtful readers.

MULTI-ASSET CLASS INVESTING

Investing in Beta Exposure and Other Factors

Many times portfolio managers are interested in acquiring a specific market exposure, and the actual holdings in the portfolio are not as important as long as the portfolio achieves the desired exposure level. For example, managers looking to acquire exposure to the general market index—SP500—have several different investment vehicles. They could invest in the underlying stocks, they could purchase an exchange traded fund, or they could construct a portfolio comprised of futures contracts on the SP500. In all of these cases, the manager's portfolio will have the same beta exposure and the same underlying growth characteristics. Therefore, from the perspective of investor utility, portfolio managers should be indifferent as to which portfolio they actually hold.

The only difference in returns across these portfolios then should be due to the implementation cost of acquiring the asset and any corresponding management fee. To achieve the highest returns possible, managers need to understand the cost structure of each asset, and choose the most cost effective path to gain the desired exposure.

Example:

A portfolio manager looking to gain exposure to the general market index can purchase either stocks, exchange traded funds, or futures. How should the manager determine the best approach to acquire the exposure?

Based on what we have observed in practice, stocks often have the lowest cost for smaller sizes, followed by exchange traded funds for slightly larger sizes, and then futures for the largest sizes. This is illustrated in Figure 12.7a. In this example, stocks are the most inexpensive option for investment dollars up to $250 million. Exchange traded funds are the most inexpensive investment vehicle for investment dollars from $250 million to $667 million. And futures are the most inexpensive for investments of $667 million and higher.

Note: These values and breakpoints are for illustration purposes only based on data from 2011. Investors need to apply techniques provided in this chapter to determine the exact break points for today's market conditions.

What causes these differences across investment vehicles?

Equities: Equities often have the lowest implementation cost for smaller dollar values but these costs increased at the fastest rates. Equity trading costs tend to increase at the fastest rates because buying or selling pressure often causes market participants to believe that the excessive transaction pressure is due to a mispricing of the stock price or due to changing company fundamentals that have not yet been fully disseminated into the market. The corresponding price change then often attracts momentum players and active managers hoping to achieve a short-term trading profit which further impacts the stock price.

Exchange Traded Funds: ETFs tend to have initial trading costs (intercept term) higher than the underlying stocks. This is primarily due to the trading cost corresponding with acquiring the position (similar to stocks) and also the management fee charged by the ETF fund manager for maintaining the appropriate ETF portfolio. ETF trading costs, however, increase at a slower rate with investment value than the underlying stocks due to shadow liquidity. Shadow liquidity refers to the potential trading volumes from market participants who stand ready to buy or sell shares in an ETF if there is a mispricing in the market. As soon as the ETF price

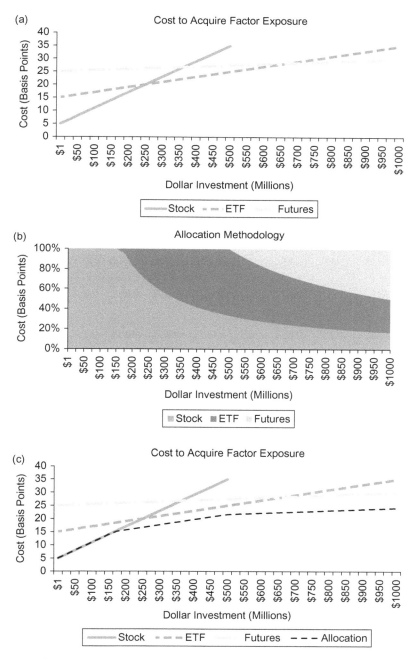

■ **Figure 12.7** Beta Investment Allocation.

is pushed too high or too low compared to the underlying securities these market participants jump in and perform a statistical arbitrage type function. If these participants can acquire the underlying stocks at a lower price than the ETF they will buy the shares and sell the ETF. If the ETF price is lower (due to selling pressure) than the price of the underlying securities these participants will buy the ETF and sell the stock. In theory, they will have a net zero risk position and will profit from the difference in prices. Since the acquired position is hedged—they have the exact same long and short exposure—they can trade out of both positions passively over time without incurring high trading costs and lock in a profit. Additionally, ETFs can often increase or decrease the number of its outstanding shares through creation and/or redemption. This differs from company stock that has a constant number of shares outstanding. In these cases, portfolio managers can purchase the underlying ETFs in the market if there are sufficient sellers, and if not, they can purchase the underlying equity shares and create the ETF. Alternatively, portfolio managers could sell the ETFs in the market if there are sufficient buyers, and if not, managers could redeem the shares and sell them in the market.

There is often a fee corresponding to the creation and redemption process but in these cases the fee would still be less than the incremental market impact cost for attracting necessary counterparties. The last point worth mentioning here is that ETFs do not suffer the same information content as stocks since ETF transactions are more likely believed to be due to a macro event rather than any specific company event. If there is a large buyer or seller of a stock it is more often believed to be due to company specific information such as a mispricing, undervaluation, or simply changing company fundamentals that have not yet been fully disseminated into the market.

How does this affect the ETF volumes in the market impact model? Since ETFs have corresponding "shadow" liquidity, analysts will often use a higher volume estimate than is reflected in the data. Some analysts may apply an adjustment factor, e.g., 1.5, and some analysts may attempt to measure total potential volume that could be used to create and redeem shares from the underlying stock volume data and corresponding weights in the ETF.

How about ETF market impact price sensitivity? The price sensitivity expression in the ETF market impact model a_1 is often higher than the stock model price sensitivity parameter. This is because the ETF volatility is lower due to diversification (e.g., market risk only) whereas stock volatility is comprised of both market risk and company specific risk. The ETF price sensitivity a_1 parameter needs to be higher to avoid a potential arbitrage opportunity. For example, suppose volumes are identical across stocks and ETF (we are ignoring shadow liquidity here for simplicity). If the manager invests the same dollar amount in the ETF or in the underlying stock portfolio (following the same weightings) the trading costs should be the same. But since the ETF volatility will be less than the weighted stock portfolio the market impact model will estimate lower costs for the ETF than the portfolio of stocks. To correct for the mispricing, a proper ETF market impact model will need to have a higher sensitivity term as follows:

$$a_{1ETF} = a_{1Stock} \cdot \left(\frac{w'\sigma}{\sigma_p}\right)^{a_3}$$

where $w'\sigma$ is the weighted volatility of the ETF portfolio and σ_p is the ETF portfolio volatility incorporating all covariance and correlation benefits.

Futures: The initial cost of trading futures is usually the highest compared to ETFs and equities. The reason is primarily due to the roll cost associated with purchasing the next futures contract at the time of the contract expiration. Managers who maintain a portfolio of futures contracts will need to continuously purchase the next contract. This creates a recurring cost. An advantage of trading futures is that the contract sizes are usually extremely large in size and trading costs increase at the slowest rate compared to ETFs and equities. Investors will transact in futures for various reason. First investors purchase futures to hedge positions. Second, managers purchase futures contracts often for speculation. Third, investors have begun investing in futures portfolios due to the cost advantage they provide for very large orders. All of these investment reasons provide a great deal of liquidity for futures contracts and result in lower incremental trading costs.

Beta Investment Allocation

A common misconception with beta allocation investment strategies is that the portfolio manager should purchase only a single investment vehicle to achieve their exposure. For example in Figure 12.7a we found that equities are the most cost efficient vehicle if dollar value is less than $250 million. ETFs are the most cost effective vehicle if dollar value is between $250 and $667 million, and futures are the most cost effective if dollar value is greater than $667 million. But this is only true if the

Table 12.1 Allocation Schedule

Investment Value	Allocation
Dollar Value < $167M	All Dollars in Equities
$167M ≤ Dollar Value ≤ $500M	$167M in Stock, (Dollar Value−$167M) in ETF
Dollars > $500M	$167M in Stock, $333M in ETF, (Dollar−$500) in Futures

portfolio manager can only invest in a single asset. Managers, however, can incur lower trading costs if they allocate investment dollars across all three alternative options (Figure 12.7b). For example, equities are the lowest trading cost up to $167 million dollars. For investment values between $167 million and $500 million managers would be $167 million in equities and the remainder in ETFs up to $500 million. And after $500 million, the allocation should be $167 million in equities, $333 million in ETFs, and the remainder in futures. Now the manager could achieve the same exposure but a lower trading cost.

Figure 12.7c shows the minimum cost allocation scheme compared to the all-or-none examples in Figure 12.7a. Notice that for values up to $167 million the costs are the same but at higher dollar values the manager is best served via an allocation schedule as shown in Figure 12.7b (Table 12.1).

MULTI-ASSET TRADING COSTS

In this section we examine multi-asset trading costs[2]. We utilize the I-Star impact model, developed in earlier chapters, to investigate differences in cost structures across the asset classes, as well as to estimate the underlying transaction costs for various order sizes. Our analysis across asset classes found that the I-Star model performed well across both global equity markets and across the different asset classes, hence, it is an important decision-making tool for portfolio managers to evaluate asset allocation, portfolio construction, and best execution trading strategies.

---

[2]Scott Wilson, Ph.D. provided much of the early direction and insight in applying the I-Star impact model to estimate trading costs across various asset classes. He performed this leading research while an intern at a large pension plan and while completing his Ph.D. in Economics. He is currently working for Cornerstone Research.

The market impact model parameters used in our analysis were calibrated using data over the period 1H2011. See Kissell Research Group (www. KissellResearch.com) for updated impact parameters[3].

Global Equity Markets

The first part of this analysis consisted of evaluating trading costs across the global equity markets. This included: US Large Cap Stocks (US-LC), US Small Cap Stocks (US-SC), Canada, Europe, Australia, Hong Kong, Japan, and China, as well as for Developed Europe, Developed Asia, Latin America, and the Frontier Markets.

The model parameters were calibrated following the techniques provided in chapter 5 and using data from 1H2011. We then estimated trading cost for an order size of 10% ADV executed via a full day VWAP strategy. We used a constant volatility rating of 25% for all groups to allow for fair comparison across all markets. As expected, US and Canada large cap stock trading costs were relatively stable over the analysis period except for a spike in Aug-Sep 2011 due to the US debt crisis. Canadian trading costs were stable throughout the period and were not as affected by the economic issues encountered in the US. Japan and Hong Kong were the markets with the next lowest trading costs. Both countries had months in 2011 where impact cost spiked due to changing price sensitivity caused by economic and political events in the region. Developed Europe experienced the greatest fluctuation in trading costs throughout 2011 with periods of spiking cost which appeared to be related to the ongoing macro-economic climate and uncertainty in Europe. Australia and China had trading costs that were consistent with US-SC stocks and relatively stable over the period. The emerging markets and Latin America countries experienced much higher trading costs over the analysis period 2011. This was primarily due to a much higher information content of the order (at least a much higher perceived information content) and a resulting higher permanent market impact cost. Costs in these markets were more than 2.5 times greater than for US large cap stocks. The Frontier markets had by far the highest trading costs in 2011. These costs were more than 4.0 times higher than for US large cap stocks. The higher costs in the Frontier markets appeared to be driven by hyper sensitivity to order flow and trade imbalance, resulting in high information

---

[3]Kissell Research Group maintains updated market impact parameters and trading cost estimates for various asset classes, see www.KissellResearch.com for most recent data sets.

contents of the trade and higher permanent impact costs. This was consistent with the findings in the other emerging markets and Latin America.

The parameters of the model for our analysis and trading costs estimates for 2011 are provided in Table 12.2a. Investors interested in current trading costs by global region and country are referred to www.KissellResearch.com (see I–Star Global Cost Index Quarterly Report and Country Trading Cost Analysis).

Multi-Asset Classes

The second part of this analysis consisted of evaluating trading costs across multi-asset classes. Our asset classes consisted of US Large Cap Stocks (US-LC), US Small Cap Stocks (US-SC), liquid and illiquid Exchange Traded Funds (ETFs), Futures, Government and Corporate Bonds, Commodities, and Exchange Rates (FX). In this analysis, we computed the trading cost for a trade value of $10 million USD executed via a strategy of POV = 10%. This was used in place of a constant order size (% ADV) which is more commonly used in the equity markets because transaction values can vary dramatically across asset classes. In our multi-asset trading cost analysis, we also placed boundaries on some of the model parameters in order to make fair comparisons of costs across different asset classes. Portfolio managers and analysts can achieve improvements in the models forecasting accuracy by allowing more freedom on the values of the parameters and eliminating the constraints using the parameter estimation phase. Our first goal, however, was to uncover an appropriate market impact model, determine the cost structure surrounding trading costs in the different asset classes, and for evaluation and comparison to the global equity markets. Further research is suggested to determine appropriate bounds on the underlying model parameters.

Why do trading costs vary across asset classes?

Our analysis found that trading costs vary across asset classes for several reasons. These include: 1) Investment Objective, 2) Trading Liquidity, and 3) Competition.

Definitions. There are many reasons why investors will select to transact different instruments. For example, the most common investment objectives include: 1) buy and hold investing, 2) risk hedging, 3) speculation, and 4) price arbitrage opportunities.

Trading Liquidity includes: 1) trade volume, 2) shadow liquidity, 3) mispricing liquidity, and 4) factor exposure liquidity. Each is described as follows. Trade volume across the asset classes is a very general term

Table 12.2a Equity Market Trading Cost Analysis (2011): Quantity Expressed in Terms of Order Size (%ADV)

Parameter	US-LC	US-SC	Canada	Developed Europe	Australia	Hong Kong	Japan	China	Emerging Europe	Emerging Asia	Latam	Frontier
a1:	1507.5	1831.7	1525.6	1772.7	1809.9	1333.4	1543.7	1351.2	1945.9	2431.9	2356.0	2756.0
a2:	0.38	0.45	0.41	0.60	0.65	0.50	0.49	0.41	0.56	0.52	0.52	0.42
a3:	0.94	0.91	0.95	0.81	0.60	0.81	0.85	0.91	0.74	0.92	1.05	1.05
a4:	1.05	1.04	0.94	1.05	0.94	0.95	0.93	0.96	1.00	1.00	0.80	0.80
b1:	0.97	0.93	0.97	0.90	0.95	0.94	0.95	0.90	0.83	0.84	0.81	0.82
Size (% ADV)	10%	10%	10%	10%	10%	10%	10%	10%	10%	10%	10%	10%
Volatility	25%	25%	25%	25%	25%	25%	25%	25%	25%	25%	25%	25%
POV Rate	9%	9%	9%	9%	9%	9%	9%	9%	9%	9%	9%	9%
I-Star (bp)	169.6	183.1	159.0	144.1	177.5	137.4	152.3	148.9	192.1	205.1	166.0	244.4
MI (bp)	18.4	27.0	21.0	24.8	26.6	21.5	23.0	28.3	46.8	48.5	51.3	73.4

Table 12.2b Multi-Asset Trading Cost Analysis (2011): Quantity Expressed in Terms of Constant $USD Value

Parameter	US-LC	US-SC	Liquid ETF	Illiquid ETF	Futures	Gov't Bond	Corp. Bond	Commodity	Currency
a1:	0.97	1.13	0.24	0.41	0.22	0.19	2.76	0.54	0.15
a2:	0.38	0.45	0.38	0.40	0.38	0.37	0.38	0.38	0.41
a3:	1.00	1.00	1.00	1.00	1.00	1.00	1.00	1.00	1.00
a4:	1.00	1.00	1.00	1.00	1.00	1.00	1.00	1.00	1.00
b1:	0.97	0.93	0.99	0.94	0.99	1.00	0.80	0.99	0.90
Dollars	$10,000,000	$10,000,000	$10,000,000	$10,000,000	$10,000,000	$10,000,000	$10,000,000	$10,000,000	$10,000,000
Volatility	25%	25%	25%	25%	25%	25%	25%	25%	25%
POV Rate	10%	10%	10%	10%	10%	10%	10%	10%	10%
I-Star (bp)	110.9	399.6	27.7	65.2	24.7	18.9	314.9	62.1	27.7
MI (bp)	14.1	65.1	3.0	10.0	2.7	1.9	88.2	6.8	5.3

used to denote actual transaction volume, transaction value (in dollars), as well as number of contracts, etc. Shadow liquidity refers to the underlying stock liquidity for a financial product where the underlying pricing scheme is a financial instrument that trades on its own in the market. The term is most commonly used to refer to the underlying stock volume for an ETF instrument. For example, investors wishing to buy an ETF can either purchase the ETF in the market or purchase the underlying stocks that comprise the ETF in the market and then create the ETF. Hence, investors wishing to purchase an ETF have two available sources of liquidity that can be used to complete the transaction. Mispricing liquidity refers to the volume that is on-standby in the market and ready to transact if there is a mispricing between two instruments or an arbitrage opportunity. The most common occurrence of mispricing liquidity is associated with statistical arbitrage traders who are standing by ready to transact in an index and its underlying stock members or in an exchange traded fund and its underlying stock members if there is a market mispricing. These traders will sell (short) shares in the over-valued instrument and buy shares in the under-valued instrument. Factor exposure liquidity refers to the investors' ability to invest in a similar financial instrument which provides the risk characteristics and expected returns stream as the desired instrument. For example, investors interested in gaining exposure to the SP500 index have numerous options available. They can purchase the stocks that comprise the SP500 index, purchase any of the large cap SP500 index exchange traded funds, an SP500 futures contract, a mini futures contract, etc. All of these instruments will provide the investor with exact same returns and risk. They all have the same risk composition and same stream of future returns.

Competition refers to the investors' ability to transact the financial instrument from various venues, broker-dealers, and/or market participants. Competition has been found to dramatically reduce trading costs in the equity markets. A marketplace with multiple venues is more competitive and cost efficient than a market with a single or relatively few dealers for the financial instrument.

Observations

The following results are based on empirical data and market observations over the period 1H2011 and trade costs estimates were based on a transaction size of $10 million USD.

Equities. A trade value of $10 million USD is equivalent to an order size of 4% ADV for a US large cap stock and an order size of 80% ADV for a US small cap stock. The reason the small cap order size is so much larger than for the large cap order size is that the price of the small cap stock is about one-half the price of large cap stocks and small cap volume is about one-tenth large cap stock. Thus, resulting in small cap order sizes that are dramatically larger than for large cap stocks for the same dollar value. The corresponding cost estimate of $10 million USD was 14.1bp for the large cap stock and 65.1bp for the small cap stock. This results in a small cap stock cost that is 4.6 times greater than for the equivalent dollar value invested in the large cap stock.

Exchange Traded Funds (ETFs). Cost estimates for the liquid ETF were approximately −80% less than costs for large cap stocks. This reduction in cost was primarily due to the shadow liquidity and factor exposure liquidity corresponding to exchange traded funds. For example, using a single stock market impact model, the estimated cost for transacting 100% ADV of a broad market ETF such as the SPY could be as high as 200bp. But the actual trading cost for this investment is closer to 10−20bp since the investor has many options to achieve this market exposure and the desired ETF. Investors could 1) purchase the desired ETF, 2) they could purchase the underlying stocks and create the ETF, 3) they could purchase a Futures contract, exchange the futures for the physical stock, and then create the ETF. Costs corresponding to the illiquid ETFs were found to be −25% to −35% less than large cap stocks.

Futures. Future trading costs (stock index futures) were found to be −80% less than large cap stocks. Future contracts were also found to be less sensitive to larger order sizes than for stocks or for ETFs. Investors could improve the model forecasting accuracy by fine tuning the impact model for each index individually. Portfolio managers wishing to invest in a futures portfolio, however, will incur an incremental cost at the time of futures expiration where they will have to settle the contract then purchase another futures contract. This is known as the "roll cost" and it plays a large part in the total trading cost of futures contracts.

Bonds. The corresponding cost of government bonds were -90% lower than the cost of large cap stocks. Most of the cost was due to the spread cost of the bonds. Unlike equities, where investors can purchase shares at the bid and sell shares at the ask (offer), investors are much more likely to pay the full spread cost when transacting government bonds. The corresponding cost of corporate bonds was dramatically higher than large cap stocks. Our analysis found that the cost of transacting corporate bonds was +526% higher than the cost of transacting large cap stocks. That is, corporate bonds were 6.26 times more expensive to transact than large cap stocks! This appears to be due to much smaller availability of corporate bonds than for equities and also the bid-ask spread. Investors seeking to purchase corporate bonds will often have to find a dealer with an existing inventory and then additionally pay the ask price. Investors seeking to sell corporate bonds will need to find a dealer willing to take on inventory and then sell the bonds at the bid price. The corresponding risk-premium of a corporate bond can be dramatically reduced by these transaction costs if the investor does not hold the bond for its remaining duration. We do expect corporate bond transaction costs to decrease with increased market transparency. We did not find a large relationship between corporate bonds and order size as has been found with equities. Transaction costs in the corporate bond market appear to be related to the competitiveness of the market and number of corresponding dealers.

Commodities. Commodity transaction costs were on average about -50% lower than large cap stocks. Much of this cost was due to the bid-ask spread rather than due to the transaction size. We also found that trading costs across different commodities varied greatly. For example, precious metals had costs that were much different than fossil fuels such as oil and natural gas, and were much different than agricultural goods such as corn, sugar, wheat, etc. Commodity prices did not appear to be strongly related to actual transaction size as it is for equities. Thus, a structural difference between equities and commodities.

Currency. Currency trading (e.g., exchange rates, FX) was -63% less than large cap stocks. The largest component of the FX trading cost was the market spread. We did not find as large a relationship between trading cost and transaction value for currencies as we observed for stocks. But investors did transact at the full spread rather than transacting within the spread as is often accomplished in the equities markets. The market structure for FX trading is much different than the equities markets.

These results are shown in Table 12.2b.

A major finding between the cost structure with equities and other asset classes is that price sensitivity to the underlying order size (e.g., price elasticity) is much more instrument specific than it is for stocks. We found the parameters a_1, a_2, a_3, a_4 and b_1 were relatively stable across stocks in our equity market grouping (e.g., US-LC, US-SC, Europe, etc.), but these parameters did vary by instruments in other asset classes. For example, the model parameters could be much different for a liquid broad market ETF compared to a specific factor ETF such as a dividend yielding ETF or a bond index ETF. Model parameters could also vary greatly across the many different commodities such as Oil, Natural Gas, Gold, Silver, Corn, Wheat, etc.

Room for Improvement. As mentioned above, in our multi-asset trading cost analysis we bounded the market impact parameters (e.g., set constraints on the potential set of solutions) in order to be able to make a fair cost comparison of trading costs across the different assets classes. But as our results above found, we can improve the forecasting accuracy of the model by allowing these parameters to vary by asset class. Many of the asset classes have relationships different from the equity markets. For example, currencies, commodities, and corporate bonds were found to have a much lower relationship with cost and size as was found for equities. Portfolio managers investing in multi-asset classes were best served by using a market impact model that was constructed specifically for that financial instrument, and allowing the parameters of the model to vary appropriately.

Finally, the cost structure of many of these asset classes has changed since 2011. Investors seeking to have the most up to date cost estimates and impact model parameters are referred to www.KissellResearch.com as well as Kissell (2013), "Multi-Asset Trading Cost Estimates," working paper available upon request.

High Frequency Trading and Black Box Models

Ayub Hanif, Ph.D.

INTRODUCTION

The growth in active investing has seen the rise of high frequency trading. Since the credit crisis and the subsequent flurry of regulations with respect to derivatives, no area of finance has been under such scrutiny as high frequency trading. High frequency trading is characterized by an enhanced turnover of capital in response to market dynamics. This usually results in enhanced trading activity coupled with smaller gains per trade. In contrast with traditional investing and trading, the holding period and thus the investment horizon is orders of magnitude smaller.

High frequency trading strategies themselves have found preponderance amongst money managers with most opting to close these strategies out daily (Aldridge, 2009). Closing out positions overnight provides three key advantages to a money manager:

- Through the proliferation of trading and execution venues, trading is now virtually a 24-hour a day activity. Thus overnight positions, given the slivers they are trying to capitalize on, could become extremely risky. Closing flat at the end of trading in one's main market removes this risk.
- Capital is not committed beyond the course of close of business, enabling transparent accounting.
- Typically, overnight positions taken out on margin are paid for at the overnight carry rate which is usually slightly above LIBOR. Overnight LIBOR volatility and the chances of hyperinflation can quickly deteriorate with overnight positions becoming increasingly expensive and diminishing any profits accrued. Closing out avoids overnight carry risk, providing substantial savings in tight lending and high-interest regimes.

Furthermore, it has been argued that high frequency trading contributes to market efficiency. There is added liquidity, reduced trading costs and

The Science of Algorithmic Trading and Portfolio Management. DOI: http://dx.doi.org/10.1016/B978-0-12-401689-7.00013-1

a general stabilization of market systems. Though they are premised in the antithesis of efficient markets, high frequency strategies remove market inefficiencies and impound information into prices faster. Liquidity is enhanced, owing to the increased trade count; however, during severe dislocations this may become scarce. In addition, automation provides numerous operational benefits, i.e., a reduced headcount and thus reduced expenses alongside removal of human error.

Frequently, the term black box trading is used to describe high frequency trading strategies. Black box trading or black box models in general are seemingly complex and mathematically sophisticated ways of detecting anomalies and inefficiencies in the market. Indeed, quantitative trading thrives through such a façade; however; the innards of such models are not beyond the understanding of most people. Quantitative trading applies a rigorous, thorough and scientific method to the trading process. These are the inner workings of the black box, which are commonly understood to be unknown and unknowable. Figure 13.1 removes the illusory veneer from the black box. It is important to note that such a setup is not the same across traders. However, it succinctly captures the key groupings within their frameworks (Narang, 2009).

Inside the black box there are three key components: the alpha model, the risk model, and the transaction cost model, which feed into the portfolio construction model, which in turn feeds into the execution model. The alpha model is the edge sought by quant traders, being the focus of this

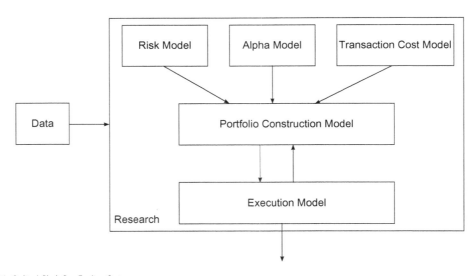

■ **Figure 13.1** Stylized Black Box Trading System

chapter, and usually involves some form of price prediction. The risk model in turn ensures checks and balances against adverse positions and exposures are in place. The transaction cost model aids the portfolio construction model to decipher what trades and/or positions are necessary to achieve the trading goal, passing these decisions onto the execution model to enact.

Transient to the core of the black box is accurate and well-curated data. High frequency, or tick-by-tick, data is vast. The number of observations for a single day in a liquid market is around 30 years of daily data (Dacorogna et al., 2001). Rigorous research on accurate data is at the very core of a successful high frequency strategy. Prior to detailing some strategies and an evaluation framework, we shall take a closer look at high frequency data.

DATA AND RESEARCH

High frequency data originates from the financial markets. By its very nature it is irregularly spaced in time, however, and with the sheer volume being reported by liquid markets can only be understood using continuous dynamics (Hanif and Protopapas, 2013). Financial data providers usually report hundreds of thousands of prices for a single market a day. Dacorogna et al. (2001) argue, correctly, that high frequency data should be the primary concern of those interested in understanding the financial markets, especially given the effect of market dynamics on everyday investors. Unfortunately this is not the case with most academic and empirical financial literature for, they argue, two reasons.

Firstly, it is costly and resource-intensive to collect, store, manipulate and curate high frequency data. This is precisely why most available data is either daily or lower frequencies. The second reason is more subtle. Most statistical modeling and machine learning tools assume time-series homogeneity; that is they assume regularly spaced data points. Little work has been done to look into irregular data with Hanif and Protopapas (2013), described above, working towards bridging the gap between regularized and irregular time series. Homogeneous financial time-series do not come from the markets but are the result of post-processing on raw tick-by-tick data.

Time-series operators such as interpolation methods are used to transform inhomogeneous time-series into a homogeneous time-series ready for analysis. Variables which are commonly used to capture intraday dynamics include price, return, realized volatility, bid-ask spread, tick frequency, realized skewness, volatility ratio between time resolutions, direction change indicators, and overlapping returns. A word of caution. A cursory glance at financial markets will display the complexity of simply inquiring about the price. Price quotes could be the bid-ask pair; transaction prices, which may or may not be former bid or ask quotes; irregular bid, ask, transaction prices; or simply the mid-price. The research process needs to be clear in its acquisition of the correct data needed, as trivial requests may lead to adverse results.

Having acquired high frequency data we are now ready for the analysis. The preferred methodology for research is the scientific method which consists of three steps (Dacorogna et al., 2001). The first of these steps is to discover and understand the fundamental, statistical properties of the data. These properties are known as "stylized facts" in the econometrics and finance literature and are quite widely reported. See, for example, Cont (2001) for an empirical analysis of asset returns and their associated stylized facts and statistical properties.

Secondly, we formulate models based on empirical facts. The focus here should be on robust models derived from empirical analysis rather than on loose conjecture. This is where our understanding of the markets and the data should coincide. There has been quite some debate on whether to take a time-series (modeling statistical properties of the data) or microstructure (modeling market behavior) approach, though with high frequency data one should be able to test microstructure models (Hasbrouck, 2007; Dacorogna et al., 2001). The third and final step is to test whether these models reproduce the stylized facts and statistical properties discovered in the first step. The focus here is on prediction of future movements, opportunities, risks and rewards. It is thus clear that to undertake this process successfully, we need good, reliable high frequency data.

STRATEGIES
Statistical Arbitrage

If you can simultaneously buy an asset low and trade it high you have pure, deterministic arbitrage. Existence of deterministic arbitrage lies around the idea that a riskless profit can be made by replicating the future payoffs of an asset with a basket of other assets, with the price of the replicating portfolio not exceeding the costs of the original asset. This can be formalized thus:

$$|\text{payoff}(S_t - R(S_t))| > \text{TransactionCost} \qquad (13.1)$$

where S_t is some asset or combination of assets, $R(S_t)$ is the cost of the replicating portfolio and where TransactionCost represents the net costs involved in buying and selling the assets. As you are not committing any capital this is also known as riskless arbitrage—however, given the fleeting nature of such opportunities it is diminished by competition amongst arbitrageurs. Statistical arbitrage can be seen as a generalization of deterministic arbitrage, relying more heavily on the statistical properties of the mispricing dynamics. Mispricing dynamics refers to the tendency for basis risk to mean-revert or fluctuate around a stable, long-run level (Dunis et al., 2004). As such, statistical arbitrage strategies are also known as mean-reverting or relative value strategies. A critical point must be noted here, as statistical arbitrage is grounded in statistical regularities, assumptions on the existence of fair value amongst assets can be relaxed and in most cases are found to be irrelevant.

A *simple* statistical arbitrage strategy is pairs trading which aims to capitalize on the imbalances between two assets in the hope of making money once the imbalance is corrected. Pairs trading is a pure relative value strategy between two (possibly more) assets: Given two securities which historically moved together, take a long-short position as they diverge and realize a profit as the spread, the mispricing dynamics, converge back to the long-run mean.

Cointegration analysis is used to identify stochastic trends between securities. This is enhanced with an error-correction model (ECM) to provide a mechanism to capture short-term dynamics describing how the long-term equilibrium can be corrected to and restored. Though not proposing the theory, we demonstrate that if the markets were indeed efficient, predictable components could be identified given the right techniques.

Following from Kakoullis (2010), a random walk (RW) with a drift component is given by:

$$y_t = a + y_{t-1} + \varepsilon_t \tag{13.2}$$

where $y_t \sim I(1), \varepsilon_t \sim I(0)$ which can be seen as a first-order autoregressive model AR(1):

$$y_t = a + \varphi y_{t-1} + \varepsilon_t$$

where $\varepsilon \sim N(0, \sigma^2)$ with $\varphi = 1$ indicative of stationarity. We use the standard notation $I(1)$, integrated order 1, to denote that the process is nonstationary and needs to be differenced once in order to become stationary. Such a process is said to have a stochastic trend. To introduce cointegration we consider two time-series v_t and m_t with common stochastic trends

$$v_t = n_{t-1} + \varepsilon_t^n + \varepsilon_t^v \qquad (13.3)$$

$$m_t = n_{t-1} + \varepsilon_t^n + \varepsilon_t^m \qquad (13.4)$$

where n_t is a RW process $\varepsilon_t^n, \varepsilon_t^v, \varepsilon_t^m \sim I(0)$ independent of each other and $v_t, m_t \sim I(1)$. Thus

$$v_t - m_t = \varepsilon_t^v - \varepsilon_t^m \qquad (13.5)$$

therefore it can be said that v_t and m_t have a stochastic trend given by n_t and are therefore cointegrated. This result has important forecasting consequences: Given a pair of RW processes with a common stochastic trend, if a combination of them produces a stationary error/disequilibrium term then theoretically they can be predicted and exploited.

Dickey and Fuller (1979) demonstrated that if Equation 13.2 has a unit root, the process itself may be cointegrated where one or more combination of variables may be stationary though the individual variables may be non-stationary. Two variables which are cointegrated do not trend far away from one another. Thus cointegration is the long-term relationship between the series, where both are $I(1)$ but a linear combination of the two series is stationary. Given this framework we can describe a standard cointegration analysis for statistical trading:

1. Test for unit roots: usually done using the augmented Dickey-Fuller (ADF) test of order q which is based on the regression:

$$\Delta S_t + a + \beta S_{t-1} + \gamma_1 \Delta S_{t-1} + \cdots + \gamma_q \Delta S_{t-q} + \varepsilon_t \qquad (13.6)$$

where Δ is the first difference operator. The lags q are used to remove any autocorrelation which could introduce bias into the residuals. Use the t-ratio test statistic on $\hat{\beta}$ with $H_0{:}\beta$ vs $H_1{:}\beta < 0$.

2. If all series are confirmed to be stationary, test for cointegration. The simplest measure uses the Engle-Granger regression which is an ordinary least squares (OLS) regression:

$$S_t^1 = \beta_1 + \beta_2 S_t^2 + \cdots + \beta_n S_t^n + \varepsilon_t \qquad (13.7)$$

If the unit roots indicate that ε_t is stationary, then the variables $S^A \dots S^n$ are cointegrated with cointegrating vector $(1, -\hat{\beta}_2, \dots, -\hat{\beta}_n)$ or in other words $W = S_1 - \hat{\beta}_2 S_2 - \cdots - \hat{\beta}_n S_n$. A portfolio is constructed off of this vector as follows (where " + " (" − ") indicates a long (short) position respectively):

Security	1	2	...	N
Position	+1	$\hat{\beta}_2$	...	$\hat{\beta}_N$

3. Employ an ECM to provide a mechanism for the spreads deviation from the long-term equilibrium to be corrected. Given the time-series S_A and S_B an ECM can be expressed:

$$\Delta S_t^A = a_1 + \sum_{i=1}^{m} \beta_{1,1}^i \Delta S_{t-i}^A + \sum_{i=1}^{m} \beta_{1,2}^i \Delta S_{t-i}^B + \gamma_1 W_{t-1} + \varepsilon_t^1 \qquad (13.8)$$

$$\Delta S_t^B = a_2 + \sum_{i=1}^{m} \beta_{2,1}^i \Delta S_{t-i}^B + \sum_{i=1}^{m} \beta_{2,2}^i \Delta S_{t-i}^A + \gamma_2 W_{t-1} + \varepsilon_t^2 \qquad (13.9)$$

where W is the mispricing term as given above. The lag lengths, as well as the coefficients, can be determined by OLS.

4. As the errors $\varepsilon_t^1, \varepsilon_t^2$ are normally distributed, the vector

$$\begin{bmatrix} \gamma_1 \\ \gamma_2 \end{bmatrix}$$

defines how the mispricing shall be corrected. The coefficients of this vector correction determine the speed of adjustment. The generalized vector error-correction model (VECM) can thus be represented:

$$\Delta S_t = a + \sum_{i=1}^{m} \beta_i \Delta S_{t-i} + \sum_{j=1}^{q} \Gamma W_{t-j} + \varepsilon_t \qquad (13.10)$$

where

$$S_t = \begin{pmatrix} s_t^1 \\ \vdots \\ s_t^n \end{pmatrix}, a = \begin{pmatrix} s_1 \\ \vdots \\ a_n \end{pmatrix}, B = \begin{pmatrix} \beta_{1,1} & \cdots & \beta_{1,n} \\ \vdots & \ddots & \vdots \\ \beta_{n,1} & \cdots & \beta_{n,n} \end{pmatrix}$$

$$\Gamma_j = \begin{pmatrix} \gamma^{j_{1,1}} & \cdots & \gamma^{j_{1,n}} \\ \vdots & \ddots & \vdots \\ \gamma^{j_{n,1}} & \cdots & \gamma^{j_{n,n}} \end{pmatrix}, W_t = \begin{pmatrix} w_t^1 \\ \vdots \\ w_t^n \end{pmatrix}, \varepsilon_t \begin{pmatrix} \varepsilon_t^1 \\ \vdots \\ \varepsilon_t^n \end{pmatrix}$$

which can be estimated via OLS applied to each equation individually.

Once a pair has been identified any deviation from the long-term equilibrium should be corrected through an investment strategy to address the disequilibrium: short the overpriced security and long the underpriced security. The ECM can be used to calculate the length of the mispricing and timings for order execution.

Applying this to real 5-minute data for the GBL and GBM EUREX futures contracts, following Kakoullis (2010), first rebase the data to aid visual inspection and test for stationarity using ADF tests. The rebased price series alongside the rebased spread series can be seen in Figure 13.2(a) and Figure 13.2(b) respectively. Tables 13.1 and 13.2 indicate we can reject stationarity at the 5% significance level (as the spread t-statistic $> 5\%$ critical significance level). It can thus be concluded that both GBL and GBM are non-stationary.

Secondly, test for cointegration by running an Engle-Granger regression of the form

$$GBL_t^{reb} = a + bGBM_t^{reb} + \varepsilon_t \qquad (13.11)$$

and test the error variable term ε_t for stationarity. Following the same t-statistic test as above it can be said that the spread is weakly stationary, thus GBL and GBM are stationary with cointegrating vector $(1, -b)$. Thirdly, build an ECM of the form of Equation 13.4. A VECM is built using a general-to-specific approach: a general model is initialized with $i = 5$ (five period lags) for both series which are then used to systematically remove insignificant variables and test the error variable term ε_t for stationarity. Following the same t-statistic test as above it can be said that the spread is weakly stationary.

Finally, as the coefficients of the disequilibrium term W_{t-1} have opposite signs and both are highly significant given their t-statistic, then an error correction mechanism from long-term deviations exists. Furthermore, the magnitude of the coefficients gives an indication of the timing until adjustment, e.g., $\gamma 1 = -0.027$ can be interpreted as a 2.7% adjustment in 5 minutes or 3 hours for mean-reversion. Corollary, as the coefficient of $\Delta GBM^{reb}(-1)$ is significant in Equation 13.11 we can say that GBM leads GBL.

Triangular Arbitrage

Triangular arbitrage exploits mispricing across at least three foreign exchange (FX) rates. Consider the scenario where you initially hold x_i dollars. If you sell these dollars to buy euros, convert these euros to pounds, and finally convert these pounds into x_f dollars then you will realize a profit if $x_f > x_i$. If the intermediate rate does not exist you can calculate the synthetic cross to complete the exchange.

In highly liquid markets, such opportunities should be limited and if they were to exist you would expect to find the difference $x_f - x_i$ to be very small (Daniel et al., 2009). Given this restriction, when such

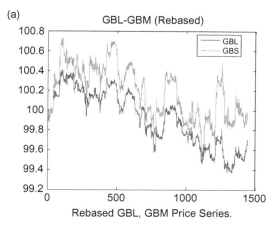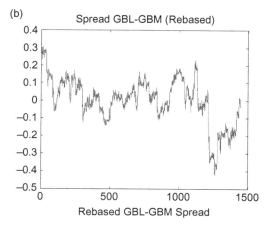

■ **Figure13.2** Pairs Trading: EUREX Contracts.

Table 13.1 Pairs Trading: ADF Tests

	ADF t-stat	5% critical
GBL	−2.692	−3.458
GBM	−1.477	−3.458
Spread(εt)	−3.128	−2.871

Source: Kakoullis, 2010
Determined by their respective t-statistic. Final parameters are given in Table 13.2

Table 13.2 Pairs Trading: VECM Spread Model

	ΔGBL^{reb}		ΔGBM^{reb}	
	Coefficients	t-stat	Coefficients	t-stat
$\gamma1$	−0.027	−3.63	—	—
$\gamma2$	—	—	0.014	−3.06
$\Delta GBL^{reb}(-1)$	−0.096	−1.76	—	—
$\Delta GBM^{reb}(-1)$	0.159	1.90	—	—

Source: Kakoullis, 2010

discrepancies are identified the differential between the identified and execution price becomes extremely important. Implementation shortfall needs to be minimized through, for instance, a highly optimized trading architecture (Hanif and Smith, 2012a).

Taking the example from Dacorogna et al. (2001), for a trader interested in yen, the interrelation between the three currencies (Japanese yen, Great British pound and the US dollar) is captured in the dynamics:

$$GBP/JPY_{bid} = \frac{USD/JPY_{bid}}{USD/GBP_{ask}}, \ GBP/JPY_{ask} = \frac{USD/JPY_{ask}}{USD/GBP_{bid}}.$$

These dynamics represent the triangular relationship between the three currencies. If a direct market between them all exists and these dynamics are strongly deviated from, the deviation can be exploited through a set of riskless transactions.

A frequently employed mechanism for identifying triangular arbitrage opportunity is the rate product:

$$\gamma(t) = \prod_{i=1}^{3} r_i(t) \tag{13.12}$$

where $r_i(t)$ is an exchange rate at time t (Aiba et al., 2002). An arbitrage opportunity is available if $\gamma > 1$ but, again crucially, this shall only be realized if the transaction is completed at the arbitrage decision prices. Continuing from the above, if you initially hold yen there are two unique rate products which can be calculated:

$$\gamma_1(t) = (JPY/GBP_{bid}(t)) \cdot (GBP/USD_{bid}(t)) \cdot \left(\frac{1}{JPY/USD_{ask}(t)}\right) \tag{13.13}$$

$$\gamma_2(t) = \left(\frac{1}{JPY/GBP_{ask}(t)}\right) \cdot \left(\frac{1}{GBP/USD_{ask}(t)}\right) \cdot (JPY/USD_{bid}(t)) \tag{13.14}$$

These two rate products represent the universe of arbitrage opportunities for this set of exchange rates (Daniel et al., 2009).

Suppose we observe the quotes $JPY/GBP_{bid} = 0.00761$, $GBP/USD_{bid} = 1.60085$ and $JPY/USD_{ask} = 0.01100$. We check the implicit cross rate by Equation 13.4.

$$JPY/USD_{ask}^{implied} = JPY/GBP_{bid} \cdot GBP/USD_{bid} \tag{13.15}$$

$$0.01208 = 0.00761 \cdot 1.60085 \tag{13.16}$$

and find that an arbitrage opportunity exists as the implicit cross rate does not equal the quoted rate:

$$JPY/USD_{ask}^{implied} \neq JPY/USD_{ask} \qquad (13.17)$$

We can capitalize on this inefficiency by buying dollars and spending pounds, spending those dollars to purchase yen and subsequently exchanging our recently acquired yen into pounds. Given a £1,000,000 we earn an arbitrage free profit of £107,497.14. All these steps are summarized in Figure 13.3.

Liquidity Trading

Liquidity trading is a particularly adept high frequency trading strategy which mimics the role of the traditional market maker. Liquidity traders, or scalpers for short, attempt to make the spread (buy the bid, sell the ask) in order to capture the spread gain. Such efforts allow for profit even if the bid or ask do not move at all. The key idea here away from traditional day traders is to establish and liquidate positions extremely quickly.

Contrary to traditional traders who would double down on winners, scalpers gain their advantage through increasing the number of winners. The aim here is to make as many short trades as possible, whilst keeping a

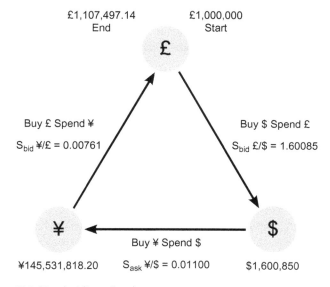

■ **Figure 13.3** Triangular Arbitrage Example.

close eye on market microstructure. Decimalization has eroded scalpers' profits so the imperative on keeping things simple and getting the market data and execution architecture in place is ever increased. Scalping is the least quantitatively involved strategy which we are covering. To ensure it works well one needs to have a good quantitative model of market direction on which there is a plethora of literature—see for instance the application of neural networks (Resta, 2006), traditional time-series analysis techniques (Dacorogna et al., 2001), and sequential Monte-Carlo methods (Hanif and Smith, 2012b; Javaheri, 2005).

Complementary to the above, there are two further types of scalping (Graifer, 2005). Volume scalping is usually achieved through purchasing a large number of shares for a small gain and betting on an extremely small price improvement. This type of scalping works only on highly liquid stocks. The third type of scalping is closest to traditional trading. The trader enters into a trade and closes out as soon as the first exit condition, usually when the 1:1 risk/reward is met.

Scalping is affected by liquidity, volatility, time horizons and risk management. As mentioned above, some traders favor liquid markets, whilst others favor illiquid markets. Volatility is a specific concern of novice scalpers; however, professional scalpers are indifferent to periods of volatility as a thoroughly calibrated directional model should enable the scalper to benefit from both upswings and downturns. Scalpers typically operate in timeframes beyond common technical analysis, and hence require bespoke tools to enable them to capitalize on microstructural changes. Finally, it is imperative that robust risk management protocols are in place to avoid any accumulations of losses and losing streaks (Graifer, 2005).

It would be prudent to note, financial advisers engaged in scalping are frequently found guilty of market manipulation. Scalping by traders is akin to front-running by advisers (buy a security and subsequently advise a purchase) and has been ruled illegal by the US Supreme Court with the SEC taking a blunter view and banning any scalping where a trust relationship exists between the trader and recommendee (Court, 1963; Commission, 2012).

Market-Neutral Arbitrage

Market-neutral arbitrage is a class of strategies based on fundamental theories of finance. An active strategy, the trader seeks to capitalize on stock selection as a long-short investment strategy (Fabozzi et al., 2008). In contrast to traditional investing, the trader here is aiming to use insights about both stocks which are expected to outperform and those expected to underperform the market. Achieving market neutrality from an investment perspective usually involves commitments of equal amounts of capital to over- and underperforming stocks, though in systematic trading market neutrality is achieved solely through exploitation of systematic risks of the stocks. The key idea here is to neutralize the portfolio against broad market moves, achieved by the offset of long positions' price sensitivity by short positions' price sensitivity (Jacobs and Levy, 2004).

Market-neutral strategies are underpinned by the capital asset pricing model (CAPM) developed by Sharpe (1964); Lintner (1965); Treynor (1961); and Mossin (1966). CAPM is given by:

$$E[R_i] = R_f + \beta_i(E[R_M] - R_f) \tag{13.18}$$

where R_i is the return on the capital asset, R_f is the risk-free rate, R_M is the return on the market portfolio and β_i is a measure of the systematic risk of the capital asset i relative to the market portfolio. CAPM is typically solved using OLS. The market portfolio is understood to be a portfolio consisting of market assets and the corresponding return is defined as the market return. Taking long-short positions with the same beta neutralizes systematic risk (having a beta of zero). Self-evidently such a construction is not risk-free though it will provide positive returns in recompense for accrued risk.

A common extension includes measures of investment managers' performance. Alpha is the intrinsic return on a capital asset and is used to measure returns in excess of the market. Taking the market regression of CAPM we have:

$$R_{i,t} - R_f = \alpha_i + \beta_i(R_{M,t} - R_f) + \varepsilon_t \tag{13.19}$$

Following Aldridge (2009), we shall proceed to describe a common market-neutral trading strategy.

Systematically trading the market-neutral pair of securities i and j involves determining the statistical significance means tests between the alphas and betas:

$$\hat{\beta} = \hat{\beta}_i - \hat{\beta}_j \tag{13.20}$$

$$\hat{\sigma}_{\Delta\beta} = \sqrt{\frac{\sigma_{\beta i}^2}{n_i} + \frac{\sigma_{\beta j}^2}{n_j}} \tag{13.21}$$

where n_i and n_j are the number of observations used in the estimation of Equation 13.19. The t-statistic is then calculated:

$$t_\beta = \frac{\Delta\hat{\beta}}{\hat{\sigma}_{\Delta\beta}} \tag{13.22}$$

The difference test and t-statistics for alpha follow the same form as Equations 13.20 to 13.22.

A typical t-test is conducted, which evaluates to statistically similar if the t-statistic falls into a one standard deviation interval:

$$t_\beta \in [\Delta\hat{\beta} - \hat{\sigma}_{\Delta\beta}, \Delta\hat{\beta} + \hat{\sigma}_{\Delta\beta}] \tag{13.23}$$

Complementary to the above, the difference in the alphas must show both economic viability (exceeding trading costs TC) and strong statistical significance usually at the 95% confidence interval:

$$\Delta\hat{\alpha} > TC \tag{13.24}$$

$$|t_\alpha| > [\Delta\hat{\alpha} + 2\hat{\sigma}_{\Delta\alpha}] \tag{13.25}$$

Given a pair of securities satisfying the tests of Equations 13.23 to 13.25, a long position is taken in the security with higher alpha and a short position is taken in the security with lower alpha, held and closed out by the time horizon determined in the forecast.

Index and Exchange Traded Fund Arbitrage

Index arbitrage is driven by relative mispricing dynamics of constituent securities. Recollecting that an index and exchange traded fund (ETF) is comprised of a basket of securities with the prevailing price being, usually, a weighted average of the constituent securities, the Law of One Price from economics dictates that in an efficient market, all identical goods must have only one price (Burdett and Judd, 1983). If the relative prices between constituents and their index/ETF trackers diverge an arbitrage opportunity exists. In what follows index and ETF are used interchangeably.

Exploitation of an index arbitrage opportunity where the price of the replicating portfolio net of transaction costs RP_{TC} is greater than the index price net of transaction costs I_{TC}

$$RP_{TC} \geq I_{TC} \tag{13.26}$$

would involve shorting the replicating portfolio and going long the index. This position would be closed once a profit had been realized. Similarly, where the price of the replicating portfolio net of transaction costs is less than the price of the index itself, again net of transaction costs, an arbitrageur would go long the replicating portfolio and short the index, closing out once profits have been realized. A preponderant form of index arbitrage is basis trading where mispricing dynamics (difference between the spot-price and related futures contract) are exploited by buying the index and selling related futures contracts. Intuitively, the gain from one of these positions will exceed the loss on the other, resulting in a net gain.

Realizing the mispricing correction is in effect mean-reversion. Alexander (1999) details a cointegration based index arbitrage strategy to deliver stable returns. Following a cointegration statistical portfolio trading process:

1. We choose to go long-short in eight different countries with the European, Asian and Far East (EAFE) Morgan Stanley index as our benchmark. We aim to beat EAFE and are tasked with finding the basket of eight countries which are currently most highly correlated with the EAFE index.
2. Perform cointegration regression of log-EAFE price index y on log price indices of the local currencies $x_1, \ldots, x_n$:

$$y_t = a + \beta_1 x_1, t + \ldots + \beta_n x_n, t + \varepsilon_t \qquad (13.27)$$

   to find optimal allocations and associated cointegrating vector.
3. Having identified which countries lead EAFE, we arbitrage by the exploitation rules detailed above.

Merger Arbitrage

Mergers and acquisitions is a mainstay of financial intermediation and the global markets. Academic effort has focused on wealth effects and associated economic reasons around merger periods which is in stark contrast to traders' efforts to capitalize on mispricing preceding deal completion (Branch and Yang, 2003). Such a trading strategy is known as merger arbitrage, though it is commonly also referred to as risk arbitrage and event arbitrage.

Merger arbitrage usually involves buying the stock that is being acquired and shorting the stock of the acquirer aiming to capture the offer premium which always persists after an announcement of intention. This premium reflects the risk inherent to the deal, hence risk arbitrage, and is the difference between the offered consideration and the target price. The long-short positions reflect the spread between the offered consideration and the target price through the merger period.

To enact this, an arbitrageur would need to calculate the rate of return implicit to the current spread compared with the event risk of the deal.

If the spread can cover and exceed the expected event risk, arbitrage trades are executed (Anson, 2008). If the merger is successful the target stock price will converge to the offered consideration, thus the arbitrageurs will earn initial spreads from the day the long-short positions were taken (another relative-value strategy). However, if the merger fails arbitrage performance is correlated to the performance of the individual long-short positions. Clearly, predicting merger success is the most important factor for merger arbitrageurs (Baker and Savaşoglu, 2002; Mitchell, 2001). The three key components of merger information for merger arbitrage success: firm/deal information, market price information and risk arbitrageur information (Branch and Yang, 2003).

Examining the purchase of MCI Corporation by Verizon Communications as detailed by Anson (2008), we can understand the dynamics of merger arbitrage. Verizon was locked in a bidding war with QWest Communications for the purchase of MCI through 2005 with Verizon finally coming out on top with a purchase price of $8.44 billion. At the announcement of its intention, the Verizon arbitrage trade was:

- Sell 1000 shares of Verizon at $36.
- Buy 1000 shares of MCI at $20.

While for QWest the arbitrage trade was:

- Sell 1000 shares of QWest at $4.20.
- Buy 1000 shares of MCI at $20.

Verizon and QWest competed for MCI with Verizon winning in October 2005. By then, MCI was trading at $25.50, Verizon had lost value and was trading at $30, and QWest was unchanged. The total return for the MCI/Verizon merger arbitrage trade:

Long MCI gain	$1000 \times (\$25.50 - \$20)$	=	$5500
Short Verizon gain	$1000 \times (\$36 - \$30)$	=	$6000
Short rebate interest	$4\% \times 1000 \times \$36 \times 240/360$	=	$960
Total		=	$12,460

The return on investment is: $\$12,460 \div \$20,000 = 62.3\%$. Similarly, for the MCI/QWest merger arbitrage trade:

Long MCI gain	$1000 \times (\$25.50 - \$20)$	=	$5500
Short QWest gain	$1000 \times (\$4.20 - \$4.20)$	=	$0
Short rebate interest	$4\% \times 1000 \times \$4.20 \times 240/360$	=	$112
Total		=	$5612

Here, the return on investment is: $5612 ÷ $20,000 = 28.06% which, when compared to the alternate bet, patently reflects an incorrect call. Again, making this call makes or breaks the arbitrage opportunity.

A simple model of evaluating completion provides the market's assessment of the likelihood (not to be confused with desirability) of completion. Brown and Raymond (1986) provide a formal model for the prediction of successful mergers and acquisitions. Given an announcement of consideration, if the merger is successful a long-short position would return a yield:

$$(P_{Tt} - P_{mt})/P_{mt} \qquad (13.28)$$

However, as the tender is not given to be accepted, the risk of falling through must be considered through the possibility of a negative return:

$$(P_F - P_{mt})/P_{mt} \qquad (13.29)$$

where P_F is the price the target stock falls to if the merger is unsuccessful. Indeed it cannot be said with certainty what value P_F will take. However, pre-announcement prices can be used as proxies to assess the overall risk-reward profile.

Given this framework, we can assess the market's prediction of the likelihood of the merger. Computing the period t merger probability x_t given the merger arbitrage payoff is zero:

$$
\begin{aligned}
E(\Pi_t) &= x_t[(P_{Tt} - P_{mt})/P_{mt}] + (1 - x_t)[(P_F - P_{mt})/P_{mt}] \\
&= x_t[(P_{Tt}/P_{mt}) - 1] + (1 - x_t)[(P_F/P_{mt}) - 1] \qquad (13.30) \\
&= 0
\end{aligned}
$$

Thus:

$$x_t = \frac{1 - (P_F/P_{mt})}{(P_{Tt}/P_{mt}) - (P_F/P_{mt})} \qquad (13.31)$$

where (P_{Tt}/P_{mt}) and (P_F/P_{mt}) are defined as tender and failure premiums respectively. Thus likelihood x_t describes the price assimilation of the tender into the market price as of period t.

Merger arbitrage is deal driven and not market driven. Consequently returns are correlated to the relative value between the two stocks when the arbitrage is executed and as such they are not, in general, correlated to returns of the general stock market (Anson, 2008). Exceptions to this rule are statistical speculators who will actively seek risk arbitrage opportunities.

EVALUATION

Back-testing is the quantitative evaluation of a model's performance, both from a statistical and trading perspective. Revisiting our modeling paradigm described in the section Data and Research above, our models are calibrated on a subset of historical data known as the in-sample set. To lend any sort of value to the model we must test performance on a far larger set of data unseen to the strategy (the out-of-sample set). This enables us to provide statistical significance measures to our model and ensure we have not over-fitted to the in-sample data set. Over-fitting is the phenomenon where you have modeled the idiosyncrasies of a particular data set rather than modeling the latent dynamics. When modeling we need to ensure we reach a balance between over- and under-fitting data.

Back-testing is the closest we come to true forecasting or trading whilst preparing our models, enabling direct comparison of forecasting accuracy and trading performance (Dunis et al., 2004). Back-tests provide three complementary purposes (Dowd, 2008). Firstly, we can assess our model's statistical validity compared with real historical data. Secondly, they enable diagnostic checks to be carried out in aid of understanding the strengths and weaknesses of the model. It shows us how the strategy can actually capture opportunities and crucially where and how it fails to do so. Finally, back-tests allow us to formally compare and rank alternative models. A good strategy should do well in all three measures.

As we are evaluating a strategy and an underlying point forecast or directional change model there are two principle out-of-sample measures we need to test for: forecasting accuracy and trading performance. To back-test a strategy we need the strategy's forecasts alongside daily profits and losses (P/L) realized by the portfolio. At first hand, this task looks quite trivial. However, we are not looking for accounting excellence; rather we are looking for data on the P/L attributable to the market risks that were taken. We require P/L data to reflect our risk taking and to remove aspects which are not directly related to current market risk taking (e.g., fee income, unrealized P/L, yields, dividends, etc.). Having assembled our data set we are ready for the analysis.

To start off with we run some straightforward statistical analysis on the data. Plotting and analyzing cumulative P/L, P/L histograms to understand the empirical distribution used in conjunction with quantile-quantile charts in aid of understanding empirical distributions vs. forecasted distributions. In addition, we calculate summary statistics off of the data, including moments and tail (extreme value) analysis.

A number of measures are used to assess forecasting accuracy. These include the traditional statistical accuracy measures: mean absolute error, root mean square error, mean absolute percentage error and Theil's inequality coefficient (Theil-U). Mean absolute percentage error and Theil-U enable evaluation of instances where forecast errors must be modeled independently of the variables. As such are they constructed to lie within the range [0,1], with zero indicating a perfect fit. The measure correct directional change assesses the ability of the strategy to predict the actual next period change of a forecast variable, an important issue in a trading strategy. Statistical performance measures used to analyze forecasting accuracy are detailed in Table 13.3. Please refer to Hanke and Reitsch (1998) and Pindyck and Rubinfeld (1998) for detailed discussion on these measures.

Forecasting accuracy measures enable us to understand the statistical underpinnings of our strategy. However, they not lend well into analysis

---

Table 13.3 Statistical Performance Measures

Performance Measure	Description	
Mean Absolute Error	$MAE = \dfrac{1}{T}\sum\limits_{t=1}^{T}|\tilde{y}_t - y_t|$	(13.32)
Mean Absolute Percentage Error	$MAPE = \dfrac{100}{T}\sum\limits_{t=1}^{T}\left|\dfrac{\tilde{y}_t - y_t}{y_t}\right|$	(13.33)
Root Mean Square Error	$RMSE = \sqrt{\dfrac{1}{T}\sum\limits_{t=1}^{T}(\tilde{y}_t - y_t)^2}$	(13.34)
Theil's Inequality Coefficient	$U = \dfrac{\sqrt{\dfrac{1}{T}\sum\limits_{t=1}^{T}(\tilde{y}_t - y_t)^2}}{\sqrt{\dfrac{1}{T}\sum\limits_{t=1}^{T}(\tilde{y}_t)^2} + \sqrt{\dfrac{1}{T}\sum\limits_{t=1}^{T}(y_t)^2}}$	(13.35)
Correct Directional Change	$CDC = \dfrac{100}{N}\sum\limits_{t=1}^{N}D_t$	(13.36)
	where $D_t = 1$ if $y_t \cdot \tilde{y}_t > 0$ else $D_t = 0$	

Source: Dunis et al. (2004)
y_t is the actual change at time t.
$\tilde{y}_t$ is the forecast change.
t = 1 to t = T for the forecast period.

of trading performance. Such measures are optimized with respect to some mathematical or statistical precedent. However, ultimately our results are analyzed financially to which these measures are not optimized. To whit: the forecast accuracy is calibrated through model estimation but the true value should be based on performance of the trading strategy (Dunis et al., 2004). Common trading performance measures are detailed in Tables 13.4−13.7. Important measures include the Sharpe ratio, maximum drawdown and the average gain/loss ratio. The Sharpe ratio is a risk-adjusted measure of return and allows for direct comparison across the industry. Maximum drawdown is a measure of downside risk, and the average gain/loss ratio is a measure of overall profit (Dunis and Jalilov, 2002; Fernandez-Rodriguez et al., 2000). Back-test measures will shed light on different aspects of the strategy, determining the overall quality of the forecasts, as in the end financial gain depends more on trading performance than forecasting accuracy.

Table 13.4 Trading Simulation Performance Measures

Performance Measure	Description	
Annualized Return	$R^A = 252 \cdot \dfrac{1}{N} \sum\limits_{t=1}^{N} R_t$	(13.37)
Cumulative Return	$R^C = \sum\limits_{t=1}^{N} R_T$	(13.38)
Annualized Volatility	$\sigma^A = \sqrt{252} \cdot \sqrt{\dfrac{1}{N-1} \sum\limits_{t=1}^{N}(R_t - \overline{R})^2}$	(13.39)
Sharpe Ratio	$SR = R^A / \sigma^A$	
Maximum Daily Profit	Maximum value of R_t over the period	(13.40)
Maximum Daily Loss	Minimum value of R_t over the period	(13.41)
Maximum DrawDown	Maximum negative value of $\sum (R_t)$ over the period	(13.42)
	$MD = \min\limits_{t=1,\ldots,N} \left(R_t^C - \max\limits_{i=1,\ldots,t} (R_i^C) \right)$	(13.43)
% Winning Trades	$WT = 100 \ldots \dfrac{\sum\limits_{t=1}^{N} F_t}{NT}$	(13.44)
	where $F_t = 1$ if Transaction Profit$_t \geq 0$	

Source: Dunis et al. (2004)

Table 13.5 Trading Simulation Performance Measures

Performance Measure	Description	
% Losing Trades	$LT = 100 \cdot \dfrac{\sum_{t=1}^{N} G_t}{NT}$ where $G_t = 1$ if Transaction Profit$_t \leq 0$	(13.45)
Number of Up Periods	N_{up} = Number of $R_t > 0$	(13.46)
Number of Down Periods	N_{down} = Number of $R_t < 0$	(13.47)
Number of Transactions	$NT = \sum_{t=1}^{N} L_t$ where $L_t = 1$ if Trading Signal$_t \neq$ Trading Signal$_{t-1}$	(13.48)
Total Trading Days	$\sum (R_t)$	(13.49)
Avg. Gain in Up Periods	$AG = \left(\sum (R_t > 0)\right)/N_{up}$	(13.50)
Avg. Loss in Down Periods	$AL = \left(\sum (R_t < 0)\right)/N_{down}$	(13.51)
Avg. Gain/Loss Ratio	$GL = AG/AL$	(13.52)

Source: Dunis et al. (2004)

Table 13.6 Trading Simulation Performance Measures

Performance Measure	Description	
Probability of 10% Loss	$PoL = \left[\dfrac{(1-P)^{\left(\frac{MaxRisk}{\Lambda}\right)}}{P} \right]$	(13.53)
	where $P = 0.5 \cdot \left(1 + \left(\dfrac{\langle (WT \cdot AG) + (LT \cdot AL)\rangle}{\sqrt{[(WT \cdot AG^2) + (LT \cdot AL^2)]}}\right)\right)$	(13.54)
	and $\Lambda = \sqrt{[(WT \cdot AG^2) + (LT \cdot AL^2)]}$ MaxRisk is the risk level defined by the user.	(13.55)
Profits T-statistics	$T\text{-statistics} = \sqrt{N} \cdot \dfrac{R^A}{\sigma^A}$	(13.56)

Source: Dunis et al. (2004)

Table 13.7 Trading Simulation Performance Measures

Performance Measure	Description	
Number of Periods Daily Returns Rise	$NPR = \sum_{t=1}^{N} Q_t$	(13.57)
	where $Q_t = 1$ if $y_t > 0$ else $Q_t = 0$	
Number of Periods Daily Returns Fall	$NPF = \sum_{t=1}^{N} S_t$	(13.58)
	where $S_t = 1$ of $y_t > 0$ else $S_t = 0$	
Number of Winning Up Periods	$NWU = \sum_{t=1}^{N} B_t$	(13.59)
	where $B_t = 1$ if $R_t > 0$ and $y_t > 0$ else $B_t = 0$	
Number of Winning Down Periods	$NWD = \sum_{t=1}^{N} E_t$	(13.60)
	where $E_t = 1$ if $R_t > 0$ and $y_t < 0$ else $E_t = 0$	
Winning Up Periods (%)	$WUP = 100 \cdot (NWU/NPR)$	(13.61)
Winning down Periods (%)	$WDP = 100 \cdot (NWD/NPF)$	(13.62)

Source: Dunis et al. (2004)

SUMMARY

We have introduced high frequency trading and demystified the black box. High frequency trading is an application of systematic and quantitative algorithms for the trading of securities. Critical to an effective strategy is soundly curated high frequency data coupled with a rigorous research framework. We described a number of practical trading strategies alongside examples. These strategies form the alpha model of the black box.

Statistical arbitrage capitalizes on statistical misrepresentations in the prices of related securities. Triangular arbitrage assesses triangular relations between currencies, taking positions when these relations are broken. Liquidity trading attempts to make the spread in the high frequency domain, focusing on the number of winners rather than on the volume per given trade. Market-neutral arbitrage builds on classical relations defined by CAPM to take long-short positions in mispriced securities. Index arbitrage exploits discrepancies in the Law of One Price, taking relative positions to capitalize. And finally, merger arbitrage aims to capitalize on mispricing dynamics prior to deal completion.

We have described a thorough back-testing framework and highlighted that it is critical to assess not only forecasting accuracy of our model but also its trading performance. Various metrics were defined to assess both statistical forecasting accuracy and statistical trading performance. We discussed how the Sharpe ratio, maximum drawdown and average gain/loss ratio are important trading performance metrics in the industry, enabling us to not only rank models but to also rate traders.

References

Admati, A.R., Pfleiderer, P., 1988. A theory of intraday trading patterns. Rev. Financ. Stud. 1, 3—40.

Agresti, A., 2002. Categorical Data Analysis. John Wiley and Sons, New Jersey.

Aiba, Y., et al., 2002. Triangular arbitrage as an interaction among foreign exchange rates. Physica A 310 (3), 467—479.

Aldridge, I., 2009. High-Frequency Trading: A Practical Guide to Algorithmic Strategies and Trading Systems, vol. 459. Wiley, NY, USA.

Alexander, C., 1999. Optimal hedging using cointegration. Philos. Trans. R. Soc. London, Ser. A 357 (1758), 2039—2058.

Almgren, R., 2003. Optimal execution with nonlinear impact functions and trading enhanced risk. Appl. Math. Finance 10, 1—18.

Almgren, R., Chriss N., 1997. Optimal liquidation, Original Working Paper.

Almgren, R., Chriss, N., 1999. Value under liquidation. Risk 12, 61—63.

Almgren, R., Chriss, N., 2000. Optimal execution of portfolio transactions. J. Risk 3, 5—39.

Almgren, R., Chriss, N., 2003. Bidding principles. Risk. 97—102.

Almgren, R., Loren, J., 2006. Bayesian adaptive trading with a daily cycle. J. Trading 1 (4), 38—46, Fall.

Almgren, R., Loren, J., 2007. Adaptive arrival price. Algorithmic Trading III: Precision, Control, Execution — 2007, 2007 (1), 59—66.

Amihud, Y., Mendelson, H., 1980. Dealership market: market-making with inventory. J. Financ. Econ. 8, 31—53.

Amihud, Y., Mendelson, H., 2000. The liquidity route to a lower cost of capital. J. Appl. Corp. Financ. 12, 8—25.

Anson, M., 2008. Hedge funds. Handb. Finance 1.

Arrow, K.J., 1971. Essays in the Theory of Risk-Bearing, North Holland, Amsterdam.

Bacidore, J.M., Battalio R., Jennings R., 2001. Order submission strategies, liquidity supply, and trading in pennies on the New York Stock Exchange. NYSE Research Paper.

Bagehot, W., Treynor, J., 1971. The only Game in town. Financ. Anal. J. 27 (22), 12—14 (March/April)

Baker, M., Savasoglu, S., 2002. Limited arbitrage in mergers and acquisitions. J. Financ. Econ. 64 (1), 91—115.

Balduzzi, P., Lynch., A.W., 1999. Transaction costs and predictability: some utility cost calculations. J. Financ. Econ. 52, 47—78.

Banks, E., Glantz, M., Siegel, P., 2006. Credit Derivatives: Techniques to Manage Credit Risk for Financial Professionals. McGraw-Hill.

Barra, 1997. Market Impact Model Handbook.

Beebower, G., Priest, W., 1980. The tricks of the trade. J. Portf. Manag. 6 (1), 36—42, Winter.

Berkowitz, S., Logue, D., Noser, E., 1988. The total cost of transactions on the NYSE. J. Finance 41, 97—112.

Bertsimas, D., Lo, A., 1998. Optimal control of liquidation costs. J. Financ. Mark. 1, 1—50.

Bessembinder, H., 2003. Trade execution costs and market quality after decimalization. J. Financ. Quant. Anal.

Bessembinder, H., Kauffman, H.M., 1997. A comparison of trade execution costs for NYSE and Nasdaq Listed stocks. J. Fin. Quant. Anal. 32, 287–310.

Black, F., Litterman, R., 1992. Global portfolio optimization. Financ. Anal. J. 48 (5), 28–43, Sep/Oct 1992.

Black, F., Scholes, M., 1973. The pricing of options and corporate liabilities. J. Polit. Econ. 81 (May–June), 1973.

Bloomfield, R., O'Hara, M., 1996. Does Order Preferencing Matter? Working Paper, Johnson Graduate School of Management, Cornell University.

Blume, L., Easley, D., O'Hara, M., 1982. Characterization of optimal plans for stochastic dynamic programs. J. Econ. Theory. 28 (2), 221–234.

Bodie, Z., Kane, A., Marcus, A., 2005. Investments. McGraw-Hill, New York.

Bollerslev, T., 1986. Generalized autoregressive conditional heteroscadisticity. J. Econom. 31, 307–327.

Boni, L., 2009. Grading broker algorithms. J. Trading 4 (4), 50–61.

Branch, B., Yang, T., 2003. Predicting successful takeovers and risk arbitrage. Q. J. Bus. Econ. 3–18.

Breen, W., Hodrick, L., Korajczyk, R., 2002. Predicting equity liquidity. Manage. Sci. 48 (4), 470–483, April 2002.

Brown, K., Raymond, M., 1986. Risk arbitrage and the prediction of successful corporate takeovers. Financ. Manag. 54–63.

Broyden, C.G., 1970. J. Inst. Math. Appl. 6, 76–90.

Burdett, K., Judd, K., 1983. Equilibrium price dispersion. Econometrica 955–969.

Campbell, J., Lo, A., Mackinlay, A., 1997. The Econometrics of Financial Markets. Princeton University Press, New Jersey.

Chan, L.K., Lakonishok, J., 1993. Institutional trades and intraday stock price behavior. J. Financ. Econ. 33, 173–199.

Chan, L.K., Lakonishok, J., 1995. The behavior of stock prices around institutional trades. J. Finance 50 (4), 1147–1174.

Chan, L.K., Lakonishok, J., 1997. Institutional equity trading costs: NYSE versus Nasdaq. J. Finance 52 (2), 713–735.

Chan, N.H., 2002. Time Series: Application to Finance. Wiley-Interscience.

Chen, N., Roll, R., Ross, S., 1986. Economic forces and the stock market. J. Bus. July 1986.

Chiang, A., 1984. Fundamental Methods of Mathematical Economics. third ed. McGraw-Hill.

Christie, W., Schultz, P., 1999. The initiation and withdrawal of odd-eighth quotes among Nasdaq stocks: an empirical analysis. J. Financ. Econ. 52, 409–442.

Coarse, R., 1937. The nature of the firm. Econometrica. 36.

Cochrane, J.H., 2005. Asset Pricing. Princeton University Press, Revised Edition.

Cohen, K.J., Maier, D., Schwartz, R., Witcomb, D., 1982. Transactions cost, order placement strategy and existence of the bid-ask spread. J. Polit. Econ. 89, 287–305.

Conrad, J., Johnson, K., Wahal, S., 2003. Institutional trading and alternative trading systems. J. Financ. Econ. 70, 99–134.

Cont, R., 2001. Empirical properties of asset returns: stylized facts and statistical issues. Quant. Finance 1, 223–236.

Copeland, T.E., Galai, D., 1983. Information effects on the bid-ask spread. J. Finance 38, 1457–1469.

Cox, B., 2000. Transaction Cost Forecasts and Optimal Trade Schedule, IMN Superbowl of Indexing Conference.

Crow, E.L., Davis, F.A., Maxfield, M.W., 1960. Statistics Manual. Dover Publications, Inc.

Cuneo, L., Wagner, W., 1975. Reducing the cost of stock trading. Financ. Anal. J. 31 (6), 35–44, Nov/Dec 1975.

Dacorogna, M., et al., 2001. An Introduction to High Frequency Finance.

Daniel, J., et al., 2009. The mirage of triangular arbitrage in the spot foreign exchange market. Int. J. Theor. Appl. Finance 12 (08), 1105–1123.

DeGroot, M.H., 1986. Probability and Statistic. second ed. Addison Wesley, New York.

Demsetz, H., 1968. The cost of transacting. Q. J. Econ. 82, 32–53.

Devore, J., 1982. Probability & Statistics for Engineering and the Sciences. Brooks/ Cole Publishing.

Dickey, D., Fuller, W., 1979. Distribution of the estimators for autoregressive time series with a unit root. J. Am. Stat. Assoc. 74 (366a), 427–431.

Domowitz, I., Steil, B., 2001. Global Equity Trading Costs, working paper, ITG.

Domowitz, I., Yegerman, H., 2006. The Cost of Algorithmic Trading. J. Trading 1 (1), 22–42.

Domowitz, I., Yegerman, H., 2011. Algorithmic Trading Usage Patterns and their Costs. ITG Publication, <http://www.itg.com/news_events/papers/Algorithmic_Trading.pdf>.

Donefer, B., 2010. Algos gone wild: risk in the world of automated trading strategies. J. Trading 5 (2), 31–34, Spring 2010.

Dowd, K., 1998. Beyond Value at Risk: The New Science of Risk Management. John Wiley & Sons.

Dowd, K., 2008. Back-testing market risk models. Handb. Finance. Wiley Online Library.

Dudewicz, E., Mishra, S., 1988. Modern Mathematical Statistics. John Wiley & Sons.

Dunis, C., Jalilov, J., 2002. Neural network regression and alternative forecasting techniques for predicting financial variables. Neural Netw. World 12 (2), 113–140.

Dunis, C., et al., 2004. Applied Quantitative Methods for Trading and Investment. Wiley, NY, USA.

Easley, D., O'Hara, M., 1987. Price, trade size, and information in securities markets. J. Financ. Econ. 19, 69–90.

Edwards, H., Wagner, W., 1993. Best execution. Financ. Anal. J. 49 (1), 65–71, Jan/Feb 1993.

Elton, E., Gruber, M., 1995. Modern Portfolio Theory. fifth ed. John Wiley & Sons, Inc., New York.

Enders, W., 1995. Applied Econometric Time Series. John Wiley & Sons.

Engle, R., Ferstenberg, R., 2006. Execution risk. J. Trading 2 (2), 10–20.

Engle, R., Ferstenberg, R., 2007. Execution risk. J. Portf. Manag. 33 (2), 34–44.

Engle, R.F., 1982. Autoregressive conditional heteroscadisticity with estimates of the variance of united kingdom inflation. Econometrica 50, 987–1008.

Fabozzi, F., et al., 2008. Overview of active common stock portfolio strategies. Handb. Finance 2.

Fama, E., French, K., 1992. The cross section of variation in expected stock returns. J. Finance 47 (2), 427–465.

Fama, E., French, K., 1993. Common risk factors in the returns on stocks and bonds. J. Financ. Econ. 33 (1), 3–56.

Fernandez-Rodriguez, F., et al., 2000. On the profitability of technical trading rules based on artificial neural networks: evidence from the Madrid stock market. Econ. Lett. 69 (1), 89–94.

Fletcher, R., 1970. A new approach to variable metric algorithms. Comput. J. 13 (3), 317–322.

Fox, J., 2002. Nonlinear Regression and Nonlinear Least Squares: Appendix to An R and S-Plus Companion to Applied Regression, <http://cran.r-project.org/doc/contrib/Fox-Companion/appendix-nonlinear-regression.pdf>.

Freeman, J., 1994. Simulating Neural Networks. Addison Wesley.

Freyre-Sanders, A., Guobuzaite, R., Byrne, K., 2004. A review of trading cost models: reducing trading costs. J. Invest. 13, 93–115.

Garmen, M., 1976. Market microstructure. J. Financ. Econ. 3, 257–275.

Gatheral, J., 2010. No-dynamic-arbitrage and market impact. Quant. Finance 10 (7), 749–759.

Gatheral, J., Schied, A., 2012. Dynamical models of market impact and algorithms for order execution. In: Fouque, J.-P., Langsam, J. (Eds.), Handbook on Systemic Risk. Cambridge University Press.

Gill, P.E., Murray, W., Wright, M.H., 1981. Practical Optimization. Academic Press, London, UK.

Glantz, M., 2000. Scientific Financial Management: Advances in Financial Intelligence Capabilities for Corporate Valuation and Risk Assessment. AMACOM, Inc.

Glantz, M., 2003. Managing Bank Risk. Academic Press, California.

Glantz, M., Kissell, R., 2013. Multi-Asset Risk Modeling: Techniques for a Global Economy in an Electronic and Algorithmic Trading Era. Elsevier.

Glosten, L., Harris, L., 1988. Estimating the components of the bid-ask spread. J. Financ. Econ. 14, 21–142.

Glosten, L., Milgrom, P., 1985. Bid, ask, and transaction prices in a specialist market with heterogeneously informed agents. J. Financ. Econ. 14, 71–100.

Goldfarb, D., 1970. Math. Comput. 24, 23.

Goldstein, M.A., Kavajeczb, K.A., 2000. Eighths, sixteenths, and market depth: changes in tick size and liquidity provision on the NYSE,". J. Financ. Econ. 56-1, 125–149.

Graifer, V., 2005. How to scalp any market. Real. Trading.

Greene, W., 2000. Econometric Analysis. Fourth ed. Prentice-Hall, Inc.

Grinold, R., Kahn, R., 1999. Active Portfolio Management. McGraw-Hill, New York.

Gujarati, D., 1988. Basic Economics. second ed. McGraw-Hill, New York.

Hamilton, J.D., 1994. Time Series Analysis. Princeton University Press.

Hanif, A., Protopapas, P., 2013. Recursive bayesian estimation of regularized and irregular astrophysical time series. Mon. Not. R. Astron. Soc.

Hanif, A., Smith, R., 2012a. Algorithmic, electronic and automated trading. J. Trading 7, 4.

Hanif, A., Smith R., 2012b. Generation path-switching in sequential Monte-Carlo methods. In: Evolutionary Computation (CEC), 2012 IEEE Congress on, pp. 1–7. IEEE.

Hanke, J., Reitsch, A., 1998. Business Forecasting. sixth ed. Prentice-Hall Inc., Englewood Cliffs, USA.

Hansch, O., Naik, N., Viswanathan, S., 1998. Do inventories matter in dealership markets: evidence from the London stock exchange. J. Finance 53, 1623–1656.

Harris, L., 1994. Minimum price variations, discrete bid/ask spreads and quotation sizes. Rev. Financ. Stud. 7, 149–178.

Harris, L., 1995. Consolidation, fragmentation, segmentation, and regulation. In: Schwarz, R.A. (Ed.), Global Equity Markets: Technological, Competitive, and Regulatory Challenges. Irwin Publishing, New York.

Harris, L., 2003. Trading and Exchanges. Oxford University Press, USA.

Harris, L., Gurel, E., 1986. Price and volume effects associated with changes in the S&P 500 list: new evidence for the existence of price pressures. J. Finance 41, 815–829.

Harvey, A., 1999. The Econometric Analysis of Time Series. second ed. The MIT Press.

Hasbrouck, J., 1991. The summary informativeness of stock trades. Rev. Financ. Stud. 4, 571–594.

Hasbrouck, J., 2007. Empirical Market Microstructure: The Institutions, Economics, and Econometrics of Securities Trading. Oxford University Press, USA.

Hastings, H.M., Kissell, R., 1998. Is the nile outflow fractal? Hurst's analysis revisited. Nat. Resour. Model. 11 (2).

Hastings, H.M., Sugihara, G., 1993. Fractals: A User's Guide for the Natural Sciences. Oxford Science Publications.

Ho, T., Macris, R., 1984. Dealer bid-ask quotes and transaction prices: an empirical study of some AMEX options. J. Finance 40, 21–42.

Ho, T., Stoll, H., 1983. The dynamics of dealer markets under competition. J. Finance 38, 1053–1074.

Holthausen, R., Leftwich, R., Mayers, D., 1987. The effect of large block transactions on security prices. J. Finan. Econ. 19, 237–267.

Holthausen, R., Leftwich, R., Mayers, D., 1990. Large-block trans-actions, the speed of response, and temporary and permanent stock-price effects. J. Financ. Econ. 26, 71–95.

Huang, R.D., Stoll, H.R., 1994. Market microstructure and stock return predictions. Rev. Financ. Stud. 7, 179–213.

Huang, R.D., Stoll, H.R., 1996. Dealer versus auction markets: apaired comparison of execution costs on NASDAQ and the NYSE. J. Financ. Econ. 41 (3), 313–357.

Huang, R.D., Stoll, H.R., 1997. The components of the bid-ask spread: a general approach. Rev. Financ. Stud.s 10 (4), 995–1034.

Huang, R.D., Stoll., H.R., 1994. Market microstructure and stock return predictions. Rev. Financ. Stud. 7, 179–213.

Huberman, G., Stanzl, W., 2001. Optimal liquidity trading. Preprint.

Hull, J., 2012. Options, Futures, and Other Derivatives. eighth ed. Prentice Hall.

Hurst, H.E., 1950. Long-term storage capacity of reservoirs. Am. Soc. Civ. Eng. Proc. 76 (April 1950), Also reprinted: Transactions of the American Society of Civil Engineers 116 pg. 770–808, Sept 1951.

Jacobs, B., Levy, K., 2004. Market Neutral Strategies, vol. 112. Wiley, USA.

Jain, P., Joh, G.H., 1988. The dependence between hourly prices and trading volume. J. Financ. Quant. Anal. 23, 269–283.

Javaheri, A., 2005. Inside Volatility Arbitrage. J. Wiley & Sons, New Jersey.

Johnson, B., 2010. Algorithmic Trading and DMA: An Introduction to Direct Access Trading Strategies, 4. Myeloma Press.

Johnson, J., DiNardo, J., 1997. Econometric Methods. fourth ed. McGRaw-Hill.

Jones, C.M., Lipson, M.L., 1999. Execution costs of institutional equity orders. J. Financ. Intermediation 8, 123−140.

JP Morgan/Reuters, 1996. Risk Metrics™ - Technical Document, fourth ed. <http://gloria-mundi.com/UploadFile/2010-2/rmtd.pdf>.

Kakoullis, A., 2010. State-space methods for statistical arbitrage. Master's thesis, University College London, UK.

Keim, D.B., Madhavan, A., 1995. Anatomy of the trading process: empirical evidence on the behavior of institutional traders. J. Financ. Econ. 37, 371−398.

Keim, D.B., Madhavan, A., 1996. The upstairs market for large-block transactions: analysis and measurement of price effects. Rev. Finan. Stud. 9, 1−36.

Keim, D.B., Madhavan, A., 1997. Transactions costs and investment style: an inter-exchange analysis of institutional equity trades. J. Financ. Econ. 46, 265−292.

Kennedy, P., 1998. A Guide to Econometrics. fourth ed. The MIT Press, Cambridge, Massachusetts.

Kissell, R., 2003. Managing Trading Risk, Working Paper, <www.kissellresearch.com>.

Kissell, R., 2003. Pricing Principal Bids, Working Paper, <www.kissellresearch.com>.

Kissell, R., 2006. Algorithmic Trading Strategies, ETD Collection for Fordham University. Paper AAI3216918, ISBN 978-0-542-67789-2, <http://fordham.bepress.com/dissertations/AAI3216918>.

Kissell, R., 2006. The expanded implementation shortfall: understanding transaction cost components. J. Trading 6−16, Summer 2006.

Kissell, R., 2007. Statistical methods to compare algorithmic performance. J. Trading Spring 2007, 53−62.

Kissell, R., 2008. A Practical Framework for Transaction Cost Analysis. J. Trading 3 (2), 29−37, Summer 2008.

Kissell, R., 2009. Introduction to Algorithmic Trading: Cornell University, Graduate School of Financial Engineering- Manhattan, Class Notes, Fall Semester 2009.

Kissell, R., 2010. Introduction to Algorithmic Trading: Cornell University, Graduate School of Financial Engineering- Manhattan, Class Notes, Fall Semester 2010.

Kissell, R., 2011. TCA in the Investment Process: An Overview. J. Index Invest. 2 (1), 60−64.

Kissell, R., 2011. Creating dynamic pretrade models: beyond the black box. J. Trading 6 (4), 8−15, Fall 2011.

Kissell, R., 2012. Intraday volatility models: methods to improve real-time forecasts. J. Trading 7 (4), 27−34, Fall 2012.

Kissell, R., 2013. Multi-Asset Trading Cost Estimates. Kissell Research Group Working Paper, <www.kissellresearch.com>.

Kissell, R., Freyre-Sanders, A., 2004. An overview of the algorithmic trading process. The Euromoney Equity Capital Markets Handbook.

Kissell, R., Glantz, M., 2003. Optimal Trading Strategies. AMACOM, Inc., New York.

Kissell, R., Lie, H., 2011. U.S. Exchange auction trends: recent opening and closing auction behavior, and the implications on order management strategies. J. Trading 6 (1), 10−30.

Kissell, R., Malamut R., 1999. Optimal Trading Models, Working Paper.

Kissell, R., Malamut, R., 2005. Understanding the profit and loss distribution of trading algorithms. Institutional Investor, Guide to Algorithmic Trading .

Kissell, R., Malamut, R., 2006a. Algorithmic Decision Making Framework. J. Trading 1 (1), 12−21, Winter 2006.

Kissell, R., Malamut, R., 2006b. Unifying the Investment and Trading Theories, Working Paper.

Kissell, R., Malamut, R., 2007. Investing and trading consistency: does vwap compromise the stock selection process? J. Trading 2 (No. 4 (Fall 2007)), 12−22.

Kissell, R., Glantz, M., Malamut., R., 2004. A Practical Framework for Estimating Transaction Costs and Developing Optimal Trading Strategies to Achieve Best Execution. Elsevier, Finance Research Letters, 1, 35−46.

Kissell, R., Tannenbaum, P., 2009. 2008: the trading year in review. J. Trading 4 (2), 10−23.

Kolb, R.W., 1993. Financial Derivatives. Institute of Finance, New York.

Kolmogorov, A.N., Fomin, S.V., 1970. Introductory Real Analysis. Dover Publisher, Inc.

Konishi, H., Makimoto, N., 2001. Optimal slice of a block trade. J. Risk 3 (4), 33−51.

Krass, A., Stoll, H., 1972. Price impacts of block trading on the New York Stock Exchange. J. Finance 27, 569−588.

Kuhn, H.W., Tucker, A.W., 1951. Nonlinear programming'. Proceedings of Second Berkeley Symposium. Berkeley: University of California Press. pp. 481−492. MR47303

Kyle, A., 1985. Continuous auctions and insider trading. Econometrica 53, 1315−1335.

Lakonishok, J., Shleifer, A., Vishny, R., 1992. The impact of institutional trading on stock price. J. Financ. Econ. 32, 23−43.

Lee, C., Ready, M., 1991. Inferring trade direction from intraday data. J. Finance 46, 733−747.

Leland, H.E., 1996. Optimal asset rebalancing in the presence of transaction costs, Research Program in Finance, Working Paper RPF 261, U.C. Berkeley.

Leland, H.E., 1999. Optimal Portfolio Management with Transaction Costs and Capital Gains Taxes, Research Program in Finance, Working Paper RPF 290, U.C. Berkeley.

Li, D., Ng, W.L., 2000. Optimal dynamic portfolio selection: multi-period mean variance formulation. Math. Finance 10 (3), July 2000.

Lillo, F., Farmer, J.D., Mantegna, R.N., 2003. Master curve for price impact function. Nature 421, 129−130.

Lintner, J., 1965. The valuation of risk assets and the selection of risky investments in stock portfolios and capital budgets. Rev. Econ. Stat. 47 (1), 13−37.

Lobo, M.S., Fazel, M., Boyd, S., 2006. Portfolio optimization with linear and fixed transaction costs. Forthcoming, Annals of Operations Research, Special Issue on Financial Optimization.

Loeb, T.F., 1983. Trading costs: the critical link between investment information and results. Financ. Anal. J. 39−43, May/June 1983.

Lorenz, J., Osterrieder, J., 2009. Simulation of a limit order driven market. J. Trading 4 (1), 23−30, Winter 2009.

Madhavan, A., 2000. Market microstructure − a survey. J. Financ. Mark. 3, 205−258.

Madhavan, A., 2002. Market microstructure: a practitioners guide. Financ. Anal. J. 58 (5), 28–42, Sep/Oct 2002.

Madhavan, A., Cheng, M., 1997. In search of liquidity: an analysis of upstairs and downstairs trades. Rev. Financ. Stud. 10, 175–204.

Madhavan, A., Smidt, S., 1993. An analysis of change in specialist quotes and inventories. J. Finance 48, 1595–1628.

Madhavan, A., Sofianos, G., 1998. An empirical analysis of NYSE specialist trading. J. Financ. Econ. 48, 189–210.

Malamut, R., 2002. Multi-Period Optimization Techniques for Trade Scheduling. QWAFAFEW, April 2002.

Malamut, R., Kissell R., 2002. Multi-period Trade Schedule Optimization Working paper, <www.KissellResearch.com>.

Malamut, R., Kissell R., 2006. Multi-period Trade Schedule Optimization with Expanded Trading Trajectory Formulation, Working paper, <http://www. KissellResearch.com>.

Malkiel, B., 1995. Returns from investing in equity mutual funds 1971-1995. J. Finance 50 (2), 549–572.

Mandel, J., 1964. The Statistical Analysis of Experimental Data. Dover Publications, Inc.

Mandelbrot, B.B., 1982. The Fractal Geometry of Nature. Freeman, San Francisco, NY.

Mansfield, E., 1994. Statistics for Business and Economics: Methods and Applications. fifth ed. W. W. Norton & Company.

Markowitz, H.M., 1952. Portfolio selection. J. Finance 7 (1), 77–91.

Markowitz, H.M., 1952. The utility of wealth. J. Polit. Econ. 60, 152–158.

Markowitz, H.M., 1956. The optimization of a quadratic function subject to linear constraints. Naval Res. Logistics Q. 3, 111–133.

Markowitz, H.M., 1959. Portfolio Selection: Efficient Diversification of Investments. John Wiley & Sons, New York.

Markowitz, H.M., 2008. CAPM Investors do not get paid for bearing risk: a linear relation does not imply payment for risk. J. Portf. Manag. 34 (2), 91–94.

Menchero, J., Wang, J., Orr, D.J., 2012. Improving risk forecasts for optimized portfolios. Financ. Anal. J. 68 (3), 40–50.

Meyer, P., 1970. Introductory Probability and Statistical Applications. second ed. Addison-Wesley Publishing Company.

Michard, R., 1998. Efficient Asset Management: A Practical Guide to Stock Portfolio Optimization and Asset Allocation. Harvard Business School Press.

Mitchell, J., Braun S., 2004. Rebalancing an Investment Portfolio in the Presence of Convex Transaction Costs, unpublished manuscript.

Mitchell, M., 2001. Characteristics of risk and return in risk arbitrage. J. Finance 56 (6), 2135–2175.

Mittelhammer, R., Judge, G., Miller, D., 2000. Econometrics Foundation. Cambridge University Press.

Morgan/Reuters, J.P., 1996. Risk Metrics™–Technical Document. fourth ed.<http:// gloria-mundi.com/UploadFile/2010-2/rmtd.pdf>.

Mossin, J., 1966. Equilibrium in a capital asset market. Econometrica 768–783.

Narang, R., 2009. Inside the Black Box: The Simple Truth about Quantitative Trading, vol. 501. Wiley, NY, USA.

Nasdaq Economic Research, 2001. The impact of decimalization on the Nasdaq stock market. Final report to the SEC.

Newmark, J., 1988. Statistics and Probability in Modern Life. fourth ed. Saunders College Publishing.

Nocedal, J., Wright, S., 1999. Numerical Optimization. Springer, New York.

Obizhaeva, A., Wang, J., 2005. Optimal Trading Strategy and Supply/Demand Dynamics. Unpublished Manuscript.

O'Hara, M., 1995. Market Microstructure Theory. Blackwell, Cambridge, MA.

O'Hara, M., Oldfield, G., 1986. The microeconomics of market making. J. Financ. Quant. Anal. 21, 361−376.

Pagano, M., Röell, A., 1990. Trading systems in european stock exchanges: current performance and policy options. Econ. Policy 10, 65−115.

Patton, A., 2011. Volatility forecast comparison using imperfect volatility proxies. J. Econom. 160 (1), 246−256.

Pearson, N., 2002. Risk Budgeting: Portfolio Problem Solving with Value-at-Risk. Wiley.

Pemberton, M., Rau, N, 2001. Mathematics for Economists: An Introductory Textbook. Manchester University Press.

Perold, A., 1988. The implementation shortfall: paper versus reality. J. Portf. Manage. 14, 4−9.

Perold, A., Sirri, E., 1993. The Cost of International Equity Trading, Working Paper, Division of Research, Harvard Business School.

Peters, E., 1989. Fractal structure in the capital markets. Financ. Analysts J. July/August 1989.

Peters, E., 1991. Chaos and Order in the Capital Markets. John Wiley & Sons.

Peters, E., 1994. Fractal Market Analysis. John Wiley & Sons.

Pfeiffer, P., 1978. Concepts of Probability Theory. Second Revised Edition Dover Publications, Inc.

Pindyck, R., Rubinfeld, D., 1998. Econometric Models and Economic Forecasts. Irwin/McGraw-Hill, USA.

Pratt, J.W., 1964. Risk aversion in the small and the large. Econometrica 122−136, Jan−Apr, 1964.

Rakhlin, D., Sofianos, G., 2006a. The Impact of an increase in volatility on trading costs. J. Trading 1 (2), 43−50, Spring 2006.

Rakhlin, D., Sofianos, G., 2006b. Choosing benchmarks vs. choosing strategies: part 2: execution strategies: VWAP or shortfall. J. Trading (Winter 2006).

Rardin, R., 1997. Optimization in Operations Research. 1st ed. Prentice Hall, NJ.

Resta, M., 2006. On the profitability of scalping strategies based on neural networks. Knowledge-Based Intelligent Information and Engineering Systems. University of Geneva, Lecture Notes in Computer Science, Springer, pp. 641−646

RiskMetrics Group, 1996. RiskMetrics − Technical Document. J.P.Morgan/Reuters, NY.

Roll, R., Ross, S., 1984. The arbitrage pricing theory approach to strategic portfolio planning. Financ. Analysts J. May-June 1984.

Ross, S., 1976. The arbitrage theory of asset pricing. J. Econ. Theory 13 (3), 341−360.

Rudd, A., Clasing Jr., H.K., 1988. Modern Portfolio Theory: The Principles of Investment Management. Andrew Rudd, Orinda, CA.

SEC Commission, 2012. US SEC litigation release no. 22240/January 26, 2012. SEC litigation releases LR(22240).

Schwartz, R.A., 1991. Reshaping the Equity Markets: A Guide for the 1990's. Business One Irwin Press.

Shanno, D.F., 1970. Conditioning of quasi-newton methods for function minimization. Math. Comput. 24, 647–656.

Shanno, D.F., 1978. Conjugate gradient methods with inexact searches. Math. Oper. Res. 3 (3), 244–256.

Sharpe, W.F., 1964. Capital asset prices: a theory of market equilibrium. J. Finance 19 (3), 425–442, Sep 1964.

Sharpe, W.F., 1966. Mutual fund performance. J. Bus. 39, 119–138, Jan 1966.

Sharpe, W.F., 1964. Capital asset prices: a theory of market equilibrium. J. Finance 19 (3), 425–442.

Shleifer, A., 1986. Do demand curves for stocks slope down? J. Finance 41, 579–590.

Sorensen, E.H., Price, L., Miller, K., Cox, D., Birnbaum, S., 1998. The Salomon Smith Barney global equity impact cost model Technical report, Salomon Smith Barney.

Stoll, H., 1978a. The supply of dealer services in securities markets. J. Finance 33, 1133–1151.

Stoll, H., 1978b. The pricing of security dealer services: An empirical study of NASDAQ stocks.

Stoll, H.R., 1989. Inferring the components of the bid-ask spread: theory and empirical tests. J. Finance 44, 115–134.

Tinic, S., 1972. The economics of liquidity service. Q. J. Econ. 86, 79–93.

Treynor, J., 1961. Toward a theory of market value of risky assets. 21. Unpublished manuscript dated 8/8/1961, No. 95–209

Treynor, J., 1981a. What does it take to win the trading game? Financ. Analysts J. 37 (1), 55–60 (Jan/Feb).

Treynor, J., 1981b. The only game in town. Financ. Analyst J. 27 (2), 12–14, Mar/Apr 1971.

Treynor, J., 1994. The invisible cost of trading. J. Portf. Manage., Fall 71–78.

Tsay, R., 2002. Analysis of Financial Time Series. John Wiley & Sons.

Tutuncu, R.H., Koenig, M., 2003. Robust Asset Allocation. Working paper.

US Court, 1963. S. E. C. v. capital gains bureau, 375 U.S. 180 (1963).

Wagner, W., 1990. Transaction costs: measurement and control. Financ. Analyst J. Sept/Oct 1990.

Wagner, W., 2003. The Iceberg of Transaction Costs. Plexus Group Publication.

Wagner, W. (Ed.), 1991. The Complete Guide to Security Transactions. John Wiley.

Wagner, W., 2004. The Transaction Process: Evaluation and Enhancement, QWAFAFEW Meeting Presentation, San Francisco (March 9, 2004).

Wagner, W., Banks, M., 1992. Increasing portfolio effectiveness via transaction cost management. J. Portf. Manage. 19, 6–11.

Wagner, W., Cuneo, L., 1975. Reducing the cost of stock trading. Financ. Analyst J. 31 (6), 35–44, Nov/Dec 1975.

Wagner, W., Edwards, H., 1993. Best execution. Financ. Analyst J. 49 (1), 65–71, Jan/Feb 1993.

Wagner, W., Glass, S., 2001. What every plan sponsor needs to know about transaction costs. Inst. Investor, Trans. Cost Guide Spring, 2000.

Wood, R., McInish, T.H., Ord, J.K., 1985. An investigation of transaction data for NYSE stocks. J. Finance 25, 723–739.

Zhi, J., Melia, A.T., Guericiolini, R., et al., 1994. Retrospective population-based analysis of the dose-response (Fecal Fat Excretion) relationship of orlistat in normal and obese volunteers. Clin. Pharmacol. Ther. 56, 82–85.

Index

Note: Page numbers followed by "*f*" and "*t*" refer to figures and tables, respectively.